Probability and Mathematical Statistics (Continued)

MUIRHEAD • Aspects of Multivariate Statistical Theory
PRESS • Bayesian Statistics: Principles, Models, and Applications
PURI and SEN • Nonparametric Methods in General Linear Models
PURI and SEN • Nonparametric Methods in Multivariate Analysis
PURI, VILAPLANA, and WERTZ • New Perspectives in Theoretical and Applied Statistics
RANDLES and WOLFE • Introduction to the Theory of Nonparametric Statistics
RAO • Asymptotic Theory of Statistical Inference
RAO • Linear Statistical Inference and Its Applications, *Second Edition*
RAO • Real and Stochastic Analysis
RAO and SEDRANSK • W.G. Cochran's Impact on Statistics
ROBERTSON, WRIGHT and DYKSTRA • Order Restricted Statistical Inference
ROGERS and WILLIAMS • Diffusions, Markov Processes, and Martingales, Volume II: Îto Calculus
ROHATGI • An Introduction to Probability Theory and Mathematical Statistics
ROHATGI • Statistical Inference
ROSS • Stochastic Processes
RUBINSTEIN • Simulation and The Monte Carlo Method
RUZSA and SZEKELY • Algebraic Probability Theory
SCHEFFE • The Analysis of Variance
SEBER • Linear Regression Analysis
SEBER • Multivariate Observations
SEBER and WILD • Nonlinear Regression
SEN • Sequential Nonparametrics: Invariance Principles and Statistical Inference
SERFLING • Approximation Theorems of Mathematical Statistics
SHORACK and WELLNER • Empirical Processes with Applications to Statistics
STOYANOV • Counterexamples in Probability

Applied Probability and Statistics

ABRAHAM and LEDOLTER • Statistical Methods for Forecasting
AGRESTI • Analysis of Ordinal Categorical Data
AGRESTI • Categorical Data Analysis
AICKIN • Linear Statistical Analysis of Discrete Data
ANDERSON and LOYNES • The Teaching of Practical Statistics
ANDERSON, AUQUIER, HAUCK, OAKES, VANDAELE, and WEISBERG • Statistical Methods for Comparative Studies
ARTHANARI and DODGE • Mathematical Programming in Statistics
ASMUSSEN • Applied Probability and Queues
BAILEY • The Elements of Stochastic Processes with Applications to the Natural Sciences
BARNETT • Interpreting Multivariate Data
BARNETT and LEWIS • Outliers in Statistical Data, *Second Edition*
BARTHOLOMEW • Stochastic Models for Social Processes, *Third Edition*
BARTHOLOMEW and FORBES • Statistical Techniques for Manpower Planning
BATES and WATTS • Nonlinear Regression Analysis and Its Applications
BECK and ARNOLD • Parameter Estimation in Engineering and Science
BELSLEY, KUH, and WELSCH • Regression Diagnostics: Identifying Influential Data and Sources of Collinearity
BHAT • Elements of Applied Stochastic Processes, *Second Edition*
BHATTACHARYA and WAYMIRE • Stochastic Processes with Applications
BLOOMFIELD • Fourier Analysis of Time Series: An Introduction
BOLLEN • Structural Equations with Latent Variables
BOX • R. A. Fisher, The Life of a Scientist

Applied Probability and Statistics (Continued)

BOX and DRAPER • Empirical Model-Building and Response Surfaces
BOX and DRAPER • Evolutionary Operation: A Statistical Method for Process Improvement
BOX, HUNTER, and HUNTER • Statistics for Experimenters: An Introduction to Design, Data Analysis, and Model Building
BROWN and HOLLANDER • Statistics: A Biomedical Introduction
BUNKE and BUNKE • Nonlinear Regression, Functional Relations and Robust Methods: Statistical Methods of Model Building
BUNKE and BUNKE • Statistical Inference in Linear Models, Volume I
CHAMBERS • Computational Methods for Data Analysis
CHATTERJEE and HADI • Sensitivity Analysis in Linear Regression
CHATTERJEE and PRICE • Regression Analysis by Example
CHOW • Econometric Analysis by Control Methods
CLARKE and DISNEY • Probability and Random Processes: A First Course with Applications, *Second Edition*
COCHRAN • Sampling Techniques, *Third Edition*
COCHRAN and COX • Experimental Designs, *Second Edition*
CONOVER • Practical Nonparametric Statistics, *Second Edition*
CONOVER and IMAN • Introduction to Modern Business Statistics
CORNELL • Experiments with Mixtures: Designs, Models and The Analysis of Mixture Data, *Second Edition*
COX • A Handbook of Introductory Statistical Methods
COX • Planning of Experiments
DANIEL • Applications of Statistics to Industrial Experimentation
DANIEL • Biostatistics: A Foundation for Analysis in the Health Sciences, *Fourth Edition*
DANIEL and WOOD • Fitting Equations to Data: Computer Analysis of Multifactor Data, *Second Edition*
DAVID • Order Statistics, *Second Edition*
DAVISON • Multidimensional Scaling
DEGROOT, FIENBERG and KADANE • Statistics and the Law
DEMING • Sample Design in Business Research
DILLON and GOLDSTEIN • Multivariate Analysis: Methods and Applications
DODGE • Analysis of Experiments with Missing Data
DODGE and ROMIG • Sampling Inspection Tables, *Second Edition*
DOWDY and WEARDEN • Statistics for Research
DRAPER and SMITH • Applied Regression Analysis, *Second Edition*
DUNN • Basic Statistics: A Primer for the Biomedical Sciences, *Second Edition*
DUNN and CLARK • Applied Statistics: Analysis of Variance and Regression, *Second Edition*
ELANDT-JOHNSON and JOHNSON • Survival Models and Data Analysis
FLEISS • The Design and Analysis of Clinical Experiments
FLEISS • Statistical Methods for Rates and Proportions, *Second Edition*
FLURY • Common Principal Components and Related Multivariate Models
FOX • Linear Statistical Models and Related Methods
FRANKEN, KÖNIG, ARNDT, and SCHMIDT • Queues and Point Processes
GALLANT • Nonlinear Statistical Models
GIBBONS, OLKIN, and SOBEL • Selecting and Ordering Populations: A New Statistical Methodology
GNANADESIKAN • Methods for Statistical Data Analysis of Multivariate Observations
GREENBERG and WEBSTER • Advanced Econometrics: A Bridge to the Literature
GROSS and HARRIS • Fundamentals of Queueing Theory, *Second Edition*
GROVES • Survey Errors and Survey Costs
GROVES, BIEMER, LYBERG, MASSEY, NICHOLLS, and WAKSBERG • Telephone Survey Methodology

(*cardplate continues in back of book*)

Applied Probability and Stochastic Processes

In Engineering and Physical Sciences

Applied Probability and Stochastic Processes

In Engineering and Physical Sciences

MICHEL K. OCHI
University of Florida

A Wiley-Interscience Publication
JOHN WILEY & SONS
New York · Chichester · Brisbane · Toronto · Singapore

Copyright © 1990 by John Wiley & Sons, Inc.

All rights reserved. Published simultaneously in Canada.

Reproduction or translation of any part of this work beyond that permitted by Section 107 or 108 of the 1976 United States Copyright Act without the permission of the copyright owner is unlawful. Requests for permission or further information should be addressed to the Permissions Department, John Wiley & Sons, Inc.

Library of Congress Cataloging in Publication Data:
Ochi, Michel K.
  Applied probability and stochastic processes in engineering and physical sciences/Michel K. Ochi.
    p. cm.—(Wiley series in probability and mathematical statistics. Applied probability and statistics section, ISSN 0271-6356)
  "A Wiley-Interscience publication."
  Bibliography: p.
  Includes index.
  1. Engineering—Statistical methods. 2. Science—Statistical methods. 3. Probabilities. 4. Stochastic process. I. Title. II. Series.
TA340.O24 1989   89-34352
620'.0072—dc20   CIP
ISBN 0-471-85742-4

Printed in the United States of America

10 9 8 7 6 5 4 3 2 1

To the Memory of
My Parents

Contents

Preface   xv

1. Elements of Probability   1

    1.1. Basic Concept, 1
    1.2. Algebra of Sets and Fields, 4
    1.3. Probability, 9
    Exercises, 12

2. Random Variables and Their Probability Distributions   13

    2.1. Random Variables, 13
    2.2. Distribution Functions and Probability Densities, 15
    2.3. Quantiles, Median, and Mode of Distribution, 21
    2.4. Random Vectors and Probability Distribution Function, 25
    2.5. Conditional Probability and Distribution, 29
        2.5.1. Conditional Probability, 29
        2.5.2. Statistical Independence, 33
        2.5.3. Truncated Distribution, 35
        2.5.4. Bayes Formula, 38
    Exercises, 40

3. Moments of Random Variables   43

    3.1. Expected Value and Moments, 43
        3.1.1. Mean and Variance of Random Variables, 43
        3.1.2. Coefficient of Variation, Skewness, and Kurtosis, 51

3.2. Moments of Random Vectors, 53
    3.2.1. Moments, Covariance, and Correlation Coefficient, 53
    3.2.2. Mean and Variance of Sum and Product of Two Random Variables, 59
3.3. Conditional Moments, 62
    3.3.1. Conditional Mean and Variance, 62
    3.3.2. Application of Conditional Mean and Variance, 65
Exercises, 70

4. Moment Generating Function, Characteristic Function, and Their Application   72

4.1. Moment Generating Function, 73
4.2. Characteristic Function, 76
4.3. Cumulants (Semiinvariants), 84
4.4. Probability Generating Function, 87
Exercises, 89

5. Discrete Random Variables and Their Distributions   91

5.1. Binomial and Related Distributions, 91
    5.1.1. Binomial Distribution, 91
    5.1.2. Negative Binomial Distribution, 94
    5.1.3. Hypergeometric Distribution, 96
    5.1.4. Multinomial Distribution, 98
5.2. Poisson and Related Distributions, 101
    5.2.1. Poisson Distribution, 101
    5.2.2. Poisson Distribution with Parameter a Random Variable, 105
5.3. Relationship between Distributions of Various Discrete-Type Random Variables, 106
Exercises, 109

6. Continuous Random Variables and Their Distributions   111

6.1. Normal and Related Distributions, 111
    6.1.1. Normal Distribution, 111
    6.1.2. Log-Normal Distribution, 117
    6.1.3. Multivariate Normal Distribution, 118

CONTENTS

- 6.2. Gamma and Related Distributions, 122
  - 6.2.1. Gamma and Exponential Distributions, 122
  - 6.2.2. Chi-Square (χ^2) Distribution, 124
- 6.3. Weibull and Related Distributions, 128
  - 6.3.1. Weibull Distribution, 128
  - 6.3.2. Rayleigh Distribution, 129
- 6.4. Some Other Distributions, 131
- 6.5. Relationship between Distributions of Various Continuous-Type Random Variables, 133
- Exercises, 134

7. Transformation of Random Variables — 138

- 7.1. Transformation of Single Random Variable, 138
- 7.2. Transformation of Several Random Variables, 144
  - 7.2.1. Function of Random Variables, 144
  - 7.2.2. Sum, Difference, Product, and Ratio of Two Random Variables, 151
- 7.3. Transformation through Characteristic Functions, 161
- Exercises, 165

8. Extreme Value Statistics — 168

- 8.1. Order Statistics and Extreme Values, 169
  - 8.1.1. Order Statistics, 169
  - 8.1.2. Evaluation of Extreme Values, 172
- 8.2. Asymptotic Distributions of Extreme Values, 178
  - 8.2.1. Type I Asymptotic Distribution, 179
  - 8.2.2. Type II Asymptotic Distribution, 186
  - 8.2.3. Type III Asymptotic Distribution, 190
- 8.3. Estimation of Extreme Values from Observed Data, 194
- Exercises, 199

9. Stochastic Processes — 201

- 9.1. Introduction, 201
- 9.2. Classification of Stochastic Processes, 203
  - 9.2.1. Stationary Process, 203
  - 9.2.2. Ergodic Process, 205
  - 9.2.3. Independent Increment Process, 207

        9.2.4. Markov Process, 210
        9.2.5. Counting Process, 211
        9.2.6. Narrow-Band Process, 212
   9.3. Some Stochastic Processes for Analysis of Physical Phenomena, 214
        9.3.1. Normal (Gaussian) Process, 214
        9.3.2. Wiener–Lévy Process, 215
        9.3.3. Poisson Process, 216
        9.3.4. Bernoulli Process, 217
        9.3.5. Shot Noise Process, 217

10. Spectral Analysis of Stochastic Processes   218

   10.1. Spectral Analysis for a Single Stochastic Process, 219
        10.1.1. Autocorrelation Function, 219
        10.1.2. Spectral Density Function, 224
        10.1.3. Wiener–Khintchine Theorem, 231
   10.2. Spectral Analysis of Two Random Processes, 239
        10.2.1. Cross-Correlation Function, 239
        10.2.2. Cross-Spectral Density Function, 242
        10.2.3. Application—Directional Spectral Analysis, 245
   10.3. Integrated and Differentiated Random Processes, 249
        10.3.1. Mean, Variance, and Covariance, 249
        10.3.2. Autocorrelation Function and Spectral Density Function of Derived Random Processes, 254
   10.4. Squared Random Processes, 257
   10.5. Higher Order Spectral Analysis, 259
   Exercises, 265

11. Amplitudes and Periods of Gaussian Random Processes   267

   11.1. Distribution of Amplitudes for Narrow-Band Processes, 267
        11.1.1. Probability Density Function of Amplitudes, 267
        11.1.2. Distribution of Crest-to-Trough Excursions, 273
        11.1.3. Envelope Process, 279
        11.1.4. Significant Value and Extreme Value, 281

CONTENTS　　　　　　　　　　　　　　　　　　　　　　　　　　　　　　　　xi

- 11.2. Distribution of Maxima for Non-Narrow-Band Processes, 283
    - 11.2.1. Expected Number of Maxima, 284
    - 11.2.2. Probability Distribution of Maxima, 290
- 11.3. Joint Probability Distribution of Amplitudes and Periods, 294
    - 11.3.1. Joint Distribution for Narrow-Band Processes, 295
    - 11.3.2. Joint Distribution for Non-Narrow-Band Processes, 300
- 11.4. Distribution of Periods, 304
    - 11.4.1. Expected Period, 304
    - 11.4.2. Probability Density Function of Periods, 306
- 11.5. Estimation of Extreme Amplitude and Maxima, 312
- Exercises, 315

12. Statistical Analysis of Time Series Data　　　317

- 12.1. Principle of Statistical Estimation, 317
- 12.2. Confidence Intervals, 325
    - 12.2.1. Confidence Interval for the Rayleigh Distribution Parameter, 326
    - 12.2.2. Confidence Interval for the Variance of the Normal Distribution, 329
    - 12.2.3. Confidence Interval for the Parameter of the Poisson Distribution, 332
- Exercises, 334

13. Wiener–Lévy and Markov Processes　　　335

- 13.1. Wiener–Lévy Process, 335
    - 13.1.1. Random Walk, 336
    - 13.1.2. Wiener–Lévy Process, 339
- 13.2. Markov Process, 341
    - 13.2.1. Chapman–Kolmogorov Equation, 341
    - 13.2.2. Two-State Markov Chain and Markov Process, 344
- 13.3. Fokker–Planck Equation, 353
    - 13.3.1. Derivation of Fokker–Planck Equation, 353
    - 13.3.2. Application to Nonlinear Vibration Systems, 362

14. Linear System and Stochastic Prediction — 366

- 14.1. Linear System and Unit Impulse Response, 366
  - 14.1.1. Linear System, 366
  - 14.1.2. Impulse Response Function, 368
  - 14.1.3. Input and Output Mean Levels, 371
- 14.2. Evaluation of Impulse and Frequency Response Functions, 372
  - 14.2.1. Theoretical Approach, 372
  - 14.2.2. Experimental Approach, 375
- 14.3. Input and Output Spectral Relationship, 377
  - 14.3.1. Autocorrelations and Spectral Density Functions, 377
  - 14.3.2. Response of a System to Dual Inputs, 379
  - 14.3.3. Coherency Function, 382
- Exercises, 384

15. Nonlinear Systems and Stochastic Prediction — 387

- 15.1. Nonlinear Systems, 387
- 15.2. Equivalent Linearization Technique, 389
- 15.3. Perturbation Technique, 398
- 15.4. Markov Process Approach—Application of Fokker–Planck Equation, 405
- 15.5. Application of Bispectral Analysis, 407
- Exercises, 409

16. Non-Gaussian Stochastic Processes — 411

- 16.1. Probability Distribution by Applying the Concept of Orthogonal Polynomials, 412
- 16.2. Probability Distribution by Applying Cumulant Generating Function, 417
- Exercises, 422

17. Counting Stochastic Processes — 423

- 17.1. Poisson Processes, 423
  - 17.1.1. Fundamentals of the Poisson Process, 423
  - 17.1.2. Some Properties of the Poisson Process, 426
  - 17.1.3. Interarrival Time and Waiting Time, 429

CONTENTS

17.2. Generalization of the Poisson Process, 437
    17.2.1. Nonhomogeneous Poisson Process, 437
    17.2.2. Compound Poisson Process, 439
    17.2.3. Superimposed Poisson Process, 443
17.3. Poisson Impulse Process and Response, 446
17.4. Renewal Counting Processes, 449
    17.4.1. Renewal Counting Processes, 449
    17.4.2. Renewal Equation, 451
    17.4.3. Asymptotic Properties of Renewal Process, 454
    17.4.4. Probability Distribution of Residual Time, 455
Exercises, 459

Appendix A. Fourier Transform   **461**
    A.1. Derivation of the Fourier Transform, 461
    A.2. Properties of the Fourier Transform, 468

Appendix B. Hilbert Transform   **472**

Appendix C. Unit Impulse Function (Delta Function)   **477**

Appendix D. Statistical Goodness of Fit Tests   **481**
    D.1. Chi-Square (χ^2) Test, 481
    D.2. Kolmogorov–Smirnov Test, 484

References   **487**

Bibliography   **493**

Index   **495**

Preface

Application of probability and stochastic process theory is playing an ever increasing role in a number of diverse fields in engineering and the physical sciences. This is due, in part, to the growing realization that many random phenomena observed in physics and engineering are described with reasonable accuracy following recent comprehensive advances in stochastic prediction methodologies. As a consequence, the probabilistic approach has become an important component of practical reasoning in physical sciences as well as an integrated part of modern design technology in engineering.

This book is designed to give senior undergraduate and graduate students and researchers in engineering and the physical sciences a thorough understanding of the modern concepts of stochastic process theory and its application for predicting statistical characteristics of random phenomena. Toward this end, emphasis is placed on clarification of basic principles supporting current prediction techniques and practical application of prediction methods.

No advanced knowledge of probability theory on the part of the reader is assumed. However, a sound knowledge of advanced calculus and functional analysis is essential in order to comprehend the mathematical analysis. For the readers' convenience a brief review of certain subjects such as the Fourier transform, the Hilbert transform, the unit impulse function, etc., which are useful in understanding the prediction techniques, are summarized in the appendices.

This book consists of two parts, although no such formal division is designated in the text. The first part consists of Chapters 1 through 8, which present probability theory relevant to probabilistic analysis of stochastic processes. Effort in these chapters is devoted to selecting subjects pertinent to predictions appearing in later chapters (the second part of the text), rather than to introducing general topics in probability theory. Needless to say, probability theory is a prerequisite for predicting the statistical as well as the quantitative properties of random phenomena.

Chapter 9 through 17 discuss principles and advanced techniques in the various subjects in stochastic processes and their application in the analysis of random phenomena observed in engineering and the physical sciences. In particular, the principles and procedures of spectral analysis and development of the probability density function derived therefrom are discussed in detail, since these provide the basis for modern probabilistic prediction techniques. Included also is material found in the recent literature but which has not been incorporated in textbooks such as higher order spectral analysis, the joint probability distribution of amplitudes and periods, and non-Gaussian random processes. Many examples are provided in order to facilitate understanding of the material.

This book is a direct result of my teaching and research in stochastic processes, and I am grateful to the College of Engineering, University of Florida, for granting me sabbatical leave during which significant progress in this undertaking was achieved.

I wish to acknowledge the encouragement and suggestions received from a number of learned scholars in the diverse fields of mathematical statistics, physics, and engineering. I am especially indebted to Professor Longuet-Higgins, University of Cambridge, and Professor Emeritus St. Denis, University of Hawaii, who inspired me to apply in depth the stochastic process approach to engineering problems. I would like to express my sincere appreciation to Mrs. Cathy Freeman and Ms. Amanda Graham for typing the manuscript, and to Ms. Lillean Pieter for drawing the illustrations. Assistance in editorial work rendered by my wife Margaret for the final preparation of the manuscript is deeply appreciated.

MICHEL K. OCHI

Gainesville, Florida
October 1989

Applied Probability and Stochastic Processes

In Engineering and Physical Sciences

CHAPTER 1

Elements of Probability

1.1 BASIC CONCEPT

The theory of probability deals with the mathematical analysis of quantities obtained from observations of random phenomena. Here, the term *random phenomena* is defined as phenomena that in repeated observations under identical circumstances do not nearly yield the same outcomes. There is no deterministic regularity in the occurrence of outcomes; instead, there is a statistical regularity in the sense that the relative frequency of occurrence of the outcome may be evaluated. That is, the relative frequency of occurrence of the event fluctuates, but the degree of the fluctuation decreases, in general, with the increase in the number of observations and therefore the frequency settles to a certain value.

To elaborate on the above statement, let us consider, as an example, the magnitude of peak-to-trough excursions of wind-generated waves in the ocean. As shown in Figure 1.1(a), the magnitude of the excursion, denoted by X in the figure, varies in random fashion from one wave to another, and hence it may be said that there is no deterministic regularity. If the observed data of X are classified in 1/2-m intervals, for example, and the relative frequency of occurrence of X is calculated for each interval, then we can obtain the relative frequencies as a function of X which is called the histogram. The shape of the histogram is inconsistent when the number of observations is small. However, the degree of inconsistency is reduced and converges to a certain shape as shown in Figure 1.1(b) with the increase in the number of observations, for example, on the order of 200. This may be called statistical regularity.

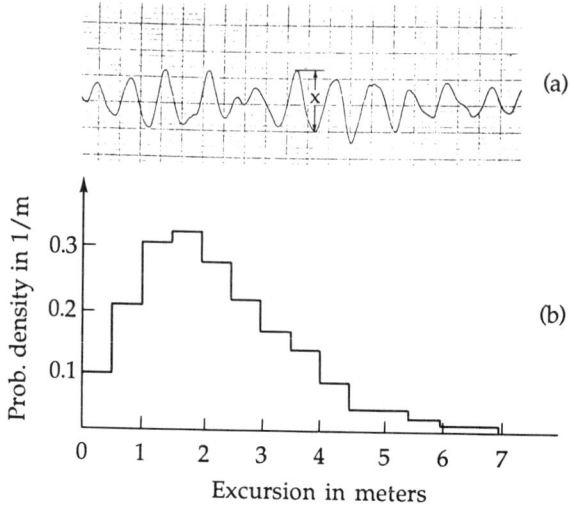

Figure 1.1 Time history of ocean waves and histogram of peak-to-trough excursions.

Prior to continuing further discussion on the fundamental concept of probability, it may be well to give the definition of sample space and a random event.

Definition 1.1. In the observation of a random phenomenon or random experiment, a set consisting of all possible outcomes that could occur is called the *sample space* and is denoted by Ω. A set belonging to the sample space for which the probability can be defined is called a *random event*.

The relative frequency of occurrence of a random event fluctuates even though observations (or experiments) are repeated under the same environment. However, it approaches a stable limit value as the number of observations becomes large, and this limit value is called the *probability* of the random event.

Example 1.1. Let us consider the simple random experiment of tossing a fair coin. The outcome of this experiment is either a head, H, or a tail, T. Hence, the sample space contains two elements, $\Omega = \{H, T\}$. Suppose we are interested in the occurrence of a head, then $\{H\}$ is a random event. Although we cannot predetermine the result of any particular toss, the frequency of occurrence of a head will converge to a certain limit value, 0.5,

BASIC CONCEPT 3

after many tosses. This limit value is called the probability of occurrence of a head. ∎

Example 1.2. Let us consider the launching of a missile from a submarine. The outcome of this random experiment is either a success, S, or a failure, F, and hence the sample space for this example is given by $\Omega = \{S, F\}$. This situation is exactly the same as shown in Example 1.1. The relative frequency of the random event, $\{S\}$, will converge to a certain limit value after many trials, but the value may not necessarily be 0.5. This is because, unlike the case of a fair coin, a success significantly depends on various factors such as performance of the launching device and the control mechanism of the missile. ∎

Example 1.3. Let us consider the launching of a missile three times from a submarine. Although the outcome is either a success, S, or a failure, F, the sample space for this case does not consist of only two elements. Note that the sample space is a set consisting of all possible outcomes; hence, for this example, we have a set consisting of eight outcomes:

$$\Omega = \{\omega_1, \omega_2, \omega_3, \omega_4, \omega_5, \omega_6, \omega_7, \omega_8\}$$

where ω_1 = S S S $\quad \omega_5$ = F S S
ω_2 = S S F $\quad \omega_6$ = F S F
ω_3 = S F S $\quad \omega_7$ = F F S
ω_4 = S F F $\quad \omega_8$ = F F F

Suppose we are interested in the possibility of hitting the target (even one hit is acceptable), then the random event is a set $\{\omega_1, \omega_2, \omega_3, \omega_4, \omega_5, \omega_6, \omega_7\}$. If we want to know the possibility of hitting the target at least twice, then the random event is a set $\{\omega_1, \omega_2, \omega_3, \omega_5\}$. The relative frequencies of occurrence of the random events for this example will be shown later. ∎

Example 1.4. The wind-generated wave profile (the deviation from the still water level) is observed at a location where the water depth is 5 m. The sample space of this example consists of elements that take any value between -5 and ∞. Suppose we are interested in the possibility that the wave profile will exceed ± 2 m, then a set of continuous ranges $\{(-5, -2)$ and $(2, \infty)\}$ is the random event. ∎

The discussion thus far briefly outlines the fundamental concept of probability in a heuristic sense. Modern probability theory, however, has been developed based on a rigorous mathematical foundation that provides a precise definition of probability, random variables, probability functions, and so on, so that the outcome of random events or experiments can be mathematically described. To discuss fundamental probability theory, it is necessary to use several definitions and terminologies from fundamental set theory. These are summarized in the following section.

1.2 ALGEBRA OF SETS AND FIELDS

A *set* is a collection of objectives. Each member, x, of a set A is called an *element* of set A, and is denoted by $x \in A$.

Definition 1.2. If every element of a set A_2 is also an element of set A_1, then A_2 is the *subset* of the set A_1 and is denoted by $A_2 \subset A_1$.

We may write the definition of a subset as follows:

$$A_2 \subset A_1 = \{x \in A_2; x \in A_2 \text{ implies } x \in A_1\}$$

It may be very convenient to illustrate various definitions concerning the algebra of sets by a pictorial sketch called a *Venn diagram*. For example, Figure 1.2 is a Venn diagram indicating the definition of a subset.

Definition 1.3. Two sets A_1 and A_2 are said to be *equal*, denoted by $A_1 = A_2$, if $A_2 \subset A_1$ and $A_1 \subset A_2$.

Definition 1.4. A set that contains no elements is called the *empty set* or *null set*, and is denoted by $A = 0$.

Definition 1.5. The set of all elements that belong to at least one of the sets $A_1, A_2, \ldots, A_n$ is called the *union* of the sets A_i, $i = 1, 2, \ldots, n$, and is denoted by $\bigcup_{i=1}^{n} A_i$.

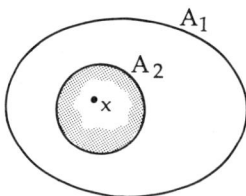

Figure 1.2 Subset $A_2 \subset A_1$.

ALGEBRA OF SETS AND FIELDS

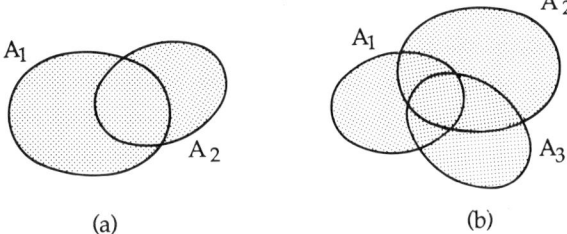

Figure 1.3 Unions (a) $A_1 \cup A_2$ and (b) $A_1 \cup A_2 \cup A_3$.

We may write the definition as

$$\bigcup_{i=1}^{n} A_i = \{x \in A_i \text{ for at least one } i = 1, 2, \ldots, n\}$$

Figures 1.3(a) and (b) show the unions of $A_1 \cup A_2$ and $A_1 \cup A_2 \cup A_3$, respectively.

Definition 1.6. The set of all elements that belong to each of the sets $A_1, A_2, \ldots, A_n$ is called the *intersection* of the sets $A_1, A_2, \ldots, A_n$, and is denoted by $\bigcap_{i=1}^{n} A_i$.

We may write

$$\bigcap_{i=1}^{n} A_i = \{x \in A_i \text{ for all } i = 1, 2, \ldots, n\}$$

Figures 1.4(a) and (b) show the intersections $A_1 \cap A_2$ and $A_1 \cap A_2 \cap A_3$, respectively.

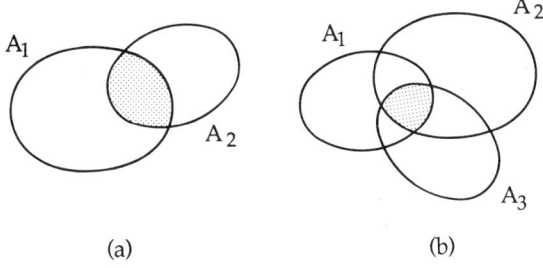

Figure 1.4 Intersections (a) $A_1 \cap A_2$ and (b) $A_1 \cap A_2 \cap A_3$.

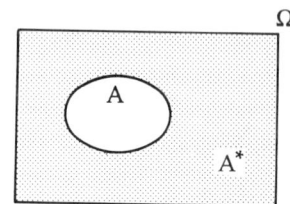

Figure 1.5 Complement A^*.

Definition 1.7. The sets A_i, $i = 1, 2, \ldots, n$ are called *mutually exclusive* if the intersection $A_i \cap A_j = 0$ for $i \neq j$. In particular, two sets A_1 and A_2 are said to be *disjoint* if $A_1 \cap A_2 = 0$.

Definition 1.8. The set that consists of all elements of the entire space, Ω, that are not elements of A is called the *complement* of A, and is denoted by A^*.

We may write

$$A^* = \{x \in \Omega; \; x \notin A\}$$

The complement A^* is given by the shaded area in the Venn diagram shown in Figure 1.5.

Definition 1.9. The aggregate of elements in A_1 that are not contained in set A_2 is called the *difference*, and is denoted by $A_1 - A_2$ (see the Venn diagram shown in Figure 1.6).

We may write

$$A_1 - A_2 = \{x \in \Omega; \; x \in A_1 \text{ and } x \notin A_2\}$$

It is noted that $A_1 - A_2 = A_1 \cap A_2^*$, and $A_1 - A_2 \neq A_2 - A_1$, in general.

Definition 1.10. The union of two differences, $A_1 - A_2$ and $A_2 - A_1$, is called the *symmetric difference*, and is denoted by $A_1 \triangle A_2$ (see the Venn diagram shown in Figure 1.7).

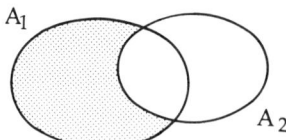

Figure 1.6 Difference $A_1 - A_2$.

ALGEBRA OF SETS AND FIELDS

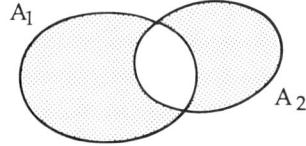

Figure 1.7 Symmetric difference $A_1 \triangle A_2$.

We may write

$$A_1 \triangle A_2 = (A_1 - A_2) \cup (A_2 - A_1)$$

Note that the symmetric difference can also be expressed as follows:

$$A_1 \triangle A_2 = (A_1 \cup A_2) - (A_1 \cap A_2)$$

By using the definitions listed above, we can write various properties of the operations on sets as follows:

1. $A \cup A^* = \Omega$

    $A \cap A^* = 0$

2. $A_1 \cup A_2 = A_1 + (A_2 - A_1)$

    $A_1 \cap A_2 = A_1 - (A_1 - A_2)$

3. $A_1 \cup (A_2 \cup A_3) = (A_1 \cup A_2) \cup A_3$

    $A_1 \cap (A_2 \cap A_3) = (A_1 \cap A_2) \cap A_3$

4. $A_1 \cup (A_2 \cap A_3) = (A_1 \cup A_2) \cap (A_1 \cup A_3)$

    $A_1 \cap (A_2 \cup A_3) = (A_1 \cap A_2) \cup (A_1 \cap A_3)$

The above relationships can be generalized as follows:

$$A \cup \left(\bigcap_{i=1}^{n} A_i \right) = \bigcap_{i=1}^{n} (A \cup A_i)$$

$$A \cap \left(\bigcup_{i=1}^{n} A_i \right) = \bigcup_{i=1}^{n} (A \cap A_i) \qquad (1.1)$$

5. $$\bigcup_{i=1}^{n} A_i = A_1 + (A_1^* \cap A_2) + (A_1^* \cap A_2^* \cap A_3) + \cdots \qquad (1.2)$$

This formula implies that the union of sets can be expressed by a sum of disjoint sets.

6.
$$\left(\bigcup_{i=1}^{n} A_i\right)^* = \bigcap_{i=1}^{n} A_i^*$$

$$\left(\bigcap_{i=1}^{n} A_i\right)^* = \bigcup_{i=1}^{n} A_i^* \tag{1.3}$$

This is called *de Morgan's law*. The proof of the first law is as follows: Let $x \in (\bigcup_{i=1}^{n} A_i)^*$. Then, $x \notin \bigcup_{i=1}^{n} A_i$, and hence $x \notin A_i$ for any i. This implies that $x \in A_i^*$ for any i, and thereby $x \in \bigcap_{i=1}^{n} A_i^*$. Inversely, let $x \in \bigcap_{i=1}^{n} A_i^*$. Then, $x \in A_i^*$ for any i, and hence $x \notin A_i$. Thus, $x \notin \bigcup_{i=1}^{n} A_i$ and thereby $x \in (\bigcup_{i=1}^{n} A_i)^*$. The second law can be proved in a similar fashion.

Example 1.5. Consider two events A_1 and A_2 and express the following events by the algebra of sets: (i) at least one event occurs, (ii) exactly one event occurs, and (iii) not more than one event occurs.

(i) At least one event is equal to the union of two; hence, we can write it as $A_1 \cup A_2$. (ii) Exactly one event is equal to the events of at least one event minus two events. Therefore, we have $(A_1 \cup A_2) - (A_1 \cap A_2) = A_1 \triangle A_2$. (iii) Not more than one event is equal to the complement of two events. Therefore, we can express it as $(A_1 \cap A_2)^*$. This can also be written as $A_1^* \cup A_2^*$ by de Morgan's law. ∎

Next, the concept of field which is necessary to define the probability is introduced in the following:

Definition 1.11. A class (collection) of sets is said to be a *field*, denoted by $\mathscr{F}$, if the following two conditions are satisfied:

(i) $A \in \mathscr{F}$, then $A^* \in \mathscr{F}$,
(ii) A_1 and $A_2 \in \mathscr{F}$, then $A_1 \cup A_2 \in \mathscr{F}$.

In consequence of the above definition of a field, a number of properties of fields follow. For example,

1. $\Omega \in \mathscr{F}$ and $\phi \in \mathscr{F}$.
2. $A_1 \cap A_2 \cap \mathscr{F}$.
3. $A_1 - A_2 \in \mathscr{F}$.

PROBABILITY 9

4. If $A_i \in \mathscr{F}$ for $i = 1, 2, \ldots, n$, then $\bigcup_{i=1}^n A_i \in \mathscr{F}$ and $\bigcap_{i=1}^n A_i \in \mathscr{F}$. Note that n is a finite number. If $n \to \infty$, then the field is called the Borel field as defined later.

Example 1.6. Consider a class of sets in which the occurrence of events ω_1 and ω_2 are involved. The set $\{\Omega, \phi, \omega_1, \omega_2\}$ is not a field, since ω_1^* and ω_2^* are not included in the set. ∎

Definition 1.12. A class of sets is said to be a *Borel field* (or σ field), denoted by $\mathscr{B}$, if the following conditions are satisfied:

(i) $\mathscr{B}$ is a field, and
(ii) $A_i \in \mathscr{B}$ for $i = 1, 2, \ldots, n$, then $\bigcup_{i=1}^\infty A_i \in \mathscr{B}$.

From the definition of the Borel field, we have the following properties:

1. If $A_i \in \mathscr{B}$, then $\bigcap_{i=1}^\infty A_i \in \mathscr{B}$.
2. A Borel field is a field, but the converse is not true.

Definition 1.13. The pair $(\Omega, \mathscr{B})$ is called a *measurable space*.

It should be noted that random events must belong to a field, and that the probability is not simply defined in a sample space; instead it is defined in a field, specifically in a Borel field. As an example, let us consider two disjoint random events, success S and failure F. Suppose we assume that these events do not belong to a field, then the probability of the union S ∪ F cannot be assigned. On the other hand, if these events belong to a field, then S ∪ F should also be a random event where the probability can be defined. In fact, the probability of S ∪ F implies the probability of occurrence of either success and/or failure and it should exist with a probability of 1.

1.3 PROBABILITY

Definition 1.14. The *probability* of occurrence of an event A can be presented by a *probability measure*, denoted by $P(A)$ or $Pr(A)$, which is a set function defined on a measurable space $(\Omega, \mathscr{B})$. $Pr(A)$ satisfies the following conditions:

(i) $0 \leq Pr(A) \leq 1$ for all $A \in \mathscr{B}$

(ii) $Pr(\Omega) = 1$

(iii) $Pr(\sum_{i=1}^{\infty} A_i) = \sum_{i=1}^{\infty} Pr(A_i)$ if A_i are mutually exclusive events. The triple $(\Omega, \mathscr{B}, P)$ is called a *probability space*.

From the definition of the probability measure, $Pr(A)$, we can derive the following various properties associated with probability:

1. $Pr(A^*) = 1 - Pr(A)$. (1.4)
2. If $A_2 \subset A_1$, then $Pr(A_2) \leq Pr(A_1)$. (1.5)

Proof. Since $A_2 \subset A_1$, we may write $A_1 = A_2 + (A_1 - A_2)$, where A_2 and $(A_1 - A_2)$ are disjoint events. Hence, $Pr(A_1) = Pr(A_2) + Pr(A_1 - A_2)$. Thus, $Pr(A_2) \leq Pr(A_1)$.

3. $Pr(A_1 \cup A_2) = Pr(A_1) + Pr(A_2) - Pr(A_1 \cap A_2)$. (1.6)

Proof. Express the sets $A_1 \cup A_2$ as a union of disjoint sets as follows:

$$A_1 \cup A_2 = A_1 \cup (A_1^* \cap A_2)$$

$$A_2 = (A_1 \cap A_2) \cup (A_1^* \cap A_2)$$

Then, from Definition 1.14 (iii), we have

$$Pr(A_1 \cup A_2) = Pr(A_1) + Pr(A_1^* \cap A_2)$$

$$Pr(A_2) = Pr(A_1 \cap A_2) + Pr(A_1^* \cap A_2)$$

The desired result can be derived from the last two formulas.

4. $Pr(\bigcup_{i=1}^{n} A_i) \leq \sum_{i=1}^{n} Pr(A_i)$. (1.7)

PROBABILITY

Proof. From Eq. (1.2) we can express $\bigcup_{i=1}^{n} A_i$ as a sum of mutually exclusive events as follows:

$$\bigcup_{i=1}^{n} A_i = A_1 + (A_1^* \cap A_2) + (A_1^* \cap A_2^* \cap A_3) + \cdots$$

Note that

$$A_1^* \cap A_2 \subset A_2$$

$$A_1^* \cap A_2^* \cap A_3 \subset A_3 \text{ etc.}$$

Then, from Eq. (1.5) we have

$$Pr(A_1^* \cap A_2) \leqslant Pr(A_2)$$

$$Pr(A_1^* \cap A_2^* \cap A_3) \leqslant Pr(A_3) \text{ etc.}$$

Thus,

$$Pr\left(\bigcup_{i=1}^{n} A_i\right) \leqslant Pr(A_1) + Pr(A_2) + Pr(A_3) + \cdots = \sum_{i=1}^{n} Pr(A_i)$$

Example 1.7. A physical device becomes inoperative when the temperature T is higher than $a°$ and/or the relative humidity H is higher than $b\%$. Then, the probability of operation can be expressed by applying de Morgan's law as follows:

$$Pr\{\text{operation}\} = Pr\{T > a \cup H > b\}^*$$

$$= Pr\{T > a\}^* \cap Pr\{H > b\}^*$$

$$= Pr\{T \leqslant a\} \cap Pr\{H \leqslant b\}$$

$$= Pr\{T \leqslant a \cap H \leqslant b\}$$

Suppose the condition of the inoperation is given that $T > a$ and $H > b$, then the probability of operation can be written as

$$Pr\{\text{operation}\} = Pr\{T > a \cap H > b\}^*$$
$$= 1 - Pr\{T > a \cap H > b\} \qquad \blacksquare$$

EXERCISES

1.1 Show that

    (a) $(A \cup B) \cap C = (A \cap C) \cup (B \cap C)$

    (b) $(A \cap B) \cup C = (A \cup C) \cap (B \cup C)$

    (c) $A \cup B = (A \cap B) \cup (A^* \cap B) \cup (A \cap B^*)$.

1.2 Let the sample space $\Omega = \{1, 2, 3, 4, 5, 6\}$ and let events $A = \{1, 2, 3\}$ and $B = \{3, 5, 6\}$. Obtain

    (a) $(A \cup B)^*$

    (b) $A^* \cup B$.

1.3 Consider the weather over 3 successive days, A_1, A_2, and A_3. Let events A_i be fine and A_i^* be rain on the ith day. Express the following events in terms of A_i and A_i^*:

    (a) At least one day is fine

    (b) Not less than two days are fine

    (c) Not more than one day is fine.

1.4 Obtain the set X that satisfies the following relationship:

$$(A^* \cap B) + (A \cup B)^* + X = \Omega.$$

CHAPTER 2

Random Variables and Their Probability Distributions

2.1 RANDOM VARIABLES

Let X be a real-valued quantity in a sequence of random experiments, and let us consider the event "X is less than or equal to x, $\{X \leq x\}$, where x is a specified real number. As an example, let X be the number of radioactive particles arriving at a counter per unit time. We are particularly interested in the event $\{X \leq x\}$, where x is a specified number. The event $\{X \leq x\}$ may not occur in every experiment; however, the relative frequency of occurrence of the event will reach a certain limiting value in a succession of experiments. Then, X is called a random variable, and the limiting value of the frequency is called the probability of occurrence of the event $\{X \leq x\}$.

As a practical example, let us consider the problem given in Example 1.3, and let X represent the number of failures that occur in the launching of a missile three times. Here, the number of failures may be 0, 1, 2, or 3. Suppose we are interested in the event of failure being one at the most (namely, either no failure or one failure in three launchings), we may write the event as $\{X \leq 1\}$. The relative frequency of occurrence of this event may change in each experiment consisting of three launchings, but the frequency will converge to a certain value if the experiment is repeated many times. As shown in Example 1.3, there is a set consisting of eight outcomes in which four outcomes (ω_1, ω_2, ω_3, and ω_5) represent the event $\{X \leq 1\}$. Therefore, the relative frequency will converge to $1/2$ after repeated trials.

RANDOM VARIABLES AND THEIR PROBABILITY DISTRIBUTIONS

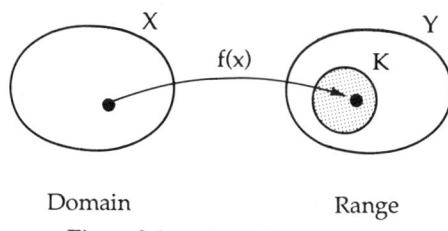

Domain   Range
Figure 2.1   Domain and range.

It should be noted that as shown in the above example, the random variable defined here is neither "random" nor is it a "variable" in the context of the usual meaning of the terms. For a precise definition of a random variable, it is necessary to apply the concept of mapping of a set.

Figure 2.1 shows the mapping of a set X into a set Y through a function f. We may write $y = f(x)$. Here y is called the image of x under f. In general, X is called the *domain* of f, while a set K that is a subset of Y consisting of all images of elements of X is called the *range* of f. That is, $f(x) \in K$ where $x \in X$. Inversely, $f^{-1}(K)$ constitutes a set of all points in the domain of f whose images are K. Then, the set $f^{-1}(K)$ is called the *inverse image* of K under f.

Now, for the definition of a random variable, let the sample space Ω be substituted for the domain X in Figure 2.1. In particular, let us consider the field $\mathscr{F}$ (which is a subset of Ω) consisting of event ω. Let the real-line R be substituted for the range Y in Figure 2.1 and consider the class of Borel set $\mathscr{B}$ in R (see Figure 2.2). Then, we can define the random variable $X(\omega)$ by letting it be a real single-valued function that maps ω in $\mathscr{F}$ into a point x in $\mathscr{B}$ in such a way that the inverse images under X or all Borel sets in R are events. The definition of the random variable can now be given as follows:

Definition 2.1. A real single-valued function $X(\omega)$ is called a *random variable* if its domain is a field $\mathscr{F}$ consisting of events in a sample space Ω, and its range is a Borel set in real numbers R such that the inverse image of $\{X(\omega) \leq x\}$ is an element of $\mathscr{F}$.

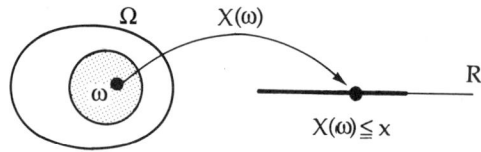

Figure 2.2   Definition of random variable $X(\omega)$.

DISTRIBUTION FUNCTIONS AND PROBABILITY DENSITIES

Random variables are usually denoted by capital letters, X, Y, etc., and their values by small letters. For example, $Pr\{X \leq x\}$ implies that the probability that a random variable X is less than or equal to a specified value "x," and $Pr\{Y = a\}$ means the probability that a random variable Y is equal to "a."

There are two types of random variables: discrete and continuous. For discrete type random variables, the outcome of an event is assigned by discrete numbers with certain probabilities. On the other hand, for continuous type random variables, the outcome of an event can be any number in the sample space. A precise definition of discrete and continuous type random variables will be given in the next section with the aid of probability distribution functions; however, the following examples may serve to clarify the difference between them.

Example 2.1. Consider the tossing of a coin in Example 1.1. For this example, the outcome of the event is either heads or tails. Let us assign to the outcome of heads the number 1 and to the outcome of tails the number 0. Then, we may say that the random variable X takes on two values $x_i = i$ ($i = 1, 0$) with the same probability $Pr\{X = x_i\} = 1/2$. A random variable of this type is called a discrete random variable. ∎

Example 2.2. Consider the case given in Example 1.4. The random variable X in this case is the deviation of the wave profile from the still water level, and X may take any value between -5 and ∞. X may be -1.02 or $+6.58$, for example. This type of random variable is called a continuous random variable. ∎

It is noted here that $Pr\{X = a\}$, where a is a number in the sample space, exists for discrete type random variables, but $Pr\{X = a\}$ is zero for continuous type random variables. In other words, the probability that the random variable X is exactly equal to a does not exist; instead, $Pr\{X < a\}$, $Pr\{X \leq a\}$, $Pr\{X > a\}$, and $Pr\{b < X < a\}$, and so on, all exist for continuous type random variables. To address this subject in detail, it is necessary to define the probability distribution function.

2.2 DISTRIBUTION FUNCTIONS AND PROBABILITY DENSITIES

Definition 2.2. The function $F(x)$ defined as

$$F(x) = Pr\{X \leq x\} \tag{2.1}$$

is called the *probability distribution function* or *cumulative distribution function* of the random variable X.

The probability distribution function satisfies the following conditions:

(i) $F(-\infty) = 0$
(ii) $F(+\infty) = 1$
(iii) $F(x + h) \geqslant F(x) \quad$ for $h > 0$ \hfill (2.2)
(iv) $F(x)$ is continuous at least from the left.

Condition (iii) implies that the probability distribution function is a monotonically increasing function. Condition (iv) is important for evaluating the probability of discrete type random variables, as will be shown later.

From the definition of the cumulative distribution function, the probability that the random variable is greater than a specified value x can be given by

$$Pr\{X > x\} = 1 - F(x) \tag{2.3}$$

By using the definition of the cumulative distribution function, we can define discrete and continuous random variables precisely.

Definition 2.3. A random variable X is said to be of the *discrete type* if every value x_i in the sample space has a positive probability $p(x_i)$, and its cumulative distribution function is given by

$$F(x) = \sum_{x \leqslant x_i} p(x_i) \tag{2.4}$$

Here, $p(x_i)$ is called the *probability mass function*.

Properties of the probability mass function of a discrete type random variable X are summarized as

(i) $Pr\{X = x_i\} = p(x_i)$
(ii) $p(x_i) \geqslant 0$ \hfill (2.5)
(iii) $\sum_{i=1}^{\infty} p(x_i) = 1.$

DISTRIBUTION FUNCTIONS AND PROBABILITY DENSITIES

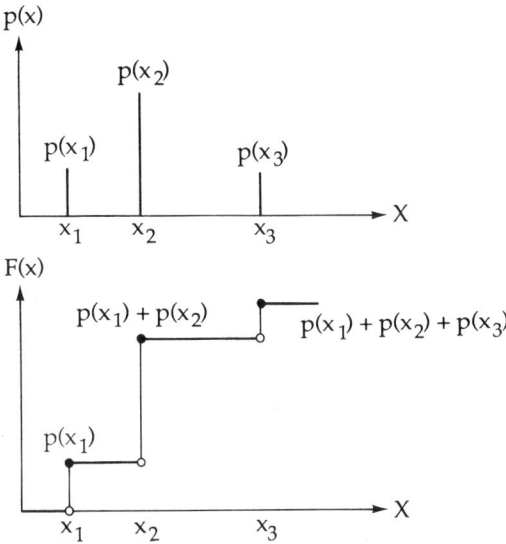

Figure 2.3 Probability mass function $p(x)$ and cumulative distribution function $F(x)$.

Figure 2.3 shows the relationship between the probability mass function and the cumulative distribution function of the discrete type random variable. As can be seen in the figure, every value x_i ($i = 1, 2, 3, \ldots$) of the random variable X has a positive probability called the probability mass function, and the cumulative distribution function is discontinuous at every x_i. The cumulative distribution function $F(x)$ is continuous from the left. For example, $Pr\{X < x_2\} = p(x_1) = F(x_2) - p(x_2)$. The cumulative distribution function increases by $p(x_2)$ as soon as X takes the value of x_2. That is, $Pr\{X \leqslant x_2\} = F(x_2) = p(x_1) + p(x_2)$.

From Eqs. (2.1) and (2.3), the following formulas are pertinent for a discrete type random variable X:

$$Pr\{X = a\} = p(a)$$

$$Pr\{X < a\} = F(a) - p(a)$$

$$Pr\{X \leqslant a\} = F(a)$$

$$Pr\{X > a\} = 1 - F(a) \qquad (2.6)$$

continues on next page

$$Pr\{X \geq a\} = 1 - F(a) + p(a)$$

$$Pr\{a < X < b\} = Pr\{X < b\} - Pr\{X \leq a\}$$
$$= \{F(b) - p(b)\} - F(a)$$

$$Pr\{a \leq X < b\} = Pr\{X < b\} - Pr\{X < a\}$$
$$= \{F(b) - p(b)\} - \{F(a) - p(a)\}$$

$$Pr\{a < X \leq b\} = Pr\{X \leq b\} - Pr\{X \leq a\}$$
$$= F(b) - F(a)$$

$$Pr\{a \leq X \leq b\} = Pr\{X \leq b\} - Pr\{X < a\}$$
$$= F(b) - \{F(a) - p(a)\}$$

Example 2.3. The cumulative distribution function of a discrete random variable X is given as shown in Figure 2.4. That is,

$$F(x) = \begin{cases} 0 & \text{for } X < -1 \\ \dfrac{x+2}{4} & \text{for } -1 \leq X < 1 \\ 1 & \text{for } 1 \leq X \end{cases}$$

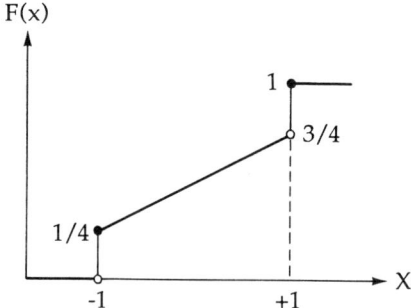

Figure 2.4 Cumulative distribution function for Example 2.3.

DISTRIBUTION FUNCTIONS AND PROBABILITY DENSITIES

We can evaluate the following probabilities from Eq. (2.6):

$$Pr\{-1/2 < X \leqslant 1/2\} = 1/4$$
$$Pr\{-1 < X < 1\} = 1/2$$
$$Pr\{-1 \leqslant X < 1\} = 3/4$$ ∎

Example 2.4. X is a discrete type random variable whose sample space is given by $(0, 1, 2, \ldots, \infty)$. Consider the function

$$p(x) = k\frac{\mu^x}{x!}, \quad \text{where } \mu > 0$$

We will determine k so that $p(x)$ is the probability mass function of X. From Property (iii) given in Eq. (2.5), $\sum_{x=0}^{\infty} p(x)$ should be equal to unity. That is,

$$\sum_{x=0}^{\infty} p(x) = k \sum_{x=0}^{\infty} \frac{\mu^x}{x!} = ke^{\mu} = 1$$

Thus, we have $k = e^{-\mu}$. ∎

Definition 2.4. A random variable X is said to be of the *continuous type* if for every value x in the sample space there exists a nonnegative function $f(x)$ such that the cumulative distribution function $F(x)$ can be expressed by

$$F(x) = \int_{-\infty}^{x} f(x)\, dx \tag{2.7}$$

Here, $f(x)$ is called the *probability density function*.

Following the definition given above, the probability density function of a continuous type random variable, $f(x)$, satisfies the following conditions:

(i) $f(x) \geqslant 0$

(ii) Integrable over every real value in sample space

(iii) $\int_{-\infty}^{\infty} f(x)\, dx = 1.$

(2.8)

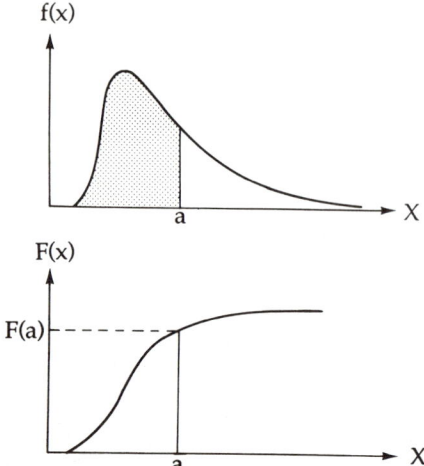

Figure 2.5 Probability density function $f(x)$ and cumulative distribution function $F(x)$.

The definition of the cumulative distribution function given in Eq. (2.7) implies that the area under the probability density function from $-\infty$ to a specified value a is equal to the probability that the random variable is less than or equal to a as shown in Figure 2.5.

It should be pointed out that for a continuous random variable X, the function $f(a)$ does not mean the probability of occurrence of the event for $X = a$. The probability of occurrence of the event for $X = a$ is zero for any value of a even though the point a belongs to the sample space. Therefore, there is no difference between $Pr\{x < a\}$ and $Pr\{x \leq a\}$ for a continuous random variable as contrasted to the case for a discrete random variable.

The following formulas are pertinent to a continuous type random variable:

$$Pr\{X = a\} = 0$$

$$Pr\{X < a\} = Pr\{X \leq a\} = F(a)$$

$$Pr\{X > a\} = Pr\{X \geq a\} = 1 - F(a) \qquad (2.9)$$

$$Pr\{a < X < b\} = Pr\{a \leq X < b\} = Pr\{a < X \leq b\}$$

$$= Pr\{a \leq X \leq b\} = F(b) - F(a)$$

Example 2.5. Consider the function $f(x) = \lambda e^{-\lambda x}$. The function $f(x)$ cannot be the probability density function unless appropriate conditions are specified. From Eq. (2.8), we have the required conditions that $\lambda > 0$ and the sample space should be $0 \leq x < \infty$. ∎

Example 2.6. The functions given by $f(x) = e^{-\lambda x^2}$ and $f(x) = \lambda x e^{-\lambda x^2}$, where $\lambda > 0$, are not probability density functions for the sample space $0 \leq X < \infty$. This is because these functions do not satisfy the third condition given in Eq. (2.8). However, $f(x) = (2\sqrt{\lambda/\pi})e^{-\lambda x^2}$ and $f(x) = 2\lambda x e^{-\lambda x^2}$ are both probability density functions for $0 \leq X < \infty$. ∎

It should be noted that the specification of sample space is extremely important for a probability function irrespective of whether it is a discrete or continuous random variable. For example, let us consider the following functions:

(a) $f(x) = e^{-\mu} \dfrac{\mu^x}{x!}$, where $\mu > 0$

(b) $f(x) = \dfrac{2x}{R} e^{-x^2/R}$, where $R > 0$

(c) $f(x) = \dfrac{1}{b-a}$, where $a > 0$ and $b > 0$

(d) $f(x) = \dfrac{1}{B(p,q)} x^{p-1}(1-x)^{q-1}$,
where $B(p,q)$ = beta function.

These functions may or may not be the probability mass (or density) function unless an appropriate sample space is specified individually. The determination of sample space for each function is left as an exercise for the reader.

2.3 QUANTILES, MEDIAN, AND MODE OF DISTRIBUTION

Definition 2.5. The value of a random variable is said to be the *quantile of order p*, denoted by x_p, if $Pr\{X < x_p\} \leq p$ and $Pr\{X \geq x_p\} \geq 1-p$, where $0 < p < 1$. In particular, x_p is called the *median* of the distribution for $p = 1/2$. For $p = 1/4$ and $3/4$, x_p is called the *lower* and *upper quartile*, respectively.

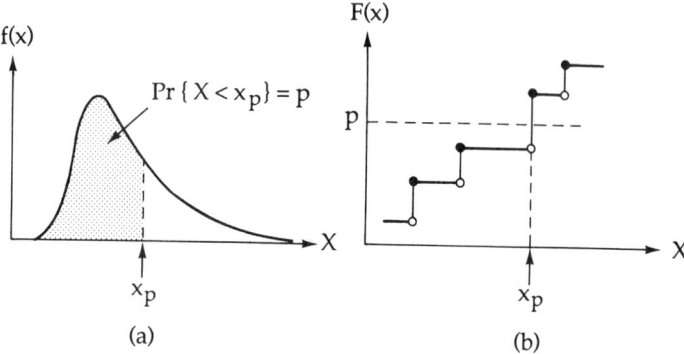

Figure 2.6 Definition of quantile of order p.

It is noted that the quantile of order p is essentially defined as the value of x_p for which $Pr\{X \leqslant x_p\} = p$. This definition is perfectly valid for a continuous type random variable as can be seen in Figure 2.6(a). However, this definition is not sufficient for a discrete type random variable, since there may be no x value that satisfies $Pr\{X \leqslant x_p\} = F(x) = p$ as shown in Figure 2.6(b). In this case, however, there exists an x value that satisfies the relationship $Pr\{X < x_p\} \leqslant p$ and $Pr\{X \geqslant x_p\} \geqslant 1 - p$ as can be seen in the figure.

The distance between the lower and upper quartiles is called the *interquartile range* and it covers 50% of the distribution (see Figure 2.7).

Example 2.7. The probability density function of a random variable X is given by

$$f(x) = \frac{1}{2(1+x)^{3/2}}, \quad -\infty < x < \infty$$

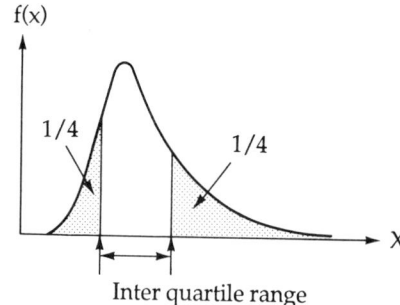

Figure 2.7 Definition of interquartile range.

QUANTILES, MEDIAN, AND MODE OF DISTRIBUTION

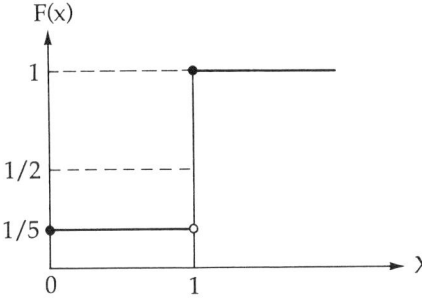

Figure 2.8 Cumulative distribution function $F(x)$ for Example 2.8.

This probability density function is symmetric with respect to 0. The upper quartile of the distribution can be obtained from

$$\int_0^{x_p} f(x)\, dx = \int_0^{x_p} \frac{1}{2(1+x)^{3/2}}\, dx = 1/4$$

which yields $x_p = 1/\sqrt{3}$. Hence the interquartile range becomes $2/\sqrt{3}$. ∎

Example 2.8. The probability mass function of a random variable X is given by $p(0) = 1/5$, $p(1) = 4/5$. Figure 2.8 shows $F(x)$. In this case the median of the distribution is $x = 1$, since it satisfies the conditions

$$Pr\{X < 1\} = Pr\{X = 0\} = 1/5 \leqslant p$$

$$Pr\{X \geqslant 1\} = Pr\{X = 1\} = 4/5 \geqslant 1 - p, \quad \text{where } p = 1/2 \quad ∎$$

Example 2.9. Let the probability mass function be $p(-1) = 1/4$, $p(0) = 1/4$, and $p(1) = 1/2$. Figure 2.9 shows $F(x)$. In this case the median is not a single value; instead, both $x = 0$ and $x = 1$ can be the median. ∎

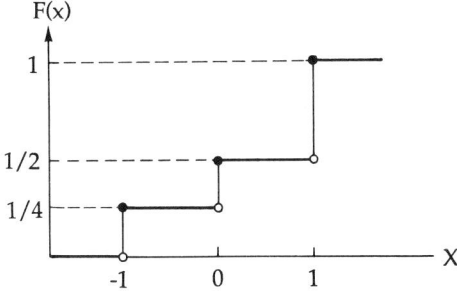

Figure 2.9 Cumulative distribution function $F(x)$ for Example 2.9.

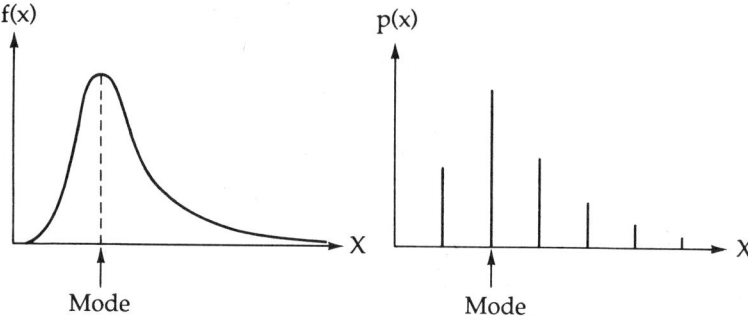

Figure 2.10 Definition of mode of distribution.

Definition 2.6. The value of a random variable X at which the probability density (or mass) function peaks, if it exists, is called the *mode* of the distribution (see Figure 2.10).

The mode can be interpreted as that value of the random variable that is most likely to occur in random experiments. This is particularly true for a random variable that has a sharply tuned density function. A typical example of the sharply tuned probability density function can be seen in the extreme value distribution that will be presented in Chapter 8.

For a continuous type random variable the mode can be obtained by letting the derivative of the probability density function be zero. For a discrete type random variable the mode can be found by taking the ratio of $p(x)$ and $p(x-1)$ and obtaining the largest x for which the ratio is greater than unity. Care must be taken in applying this method since there may be two modes in the distribution as will be shown in the following example.

Example 2.10. Let the probability mass function of a random variable X be

$$p(x) = e^{-\lambda} \frac{\lambda^x}{x!}, \quad \text{where } x = 0, 1, 2, \ldots, \infty, \lambda > 0$$

In order to determine the mode let us take the ratio of $p(x)/p(x-1)$. For this example, we have

$$\frac{p(x)}{p(x-1)} = \frac{\lambda}{x}$$

Thus, it can be seen that $p(x) > p(x - 1)$ if $\lambda > x$. Note that x is an integer but λ is not necessarily an integer. If λ is an integer, then the maximum occurs at $x = \lambda$. In this case, however, $p(x) = p(x - 1)$, and thereby $x = \lambda - 1$ is another point that yields the maximum. If λ is not an integer, then the maximum occurs at an x that is an integer smaller than (but closest to) λ. ∎

Example 2.11. Obtain the mode for the probability density function given by

$$f(x) = \frac{2x}{R} e^{-x^2/R}, \quad \text{where } 0 \leqslant x < \infty, 0 < R$$

$f(x)$ is twice differentiable. Hence from $(d/dx)f(x) = 0$, we have the mode $x = \sqrt{R/2}$. The magnitude of the probability density function for the mode is given by $f(x) = \sqrt{2/Re}$. ∎

2.4 RANDOM VECTORS AND PROBABILITY DISTRIBUTION FUNCTION

The concept of random variables and probability distribution functions discussed in the preceding three sections can be extended to more than one random variable. For brevity, let us limit our discussion to the case of two random variables. As an example, we may consider the statistical characteristics of excursion, X, and period, Y, of waves as illustrated in Figure 2.11. Here, the magnitude of X and Y varies during every cycle and they are a pair of random variables.

The individual random variables X and Y may be statistically independent or they may have some functional relationship. It is often necessary to obtain statistical characteristics of X and Y concurrently, or sometimes it is important to evaluate the probability of X given that Y has a certain

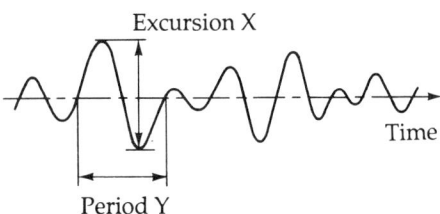

Figure 2.11 Example of two random variables X and Y.

specified value. These problems can be solved from a knowledge of the joint probability distribution of the two random variables X and Y.

In the following, the discussion will be limited to two random variables; however, the concept and formulas can easily be extended to multidimensional random variables, which may be called, in general, a random vector.

Definition 2.7. The function $F(x, y)$ defined by

$$F(x, y) = Pr\{X \leq x, Y \leq y\} \tag{2.10}$$

is called the *joint probability distribution function* of the random variables X and Y.

The joint probability distribution function satisfies the following conditions:

(i) $F(-\infty, y) = F(x, -\infty) = 0$
(ii) $F(-\infty, \infty) = 1$ (2.11)
(iii) $F(x + h, y + k) \geq F(x, y)$ for $h, k > 0$
(iv) $F(x, y)$ is continuous at least from the left.

Definition 2.8. The two-dimensional random variable (X, Y) is said to be of the *discrete type* if every values of x_i and y_j in the sample space has a positive probability $p(x_i, y_j)$, and its probability distribution function is given by

$$F(x, y) = \sum_{y_j \leq y} \sum_{x_i \leq x} p(x_i, y_j) \tag{2.12}$$

where $p(x_i, y_j)$ is called the *joint probability mass function*.

Analogous to the properties for a discrete type random variable given in Eq. (2.5), two-dimensional discrete-type random variables have the following properties:

(i) $Pr\{X = x_i, Y = y_j\} = p(x_i, y_j)$
(ii) $p(x_i, y_j) > 0$ (2.13)
(iii) $\sum_{j=1}^{\infty} \sum_{i=1}^{\infty} p(x_i, y_j) = 1.$

RANDOM VECTORS AND PROBABILITY DISTRIBUTION FUNCTION

Definition 2.9. A two-dimensional random variable (X, Y) is said to be of the *continuous type* if there exists a nonnegative function $f(x, y)$ such that the cumulative distribution function $F(x, y)$ can be expressed by

$$F(x, y) = \int_{-\infty}^{y} \int_{-\infty}^{x} f(x, y) \, dx \, dy \tag{2.14}$$

where $f(x, y)$ is called the *joint probability density function*.

Analogous to the properties for a continuous random variable given in Eq. (2.8), the joint probability density function $f(x, y)$ satisfies the following conditions:

(i) $f(x, y) \geq 0$

(ii) Integrable over all real values in the sample space

(iii) $\int_{-\infty}^{\infty} \int_{-\infty}^{\infty} f(x, y) \, dx \, dy = 1.$ \hfill (2.15)

It should be noted that evaluation of the probability associated with two random variables must be carried out with care. For example, if there is a restriction in the sample space such as $X < Y$, $X > Y + c$, and so on, the integration must be performed in an appropriate domain reflecting the restriction as demonstrated in the following examples.

Example 2.12. The joint probability density function is given by

$$f(x, y) = 4xy, \quad \text{where } 0 \leq x \leq 1, 0 \leq y \leq 1$$

Let us evaluate $\Pr\{X < 2Y\}$. The integration domain is shown in Figure 2.12.

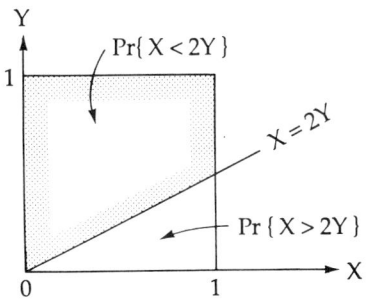

Figure 2.12 Integration domain for Example 2.12.

$$Pr\{X < 2Y\} = 1 - Pr\{X \geqslant 2Y\}$$

$$= 1 - \int_0^{1/2} \int_{2y}^1 4xy\, dx\, dy = 7/8 \qquad \blacksquare$$

Example 2.13. Let us evaluate $Pr\{X < 1, Y < 1\}$ for the joint probability density function $f(x, y) = e^{-y}$, where $0 < x < y < \infty$. First of all, it may be of interest to show that e^{-y} is a function of y only, but it is the joint probability density function of X and Y for the sample space $0 < x < y < \infty$. That is, by integrating $f(x, y)$ over the sample space, we have

$$\int_0^\infty \int_x^\infty e^{-y}\, dy\, dx = \int_0^\infty e^{-x}\, dx = 1$$

The probability $Pr\{X < 1, Y < 1\}$ can be evaluated by integration in the domain shown in Figure 2.13. We have

$$\int_0^1 \int_x^1 e^{-y}\, dy\, dx = \int_0^1 (e^{-x} - e^{-1})\, dx = 1 - 2e^{-1} \qquad \blacksquare$$

Definition 2.10. Let $F(x, y)$ be the joint distribution function of two random variables X and Y. The *marginal distribution function* of X and Y is defined as follows:

$$F_x(x) = F(x, \infty) = Pr\{X \leqslant x, Y < \infty\}$$
$$F_y(y) = F(\infty, y) = Pr\{X < \infty, Y \leqslant y\} \qquad (2.16)$$

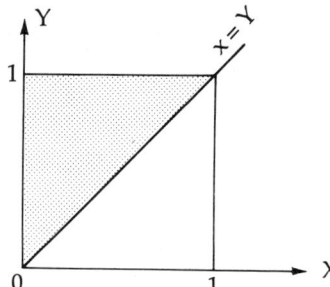

Figure 2.13 Integration domain for Example 2.13.

CONDITIONAL PROBABILITY AND DISTRIBUTION

By using the joint probability mass function or density function, the marginal distribution function of X can be written as follows:

$$F_x(x) = \begin{cases} \sum_{x_i \leq x} \left[\sum_{y_j < \infty} p(x_i, y_j) \right] = \sum_{x_i \leq x} p_x(x_i) \\ \int_{-\infty}^{x} \left[\int_{-\infty}^{\infty} f(x, y) \, dy \right] dx = \int_{-\infty}^{x} f_x(x) \, dx \end{cases} \quad (2.17)$$

The summation (or integration) inside the bracket of the above equation, namely $p_x(x_i)$ [or $f_x(x)$], is called the *marginal mass* (or *density function*), respectively, of the random variable X. Similarly, the marginal distribution function of Y can be written as

$$F_y(y) = \begin{cases} \sum_{y_j \leq y} \left[\sum_{x_i \leq x} p(x_i, y_j) \right] = \sum_{y_j \leq y} p_y(y_i) \\ \int_{-\infty}^{y} \left[\int_{-\infty}^{\infty} f(x, y) \, dx \right] dy = \int_{-\infty}^{y} f_y(y) \, dy \end{cases} \quad (2.18)$$

Example 2.14. Two random variables X and Y have the joint probability density function

$$f(x, y) = 2/a^2, \quad 0 \leq x < y < a$$

The marginal probability density function of X and Y can be evaluated as follows:

$$f_x(x) = \int_x^a f(x, y) \, dy = \int_x^a \frac{2}{a^2} \, dy = \frac{2}{a^2}(a - x), \quad \text{where } 0 \leq x < a$$

$$f_y(y) = \int_0^y f(x, y) \, dx = \int_0^y \frac{2}{a^2} \, dx = \frac{2}{a^2} y, \quad \text{where } 0 \leq y < a \quad ■$$

2.5 CONDITIONAL PROBABILITY AND DISTRIBUTION

2.5.1 *Conditional Probability*

It is often necessary to evaluate the probability that an event E_2 occurs given that (under the condition that) an event E_1 has occurred. Here, two events E_1 and E_2 may or may not have a relationship in a statistical sense.

Probability of this type is called conditional probability and is denoted by $Pr\{E_2|E_1\}$.

The concept of conditional probability plays a significant role in the solution of many practical problems in engineering and the physical sciences. As an example, let us consider the water level of a dam that may cause flooding. Suppose we assume that the annual highest water level of a dam is not influenced by past history. If we write the highest water level of the nth year as X_n, then the probability that the highest water level exceeds a specified level h in the nth year can be simply written as $Pr\{X_n > h\}$.

However, if we consider the probability that the maximum water level exceeds a specified level h for the first time in the nth year, the probability cannot be expressed without the concept of conditional probability. Note that "the first time in the nth year" implies that the water level exceeds h in the nth year given that the water did not reach the level h for the preceding $(n-1)$ years. Therefore, the probability of a first time occurrence in the nth year should be expressed in the following form:

$$Pr\left\{X_n > h \mid \bigcap_{j=1}^{n-1} X_j < h\right\}$$

Definition 2.11. Let E_1 and E_2 be two events in a sample space. The *conditional probability* of event E_2 given that event E_1 has occurred, denoted by $Pr\{E_2|E_1\}$, is defined by

$$Pr\{E_2|E_1\} = \frac{Pr\{E_1 E_2\}}{Pr\{E_1\}} \tag{2.19}$$

In order to comprehend the conditional probability given in Eq. (2.19), we first consider the probability of occurrence of event E_1, and then consider the probability of event E_2 for a reduced sample space E_1 as illustrated in Figure 2.14. We may write the probability of occurrence of event E_2 for a reduced sample space E_1 as $Pr\{E_2|E_1\}$. If we multiply the probability of occurrence of E_1 by $Pr\{E_2|E_1\}$, then it represents the simultaneous occurrence of two events $E_1 E_2$ in the sample space. That is,

$$Pr\{E_1 E_2\} = Pr\{E_1\} Pr\{E_2|E_1\} \tag{2.20}$$

Analogous to Eq. (2.19), if we consider three events E_1, E_2, and E_3, the conditional probability of E_3, given that E_1 and E_2 have occurred, is given by

$$Pr\{E_3|E_1 E_2\} = \frac{Pr\{E_1 E_2 E_3\}}{Pr\{E_1 E_2\}} \tag{2.21}$$

CONDITIONAL PROBABILITY AND DISTRIBUTION 31

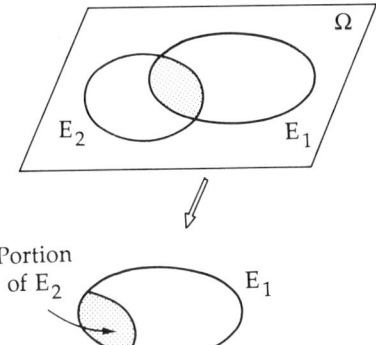

Figure 2.14 Reduced sample space for conditional probability of E_2 given E_1.

Hence, we can write the probability of the simultaneous occurrence of three events as

$$Pr\{E_1 E_2 E_3\} = Pr\{E_1 E_2\} Pr\{E_3 | E_1 E_2\}$$
$$= Pr\{E_1\} Pr\{E_2 | E_1\} Pr\{E_3 | E_1 E_2\} \quad (2.22)$$

We may generalize the joint probability given in Eq. (2.22) to that of n-random variables. That is,

$$Pr\{E_1 E_2 E_3, \ldots, E_n\} = Pr\{E_1\} Pr\{E_2 | E_1\} Pr\{E_3 | E_1 E_2\}, \ldots,$$
$$Pr\{E_n | E_1 E_2, \ldots, E_{n-1}\} \quad (2.23)$$

Example 2.15. The foundation of a nearshore building may fail when either a hurricane wind force exceeds 80 knots or a storm surge exceeds 4 m. We have the following data:

$$Pr\{\text{Wind force} > 80 \text{ kts}\} = \alpha$$
$$Pr\{\text{Storm surge} > 4 \text{ m}\} = \beta$$
$$Pr\{\text{Storm surge} > 4 \text{ m} | \text{Wind force} > 80 \text{ kts}\} = \gamma.$$

The probability of failure can be evaluated by

$$Pr\{\text{Failure}\} = Pr\{\text{Wind force} > 80 \text{ kts} \cup \text{Storm surge} > 4 \text{ m}\}$$
$$= Pr\{\text{Wind force} > 80 \text{ kts}\} + Pr\{\text{Storm surge} > 4 \text{ m}\}$$
$$- Pr\{\text{Wind force} > 80 \text{ kts} \cap \text{Storm surge} > 4 \text{ m}\}$$
$$= \alpha + \beta - \alpha\gamma \quad \blacksquare$$

Definition 2.12. The *conditional probability distribution function* of a random variable Y, given that a random variable X has the value x, is defined by

$$F(y|x) = Pr\{Y \leqslant y | X = x\} = \frac{Pr\{Y \leqslant y \text{ and } X = x\}}{Pr\{X = x\}} \quad (2.24)$$

For a continuous random variable, Eq. (2.24) can be expressed in terms of the joint probability density function as follows:

$$F(y|x) = \lim_{h \to 0} Pr\{Y \leqslant y | x \leqslant X \leqslant x + h\} = \frac{\int_{-\infty}^{y} f(x, y) \, dy}{f_x(x)}$$

$$= \frac{\int_{-\infty}^{y} f(x, y) \, dy}{\int_{-\infty}^{\infty} f(x, y) \, dy} \quad (2.25)$$

Thus, the *conditional probability density function* of a random variable Y, given that X has the value x, can be written as

$$f(y|x) = \frac{f(x, y)}{f_x(x)}$$

$$= \frac{f(x, y)}{\int_{-\infty}^{\infty} f(x, y) \, dy} \quad (2.26a)$$

Similarly, the conditional probability density function of a random variable X given that Y has the value y can be written as

$$f(x|y) = \frac{f(x, y)}{f_y(y)}$$

$$= \frac{f(x, y)}{\int_{-\infty}^{\infty} f(x, y) \, dx} \quad (2.26b)$$

CONDITIONAL PROBABILITY AND DISTRIBUTION

Example 2.16. Consider the joint probability density function of three random variables, X, Y, and Z. The conditional probability density functions $f(z|x, y)$ and $F(y, z|x)$ are given as follows:

$$f(z|x, y) = \frac{f(x, y, z)}{\int_{-\infty}^{\infty} f(x, y, z) \, dz}$$

$$f(y, z|x) = \frac{f(x, y, z)}{\int_{-\infty}^{\infty} \int_{-\infty}^{\infty} f(x, y, z) \, dy \, dz}$$ ■

2.5.2 Statistical Independence

Definition 2.13. Two events E_1 and E_2 are said to be *statistically independent* if

$$Pr\{E_1 E_2\} = Pr\{E_1\} Pr\{E_2\} \tag{2.27}$$

From the definition given in Eqs. (2.27) and (2.20), it is understood that if the events E_1 and E_2 are statistically independent, then we have

$$Pr\{E_2|E_1\} = Pr\{E_2\} \tag{2.28}$$

The concept of statistical independence can be extended to n events. That is, the events $E_1, E_2, \ldots, E_n$ are said to be statistically independent if

$$Pr\{E_1 E_2, \ldots, E_n\} = Pr\{E_1\} Pr\{E_2\}, \ldots, Pr\{E_n\} \tag{2.29}$$

It is noted that statistically independent events should not be confused with mutually exclusive events. Two events E_1 and E_2 are said to be *mutually exclusive* if the two events do not occur simultaneously. This implies that $Pr\{E_1 E_2\} = 0$.

The probability that at least one of the two events E_1 and E_2 will occur can be expressed as follows:

(i) If E_1 and E_2 are mutually exclusive events,

$$Pr\{E_1 \cup E_2\} = Pr\{E_1\} + Pr\{E_2\}$$

(ii) If E_1 and E_2 are statistically independent events,

$$Pr\{E_1 \cup E_2\} = Pr\{E_1\} + Pr\{E_2\} - Pr\{E_1\} Pr\{E_2\}$$

We now define the statistical independence of two random variables.

Definition 2.14. The random variables X and Y are said to be *statistically independent* if the joint probability distribution function $F(x, y)$ is equal to the product of the individual marginal distribution functions. That is,

$$F(x, y) = F(x)F(y) \qquad (2.30)$$

If X and Y are random variables of discrete type, the joint probability mass function can be written as the product of the marginal mass functions. That is,

$$p_{ij} = Pr\{X = x_i, Y = y_j\} = Pr\{X = x_i\}Pr\{Y = y_j\} = p_i p_j \qquad (2.31a)$$

Similarly, for continuous type random variables, we have

$$f(x, y) = f(x)f(y) \qquad (2.31b)$$

The definition of statistical independence given in Eq. (2.30) can be generalized to n-random variables. For example, the continuous type random variables $X_1, X_2, X_3, \ldots, X_n$ are statistically independent if

$$F(x_1, x_2, x_3, \ldots, x_n) = F(x_1)F(x_2)F(x_3), \ldots, F(x_n)$$

or

$$f(x_1, x_2, x_3, \ldots, x_n) = f(x_1)f(x_2)f(x_3), \ldots, f(x_n) \qquad (2.32)$$

Example 2.17. The random variables X and Y have the joint probability density function $f(x, y) = 3\exp\{-2x - y\}$, where $0 < x < y < \infty$. Then, it can be proved that X and Y are not statistically independent. This is because the marginal distribution becomes

$$f(x) = \int_x^\infty f(x, y)\, dy = 3e^{-3x}$$

and

$$f(y) = \int_0^y f(x, y)\, dx = \tfrac{3}{2}e^{-y}(1 - e^{-2y})$$

Thus, $f(x, y) \neq f(x)f(y)$, and thereby X and Y are not statistically independent. ∎

CONDITIONAL PROBABILITY AND DISTRIBUTION

Example 2.18. Consider three discrete random variables X, Y, and Z, and let $Pr\{X = 1, Y = 0, Z = 0\} = 1/4$, $Pr\{X = 0, Y = 1, Z = 0\} = 1/4$ $Pr\{X = 0, Y = 0, Z = 1\} = 1/4$, and $Pr\{X = 1, Y = 1, Z = 1\} = 1/4$. Probabilities of all other combinations of X, Y, and Z are zero. It can be proved that any two of these random variables are statistically independent, but the three random variables are not independent. As an example, let us consider the marginal joint distribution of the two random variables X and Y. The sample space for the marginal distribution of X and Y is $(0,0)$, $(1,0)$, $(0,1)$, and $(1,1)$, and the marginal probability at each point is $1/4$. Next, consider the marginal distribution of X and Y individually. The sample space is 0 and 1, and the probability at each point is $1/2$. Hence, we can prove that

$$Pr\{X = 1, Y = 1\} = Pr\{X = 1\} Pr\{Y = 1\}$$

and

$$Pr\{X = 1, Y = 1, Z = 1\} \neq Pr\{X = 1\} Pr\{Y = 1\} Pr\{Z = 1\}$$

Thus, X and Y are independent, but X, Y, and Z are not independent. Similar arguments can be made for the other pairs of random variables (Y, Z) and (Z, X). ∎

2.5.3 *Truncated Distribution*

Definition 2.15. Let S_* be a subset of the sample space of a random variable X. The conditional distribution of X within the sample space, S_*, $F(X \leq x | X \in S_*)$, is called the *truncated distribution* of X.

For a discrete type random variable X with a probability mass function $p(x_i)$, the truncated distribution is given by

$$F_*(x) = \sum_{x_i \leq x} Pr\{X = x_i | X \in S_*\} = \sum_{x_i \leq x} \frac{Pr\{X = x_i, X \in S_*\}}{Pr\{X \in S_*\}}$$

$$= \frac{\sum_{x_i \leq x \in S_*} p(x_i)}{\sum_{x_i \in S_*} p(x_i)} \tag{2.33}$$

The probability mass function of a truncated distribution X, denoted by $p_*(x_i)$, is called the *truncated probability mass function*, and it is given by

$$p_*(x_i) = Pr\{X = x_i | X \in S_*\} = \frac{p(x_i)}{\sum_{x_i \in S_*} p(x_i)}, \quad \text{where } x_i \in S_* \quad (2.34)$$

For a continuous type random variable X with a probability density function $f(x)$, the truncated distribution becomes

$$F_*(x) = Pr\{X \leq x | X \in S_*\} = \frac{Pr\{X \leq x, X \in S_*\}}{Pr\{X \in S_*\}}$$

$$= \frac{\int_{(-\infty, x) \cap S_*} f(x)\, dx}{\int_{S_*} f(x)\, dx} \quad (2.35)$$

The probability density function derived from Eq. (2.35), denoted by $f_*(x)$, is called the *truncated probability density function* and it is given by

$$f_*(x) = \frac{f(x)}{\int_{S_*} f(x)\, dx}, \quad \text{where } x \in S_* \quad (2.36)$$

Figure 2.15 shows a pictorial sketch of the truncated probability density function $f_*(x)$ whose sample space is (x_*, ∞). That is, the original probability density function $f(x)$ is truncated at $x = x_*$. Hence, $f(x)$ is no

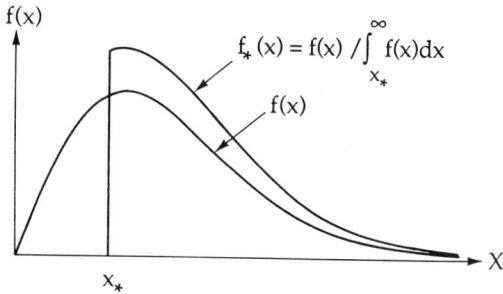

Figure 2.15 Pictorial sketch of truncated probability density function $f_*(x)$.

CONDITIONAL PROBABILITY AND DISTRIBUTION

longer the probability density function in the truncated sample space, since the integrated area under $f(x)$ is less than unity. Therefore, $f(x)$ should be normalized in the truncated sample space so that it satisfies the condition required for a probability density function.

Example 2.19. The probability density function of a random variable X is given by

$$f(x) = \frac{2x}{R} e^{-x^2/R}, \qquad \text{where } R > 0, 0 < x < \infty$$

Suppose we are only interested in the distribution of X greater than a specific value a, the probability density function to be considered is given by

$$f_*(x) = \frac{f(x)}{\int_a^\infty f(x)\,dx}$$

$$= \frac{(2x/R)e^{-x^2/R}}{e^{-a^2/R}} = \frac{2x}{R} e^{-(1/R)(x^2 - a^2)}, \qquad a \leqslant x < \infty \qquad \blacksquare$$

Example 2.20. Conditional probability plays a significant role in reliability analysis of a system and it has been increasingly used in engineering design. Let the random variable T represent the length of time in which a certain system has been operating without a breakdown. By letting $F(t)$ be the cumulative distribution function of T, we can express the probability that no failure takes place during the time interval $(0, t)$ as

$$R(t) = Pr\{T > t\} = 1 - F(t)$$

The function $R(t)$ is called the *reliability function*.

Next, let us consider that failure will occur in the time interval $(t, t + \Delta t)$ given that a failure has not occurred prior to t. This probability can be written by

$$Pr\{t < T < t + \Delta t | T > t\} = \frac{R(t) - R(t + \Delta t)}{R(t)}$$

Then, the probability of failure per unit time is given by

$$h(t) = \lim_{\Delta t \to 0} \frac{1}{\Delta t} \frac{R(t) - R(t + \Delta t)}{R(t)} = \lim_{\Delta t \to 0} \frac{1}{R(t)} \frac{R(t) - R(t + \Delta t)}{\Delta t}$$

$$= \frac{1}{R(t)} \left\{ -\frac{d}{dt} R(t) \right\} = \frac{f(t)}{R(t)} = \frac{f(t)}{1 - F(t)}$$

where $f(t)$ = probability density function of T.

The function $h(t)$ is called the *hazard function* or *hazard rate*. Note that $h(t)$ is essentially equal to the truncated probability density function $f_*(x)$ shown in Figure 2.15. It is also noted that the function $1/\{1 - F(t)\}$ is called the *return period* in extreme value statistics, which will be discussed in detail in Chapter 8. ∎

Example 2.21. Let the random variable T represent the time interval between successive breakdown of a system and let its probability density function be given by

$$f(t) = \lambda \exp\{-\lambda t\}$$

The hazard function then becomes

$$h(t) = \frac{f(t)}{1 - F(t)} = \frac{\lambda \exp\{-\lambda t\}}{\exp\{-t\}} = \lambda$$

The constant hazard rate implies that the failure process has no memory and hence the system is supposed not to be subject to wear. Failure, however, may occur by a sudden change of environment, extreme conditions, and so on. The hazard function is constant only if the time interval between successive breakdown of a system obeys the exponential probability law. If $f(t)$ does not follow the exponential distribution, the hazard rate is then a function of time. ∎

2.5.4 Bayes Formula

Let $A_1, A_2, \ldots, A_n$ be mutually exclusive component events, the union of which constitutes a certain larger scope event, and let B be another event which is contained in the union $A_1 \cup A_2 \cup \cdots \cup A_n$. Since A_i are mutually exclusive events, we have

$$Pr\{B\} = \sum_{i=1}^{n} Pr\{A_i B\} \qquad (2.37)$$

By applying Eq. (2.20), $Pr\{A_i B\}$ may be expressed in terms of conditional probability $Pr\{B|A_i\}$, and thereby we have

$$Pr\{B\} = \sum_{i=1}^{n} Pr\{A_i\} Pr\{B|A_i\} \qquad (2.38)$$

Next, consider the conditional probability $Pr\{A_i|B\}$ for any A_i, given B. It can be written as

$$Pr\{A_i|B\} = \frac{Pr\{A_i B\}}{Pr\{B\}} = \frac{Pr\{A_i\} Pr\{B|A_i\}}{Pr\{B\}} \qquad (2.39)$$

Hence, from Eqs. (2.38) and (2.39), the following formula can be derived:

$$Pr\{A_i|B\} = \frac{Pr\{A_i\} Pr\{B|A_i\}}{\sum_{i=1}^{n} Pr\{A_i\} Pr\{B|A_i\}} \qquad (2.40)$$

Equation (2.40) is called the *Bayes formula*. The simplest case is for a single event A instead of events A_i for which the Bayes formula becomes

$$Pr\{A|B\} = \frac{Pr\{A\} Pr\{B|A\}}{Pr\{B\}} \qquad (2.41)$$

The implication of the Bayes formula is as follows: $Pr\{A_i|B\}$ can be interpreted as the probability of the event A_i based on the information of event B, namely after B has occurred. Hence, it is called *posterior probability*. On the other hand, $Pr\{A_i\}$ is the probability of the event A_i before the information about B is available, and therefore it is called *priori probability*.

Example 2.22. Products are manufactured by two different processes, I and II. The number of products through Process I is 85% of that by Process II; however, the probability of defective products is 0.03 for Process I versus 0.04 for Process II. (1) What is the probability that an arbitrarily chosen product is defective? (2) When we find that an arbitrarily chosen product is defective what is the probability that the product was manufactured through Process II?

Let events A_1 and A_2 be products manufactured through Processes I and II, respectively, and let event B be the product being defective. Since the ratio of number of products through Process I to those of Process II is 0.85, we have $Pr\{A_1\} = 0.459$ and $Pr\{A_2\} = 0.541$. We also have $Pr\{B|A_1\} = 0.03$ and $Pr\{B|A_2\} = 0.04$. Hence,

(i) $Pr\{\text{Product is defective}\} = Pr\{B\} = \sum_{i=1}^{n} Pr\{A_i\} Pr\{B|A_i\} = 0.035$

(ii) $Pr\{\text{Defective product was manufactured through Process II}\}$

$$= Pr\{A_2|B\} = \frac{Pr\{A_2\} Pr\{B|A_2\}}{Pr\{B\}} = 0.618.$$ ∎

EXERCISES

2.1 Determine the sample space for each of the following functions to be a probability mass (or density) function.

(a) $f(x) = e^{-\mu} \dfrac{\mu^x}{x!}$, where $\mu > 0$

(b) $f(x) = \dfrac{2x}{R} e^{-x^2/R}$, where $R > 0$

(c) $f(x) = \dfrac{1}{b-a}$, where $a > 0$ and $b > 0$

(d) $f(x) = \dfrac{1}{B(p,q)} x^{p-1}(1-x)^{q-1}$,
where $B(p, q) = $ beta function.

2.2 The joint probability density function of two random variables X and Y is given by

$$f(x, y) = 2e^{-(x+y)}, \qquad 0 \leq x < y < \infty$$

Show that the random variables X and Y are not statistically independent.

2.3 The quality of a product is classified into three categories A, B, and C. If we select three products at random, what is the probability that at least two are of the same category?

EXERCISES

2.4 Figure 2.16 shows an assembly diagram of a system consisting of four components. The system fails if the path from A to B is broken. Let us assume that failure of each component will occur independently and let the probability of occurrence of failure be α for each component. What is the probability that the assembly of the system is completed without any failure?

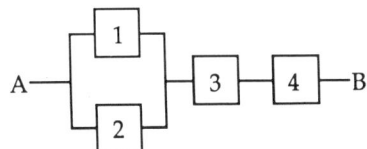

Figure 2.16 Assembly diagram of a system in Exercise 2.4.

2.5 Let us assume that the payload of a ship is proportional to the square of her displacement. We have a ship of displacement W with payload A. We will build two more ships the sum of whose displacements is W and the displacement of the smaller ship is not less than $W/4$. What is the mean value of the sum of the payloads of these ships in comparison with the ship of displacement W with payload A?

2.6 Buffon's needle problem. Parallel straight lines are drawn on the table at a distance, d, apart. A needle of length l, where $l < d$, is dropped on the table at random. Show that the probability of the needle crossing one of the lines is $2l/\pi d$. [Hint] The condition required for crossing is given by $x \leqslant (1/2)\cos\theta$, where $0 \leqslant \theta \leqslant 2\pi$, as illustrated in Figure 2.17.

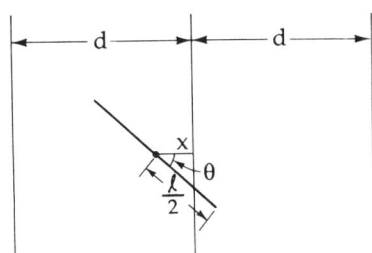

Figure 2.17 Buffon's needle problem in Exercise 2.6.

2.7 Show that the mode of the following probability mass function is an integer smaller than but closest to $(n+1)p$:

$$p(x) = \binom{n}{x} p^x q^{n-x},$$

where $x = 0, 1, 2, \ldots, n$, $0 < p < 1$, and $q = 1 - p$

2.8 The hazard function $h(t)$ is defined in Example 2.20. Assume that $h(t) = 0$ at $t = t_0$. Let T be a random variable representing the length of time a system has been operating without breakdown. Prove that the cumulative distribution function of T can be given by

$$F(t) = 1 - \{1 - F(t_0)\} \exp\left\{-\int_{t_0}^{t} h(t)\, dt\right\}$$

CHAPTER 3

Moments of Random Variables

Various statistical properties of a random variable can be obtained by evaluating moments and functions of moments of the probability distribution. For example, whether or not the probability density function has a symmetric shape can be examined by evaluating the parameter called skewness of the distribution. The moments not only provide information on properties of a random variable, but also play a significant role in the derivation of the probability distribution of quantities associated with a random process as will be shown later in Section 11.1, for example. The moments and functions of moments of a random variable and random vector, as well as the conditional moments, will be discussed in sequence.

3.1 EXPECTED VALUE AND MOMENTS

3.1.1 Mean and Variance of Random Variables

Definition 3.1. Let X be a random variable with the probability mass function $p(x_i)$ or the probability density function $f(x)$. The *expected value* (or *expection*) of a function $u(x)$ is defined as follows:

$$E[u(x)] = \begin{cases} \sum_{i \in S} u(x_i) p(x_i) & \text{for a discrete-type random variable} \\ \int_S u(x) f(x) \, dx & \text{for a continuous-type random variable} \end{cases}$$

(3.1)

where S stands for the sample space.

The expected value does not always exist (see Example 3.5, for instance). The condition required for the existence of the expected value is given by

$$\sum_{k \in S} |u(x_k)| p(x_k) < \infty \quad \text{or} \quad \int_S |u(x)| f(x) \, dx < \infty \quad (3.2)$$

Let us consider a particular case of the expected value, $u(x) = x^k$. Then, $E[x^k]$ is called the kth moment of the random variable X.

Definition 3.2. The expected value of the function $u(x) = x^k$ is called the *k*th *moment* of the random variable X. In particular, for $k = 1$, $E[x]$ is called the *mean* or *average value* of the random variable X, and is usually denoted by μ or simply $E[x]$.

Properties of the mean value of the random variable X are as follows:

(i) $E[a] = a$, where $a = $ constant
(ii) $E[ax] = aE[x]$
(iii) $E[x + a] = E[x] + a$ (3.3)
(iv) $E[x + y] = E[x] + E[y]$
(v) $E[xy] = E[x]E[y]$, where X and Y are statistically independent random variables.

Definition 3.3. The expected value of the function $u(x) = (x - \mu)^k$, where $\mu = $ mean value, is called the *k*th *central moment* of the random variable X. In particular, for $k = 2$, $E[(x - \mu)^2]$ is called the *variance* of the random variable X, and is usually denoted by σ^2 or $\text{Var}[x]$.

Let us denote the kth moment and the kth central moment of the random variable X by m_k and μ_k, respectively. The relationship between them is summarized below.

Moments:

$$m_1 = E[x] = \text{mean} = \mu$$
$$m_2 = E[x^2] = \mu_2 + \mu^2$$
$$m_3 = E[x^3] = \mu_3 + 3\mu\mu_2 + \mu^3 \quad (3.4)$$
$$m_4 = E[x^4] = \mu_4 + 4\mu\mu_3 + 6\mu^2\mu_2 + \mu^4$$

EXPECTED VALUE AND MOMENTS

Central moments:

$$\mu_1 = E[(x - \mu)] = 0$$

$$\mu_2 = E[(x - \mu)^2] = \text{variance} = m_2 - m_1^2$$

$$\mu_3 = E[(x - \mu)^3] = m_3 - 3m_1 m_2 + 2m_1^3 \quad (3.5)$$

$$\mu_4 = E[(x - \mu)^4] = m_4 - 4m_1 m_3 + 6m_1^2 m_2 - 3m_1^4$$

Definition 3.4. The square root of the variance is called the *standard deviation* of a random variable.

Properties of the variance of a random variable X is as follows:

(i) $\text{Var}[x] = E[x^2] - (E[x])^2$
   Proof. $\text{Var}[x] = E[(x - \mu)^2] = E[x^2] - 2\mu E[x] + \mu^2$
   $= E[x^2] - (E[x])^2$

(ii) $\text{Var}[a] = 0$, where $a = $ constant

(iii) $\text{Var}[ax] = a^2 \text{Var}[x]$ \quad (3.6)
   Proof. $\text{Var}[ax] = E[(ax)^2] - (E[ax])^2$
   $= a^2 E[x^2] - a^2 (E[x])^2 = a^2 \text{Var}[x]$

(iv) $\text{Var}[x + a] = \text{Var}[x]$

(v) $\text{Var}[ax + by] = a^2 \text{Var}[x] + b^2 \text{Var}[y]$, where X and Y are statistically independent random variables.

The mean and variance play a significant role in probability distribution. In particular, the variance is a measure of dispersion of the random variable around its mean. The smaller the variance, the more sharply concentrated is the probability density function around its mean value [see Figure 3.1(*a*)]. In case the variance is zero, the distribution has a point distribution at the mean value as shown in Figure 3.1(*b*).

The mean and variance of the probability density function are analogous to the center of gravity and moment of inertia in mechanics. Let us consider the center of gravity of a body that has a mass distribution $m(x)$ as shown in Figure 3.2. The location of the center of gravity from the origin, μ, is

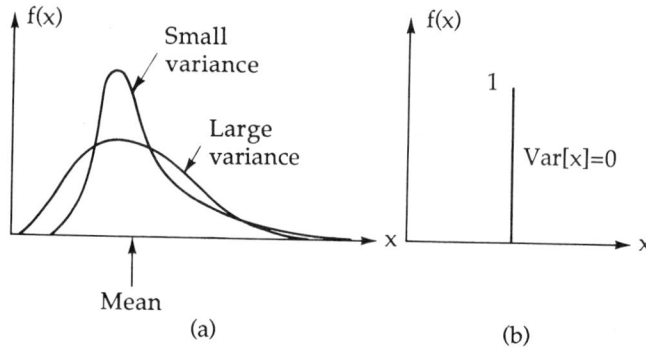

Figure 3.1 Probability density functions for (*a*) large and small variances, and (*b*) variance = 0.

simply given by

$$\mu = \frac{\int_0^L x m(x)\, dx}{\int_0^L m(x)\, dx}$$

where the denominator is the total mass of the body. This same formulation can also be applied to the mean of the probability distribution. However, in the case of the probability distribution, the denominator (total mass) is equal to 1 as was defined in Chapter 2. Thus, the mean value implies the location of the center of gravity of the probability density function.

On the other hand, the moment of inertia of a body is given by

$$\int_0^L (x - \mu)^2 m(x)\, dx$$

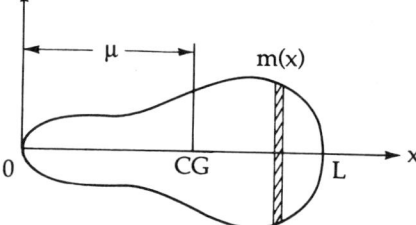

Figure 3.2 Location of center of gravity of a body with mass distribution $m(x)$.

This is exactly the same formula as for the variance defined in the probability distribution.

Example 3.1. Evaluate the mean and variance of the following distribution called the Poisson distribution:

$$p(x) = e^{-\lambda}\frac{\lambda^x}{x!}, \qquad x = 0, 1, 2, \ldots$$

$$E[x] = \sum_{x=0}^{\infty} xe^{-\lambda}\frac{\lambda^x}{x!} = e^{-\lambda} \sum_{x=1}^{\infty} \frac{\lambda^x}{(x-1)!}$$

$$= \lambda e^{-\lambda} \sum_{x=1}^{\infty} \frac{\lambda^{x-1}}{(x-1)!} = \lambda e^{-\lambda} \cdot e^{\lambda} = \lambda$$

$$E[x^2] = \sum_{x=0}^{\infty} x^2 e^{-\lambda}\frac{\lambda^x}{x!} = e^{-\lambda} \sum_{x=1}^{\infty} \frac{(x-1)\lambda^x + \lambda^x}{(x-1)!}$$

$$= e^{-\lambda}\left\{ \lambda^2 \sum_{x=2}^{\infty} \frac{\lambda^{x-2}}{(x-2)!} + \lambda \sum_{x=1}^{\infty} \frac{\lambda^{x-1}}{(x-1)!} \right\} = \lambda^2 + \lambda$$

Hence, $\text{Var}[x] = \lambda$. ∎

Example 3.2. Evaluate the mean and variance of the random variable X whose probability density function is given by

$$f(x) = \begin{cases} x & \text{for } 0 \leqslant x \leqslant 1 \\ (2-x) & \text{for } 1 \leqslant x \leqslant 2 \end{cases}$$

$$E[x] = \int_0^1 x^2\, dx + \int_1^2 x(2-x)\, dx = 1$$

$$E[x^2] = \int_0^1 x^3\, dx + \int_1^2 x^2(2-x)\, dx = 7/6$$

Thus, $\text{Var}[x] = 1/6$. ∎

Example 3.3. Let us find the mean and variance of the following probability density function called Rayleigh distribution:

$$f(x) = \frac{2x}{R} e^{-x^2/R}, \qquad 0 < x < \infty$$

$$E[x] = \int_0^\infty \frac{2x^2}{R} e^{-(x^2/R)}\, dx = \frac{\sqrt{\pi R}}{2}$$

$$\mathrm{Var}[x] = E[x^2] - (E[x])^2$$

$$= \int_0^\infty \frac{2x^3}{R} e^{-x^2/R}\, dx - \left(\frac{\sqrt{\pi R}}{2}\right)^2$$

$$= \left(1 - \frac{\pi}{4}\right) R \qquad \blacksquare$$

Example 3.4. Evaluate the mean of the following probability distribution called binomial distribution:

$$p(x) = \binom{n}{x} p^x (1-p)^{n-x}, \qquad x = 0, 1, 2, \ldots n$$

$$E[x] = \sum_{x=0}^{n} x \binom{n}{x} p^x (1-p)^{n-x}$$

$$= \sum_{x=0}^{n} n \binom{n-1}{x-1} p^x (1-p)^{n-x}$$

$$= np \sum_{x=0}^{n} \binom{n-1}{x-1} p^{x-1} (1-p)^{(n-1)-(x-1)}$$

$$= np\{(1-p) + p\}^{n-1} = np \qquad \blacksquare$$

As can be seen in Example 3.4, the evaluation of mean (or the variance) is not always as simple as one may expect. However, the difficulty of evaluating the mean and variance can be significantly extenuated by using either the moment generating function or the characteristic function, which will be discussed in Sections 4.1 and 4.2, respectively.

EXPECTED VALUE AND MOMENTS

Example 3.5. Evaluate the mean of the following probability density function:

$$f(x) = \frac{1}{\pi} \frac{1}{1 + (x - \mu)^2}, \quad -\infty < x < \infty$$

This probability distribution is called the Cauchy distribution. The probability density function is symmetric with respect to μ, as shown in Figure 3.3; hence, it appears that the mean value of the distribution is μ. However, the mean value does not exist mathematically as shown below:

$$E[x] = \int_{-\infty}^{\infty} x f(x) \, dx$$

$$= \frac{1}{\pi} \left[\int_{-\infty}^{\infty} \frac{\mu}{1 + (x - \mu)^2} \, dx + \int_{-\infty}^{\infty} \frac{x - \mu}{1 + (x - \mu)^2} \, dx \right].$$

By letting $x - \mu = y$, we have,

$$E[x] = \frac{\mu}{\pi} \int_{-\infty}^{\infty} \frac{1}{1 + y^2} \, dy + \frac{1}{\pi} \int_{-\infty}^{\infty} \frac{y}{1 + y^2} \, dy$$

$$= \mu + \frac{1}{2\pi} \lim_{\substack{a \to -\infty \\ b \to \infty}} \left[\ln(1 + y^2) \right]_a^b$$

$$= \mu + \frac{1}{2\pi} \lim_{\substack{a \to -\infty \\ b \to \infty}} \left[\ln(1 + b^2) - \ln(1 + a^2) \right]$$

Since the right side of the equation is indeterminate, $E[x]$ does not exist. ∎

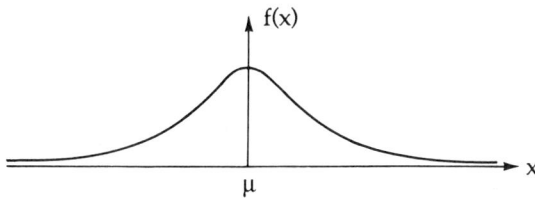

Figure 3.3 Cauchy probability distribution.

Definition 3.5. A random variable X that has a zero mean, $E[x] = 0$, and unit variance, $\text{Var}[x] = 1$, is called a *standardized random variable*.

Let the mean and variance of the random variable X be μ and σ^2, respectively. Then a random variable Z defined as

$$Z = \frac{X - \mu}{\sigma} \tag{3.7}$$

is a standardized random variable of X. It can easily be proved that Z has zero mean and unit variance. That is, from the properties of the mean and variance given in Eqs. (3.3) and (3.6), respectively, we have

$$E[z] = \frac{1}{\sigma} E[x] - \frac{\mu}{\sigma} = 0$$

and

$$\text{Var}[z] = \frac{1}{\sigma^2} \text{Var}[x] = 1$$

Example 3.6. Let us standardize the following probability density function called the normal probability distribution with mean μ and variance σ^2:

$$f(x) = \frac{1}{\sqrt{2\pi}\,\sigma} \exp\left[-\frac{1}{2}\left(\frac{x-\mu}{\sigma}\right)^2\right], \quad -\infty < x < \infty$$

Let $Z = (X - \mu)/\sigma$, then

$$Pr\{Z > z\} = Pr\left\{\frac{X - \mu}{\sigma} > z\right\} = Pr\{X > \mu + \sigma z\}$$

$$= \int_{\mu + \sigma z}^{\infty} \frac{1}{\sqrt{2\pi}\,\sigma} \exp\left[-\frac{1}{2}\left(\frac{x-\mu}{\sigma}\right)^2\right] dx = \int_{z}^{\infty} \frac{1}{\sqrt{2\pi}} e^{-z^2/2}\, dz$$

Thus, we have $f(z) = (1/\sqrt{2\pi})\exp\{-z^2/2\}$. This is called a *standardized normal probability density function*. ∎

EXPECTED VALUE AND MOMENTS

3.1.2 Coefficient of Variation, Skewness, and Kurtosis

Sometimes it is extremely important to evaluate various parameters of the distribution such as those representing the measure of asymmetry and peakedness. These parameters are summarized in the following definitions:

Definition 3.6. Let X be a random variable that has a mean μ and variance σ^2. The ratio of σ to μ is called the *coefficient of variation* denoted by v. The dimensionless ratio of the central third moment, μ_3, to σ^3 is called the *skewness* of the distribution, denoted by γ_1, and the ratio of the central fourth moment μ_4 to σ^4 is called the *kurtosis*, denoted by γ_2. That is,

$$\text{Coefficient of variation:} \quad v = \sigma/\mu$$
$$\text{Skewness:} \quad \gamma_1 = \mu_3/\sigma^3 \quad (3.8)$$
$$\text{Kurtosis:} \quad \gamma_2 = \mu_4/\sigma^4.$$

The skewness γ_1 is a measure of asymmetry of the distribution. If $\gamma_1 = 0$, then the shape of the probability density function is symmetric with respect to its mean value. γ_1 may be positive or negative depending on the shape of the distribution as illustrated in Figure 3.4. If $\gamma_1 > 0$, then the mode of the distribution is located at a value smaller than the mean, while if $\gamma_1 < 0$ then the mode is located at a value greater than the mean.

The kurtosis represents a degree of peakedness of the distribution using the normal distribution as a measure of reference. The kurtosis γ_2 for the normal distribution is equal to 3 (see Example 3.7). If $\gamma_2 < 3$, the distribution is said to be *platykurtic* (mild peak), and if $\gamma_2 > 3$, it is said to be *leptokurtic* (sharp peak).

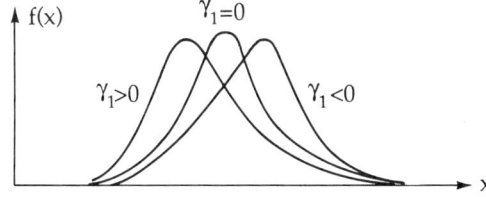

Figure 3.4 Skewness γ_1 and shape of probability density functions.

Example 3.7. The random variable X has a normal distribution shown in Example 3.6 with mean $\mu = 0$ and variance σ^2. Since the mean of the distribution is zero, the fourth central moment μ_4 is given by (see Exercise 3.4)

$$\mu_4 = \frac{(\sigma^2)^2}{2^2}\frac{4!}{2!} = 3\sigma^4$$

Hence, from Eq. (3.8), the kurtosis γ_2 is 3. ∎

Example 3.8. The random variable X has a Rayleigh distribution given by

$$f(x) = \frac{2x}{R}e^{-x^2/R}, \qquad R > 0, 0 \leqslant x < \infty$$

Evaluate the skewness and kurtosis of the distribution. The mean and variance of the distribution are obtained in Example 3.3 as $\mu = \sqrt{\pi R}/2$, and $\sigma^2 = [1 - (\pi/4)]R$, respectively. Then, we can evaluate the third and fourth central moments as

$$\mu_3 = \int_0^\infty (x - \mu)^3 f(x)\, dx = \frac{\sqrt{\pi}}{4}(\pi - 3)R\sqrt{R}$$

and

$$\mu_4 = \int_0^\infty (x - \mu)^4 f(x)\, dx = \left(2 - \tfrac{3}{16}\pi^2\right)R^2$$

Thus, we have

$$\text{Skewness} = \mu_3/\sigma^3 = \frac{\sqrt{\pi}}{4}(\pi - 3)\bigg/\left(1 - \frac{\pi}{4}\right)^{1.5} = 0.631$$

and

$$\text{Kurtosis} = \mu_4/\sigma^4 = \left(2 - \tfrac{3}{16}\pi^2\right)\bigg/\left(1 - \frac{\pi}{4}\right)^2 = 3.245 \qquad \blacksquare$$

3.2 MOMENTS OF RANDOM VECTORS

3.2.1 Moments, Covariance, and Correlation Coefficient

Definition 3.7. Let (X, Y) be a pair of random variable with the joint probability mass function $p(x_i, y_j)$ or joint probability density function $f(x, y)$. The *expected value* of an arbitrary function $u(x, y)$ is defined as follows:

$$E[u(x, y)] = \begin{cases} \sum_{i,j \in S} u(x_i, y_j) p(x_i, y_j) & \text{for a discrete random variable} \\ \int_S \int u(x, y) f(x, y) \, dx \, dy & \text{for a continuous random variable} \end{cases}$$

(3.9)

where S stands for sample space.

The above definition can be generalized to a random vector consisting of a finite number of elements. In particular, if the function $u(x, y)$ is given as $x^k y^l$ (or $x_i^k y_j^l$ for discrete type random variables) where $k \geq 0$ and $l \geq 0$, then the moment of two random variables is defined as follows:

Definition 3.8. The expected value of the function $u(x, y) = x^k y^l$, where $k \geq 0$ and $l \geq 0$, is called the *moment of order* $k + l$ of the random variables X and Y and is denoted by m_{kl}. That is,

$$m_{kl} = \begin{cases} \sum_{i,j \in S} x_i^k y_j^l p(x_i, y_j) & \text{for a discrete random variable} \\ \int_S \int x^k y^l f(x, y) \, dx \, dy & \text{for a continuous random variable} \end{cases}$$

(3.10)

In particular, $k = 1$, $l = 0$ in Eq. (3.10) yields the mean value of the random variable X, while $k = 0$, $l = 1$ yields the mean value of the random variable Y. By using the notation given in Eq. (3.10), we may write these mean values as

$$m_{10} = E[x] = \mu_x$$
$$m_{01} = E[y] = \mu_y$$

(3.11)

Next, let us consider the central moment. Similar to the definition of the central moment for one random variable given in Definition 3.3, we may define the central moment of two random variables X and Y as follows:

Definition 3.9. The *central moment* of two random variables denoted by μ_{kl}, is defined as

$$\mu_{kl} = E\left[(x - \mu_x)^k (y - \mu_y)^l\right] \quad (3.12)$$

where μ_x and μ_y are the mean values of the random variables X and Y, respectively.

By using the notation given in Eq. (3.12), we have,

$$\mu_{10} = E\left[(x - \mu_x)\right] = 0$$
$$\mu_{01} = E\left[(y - \mu_y)\right] = 0$$
$$\mu_{20} = E\left[(x - \mu_x)^2\right] = \text{Var}[x] = \sigma_x^2$$
$$\mu_{02} = E\left[(y - \mu_y)^2\right] = \text{Var}[y] = \sigma_y^2$$

(3.13)

In particular, the central moment μ_{11} is called the *covariance* of two random variables X and Y and is denoted by $\text{Cov}[x, y]$ or σ_{xy}. That is,

$$\text{Cov}[x, y] = \mu_{11} = E\left[(x - \mu_x)(y - \mu_y)\right]$$
$$= E[xy] - E[x]E[y] = m_{11} - m_{10}m_{01} \quad (3.14)$$

The covariance $\text{Cov}[x, y]$ defined above is an important parameter characterizing the distribution of two random variables. For convenience, we may consider the continuous type random variables in the following discussion. First, let us assume that X and Y are statistically independent. Then, from Eq. (2.31b), we have

$$E[xy] = \int_{-\infty}^{\infty}\int_{-\infty}^{\infty} xy f(x, y)\, dx\, dy = \int_{-\infty}^{\infty} x f(x)\, dx \int_{-\infty}^{\infty} y f(y)\, dy$$
$$= E[x]E[y] \quad (3.15)$$

Thus, from Eq. (3.14), $\text{Cov}[x, y]$ becomes zero and it is said that the two

MOMENTS OF RANDOM VECTORS 55

random variables are *uncorrelated*. In other words, if two random variables are independent, then they are uncorrelated. However, the reverse is not always true as shown in the following example.

Example 3.9. Let the joint probability density function of random variables X and Y be

$$f(x, y) = \tfrac{1}{4}(1 - x^{2n} + y^{2n}), \quad -1 \leqslant x \leqslant 1, \; -1 \leqslant y \leqslant 1,$$

$$n = \text{positive integer}$$

We can prove that X and Y are statistically uncorrelated but not statistically independent. That is, the covariance of X and Y can be evaluated as

$$\text{Cov}[x, y] = \frac{1}{4} \int_{-1}^{1} \int_{-1}^{1} xy(1 - x^{2n} + y^{2n}) \, dx \, dy = 0$$

Hence, X and Y are statistically uncorrelated. While the marginal probability density functions of X and Y become

$$f(x) = \int_{-1}^{1} f(x, y) \, dy = \frac{1}{2}\left(\frac{2n + 2}{2n + 1} - x^{2n}\right)$$

and

$$f(y) = \int_{-1}^{1} f(x, y) \, dx = \frac{1}{2}\left(\frac{2n}{2n + 1} + y^{2n}\right)$$

Thus we have $f(x, y) \neq f(x)f(y)$, and hence x and y are not statistically independent. ∎

There are many joint probability density functions of two random variables for which, if the two random variables are statistically uncorrelated, then they are independent. For instance, the bivariate normal distribution presented in Example 3.13 is a typical example of this case.

From the knowledge of the variance and covariance, the correlation coefficient of two random variables can be defined.

Definition 3.10. The *correlation coefficient* of two random variables X and Y, denoted by ρ_{xy}, is defined by

$$\rho_{xy} = \frac{\text{Cov}[x, y]}{\sqrt{\text{Var}[x] \cdot \text{Var}[y]}} = \frac{\sigma_{xy}}{\sigma_x \sigma_y} \qquad (3.16)$$

An important property of the correlation coefficient is that it is bounded as follows:

$$-1 \leq \rho_{xy} \leq 1 \qquad (3.17)$$

In order to prove this property, the following theorem referred to as *Schwarz's inequality* is introduced:

Theorem 3.1. Let X and Y be random variables with finite second moments. Then, we have

$$(E[xy])^2 \leq E[x^2]E[y^2] \qquad (3.18)$$

Proof. For any real number a, we have

$$E[(ax - y)^2] = a^2 E[x^2] - 2aE[xy] + E[y^2] \geq 0$$

The above equation is a quadratic equation with respect to a and is nonnegative; hence, it has either no solution or one solution. This implies that the discriminant should be nonpositive. That is,

$$(E[xy])^2 - E[x^2]E[y^2] \leq 0$$

Let us replace the random variable X and Y in Eq. (3.18) by $X - \mu_x$ and $Y - \mu_y$, respectively, where μ_x and μ_y are the means of X and Y. Then, we can write

$$E[(x - \mu_x)(y - \mu_y)]^2 \leq E[(x - \mu_x)^2] E[(y - \mu_y)^2] \qquad (3.19)$$

This yields the following inequality that is given in Eq. (3.17):

$$\frac{\text{Cov}[x, y]}{\sqrt{\text{Var}[x]\text{Var}[y]}} = |\rho_{xy}| \leq 1$$

The correlation coefficient ρ_{xy} is a measure indicating the extent of a functional relationship between two random variables. We may write

$$\rho_{xy} = \frac{\text{Cov}[x, y]}{\sqrt{\text{Var}[x]\text{Var}[y]}} = E\left[\left(\frac{x - \mu_x}{\sigma_x}\right)\left(\frac{y - \mu_y}{\sigma_y}\right)\right] \qquad (3.20)$$

Then, $\rho_{xy} = \pm 1$ implies that $E[(x - \mu_x)(y - \mu_y)] = \pm \sigma_x \sigma_y$. This equation can be satisfied under the following condition:

$$\frac{y - \mu_y}{\sigma_y} = \pm \frac{x - \mu_x}{\sigma_x} \tag{3.21}$$

Thus, for $\rho = \pm 1$, we have

$$y = \mu_y \pm \frac{\sigma_y}{\sigma_x}(x - \mu_x) \tag{3.22}$$

and thereby the two random variables X and Y have a functional relationship. Consequently, the statistical characteristics of one random variable can be completely evaluated from those of the other random variable. On the other hand if $\rho_{xy} = 0$, the two random variables are uncorrelated and have no functional relationship. For positive values of ρ_{xy} (namely, $0 < \rho_{xy} < 1$), it is a general tendency that large values of X are associated with large values of Y, as are small values of X associated with small values of Y. Negative values of ρ_{xy} indicate that there is a general tendency that large values of X are associated with small values of Y and that small values of X are associated with large values of Y. A value of ρ_{xy} close to ± 1 indicates that the correlation tendency is strong, while a value of ρ_{xy} close to 0 indicates that the tendency is weak.

Example 3.10. The joint probability mass function of two random variables is given by $P\{X = 0, Y = 0\} = 1/2$, $P\{X = 1, Y = 0\} = 1/8$, $P\{X = 0, Y = 1\} = 1/8$, and $P\{X = 1, Y = 1\} = 1/4$. The covariance, Cov[x, y], can be evaluated as follows: The marginal probability function of X becomes $P\{X = 0\} = 5/8$ and $P\{X = 1\} = 3/8$. Hence, the mean value of X can be evaluated as

$$E[x] = 0 \cdot (5/8) + 1 \cdot (3/8) = 3/8$$

Similarly, the mean value of Y becomes $E[y] = 3/8$. On the other hand, the expected value $E[xy]$ becomes

$$E[xy] = (1/2)P\{X = 0, Y = 0\} + (1/8)P\{X = 1, Y = 0\}$$
$$+ (1/8)P\{X = 0, Y = 1\} + (1/4)P\{X = 1, Y = 1\} = 1/4$$

Thus, Cov[x, y] = $E[xy] - E[x]E[y] = 7/64$. ∎

Example 3.11. The random variables X and Y have the joint probability density function given by

$$f(x, y) = \frac{1}{2\pi} e^{-(1/2)(x^2 - 2\sqrt{3}xy + 4y^2)} \qquad -\infty < x < \infty, \ -\infty < y < \infty$$

We first evaluate the marginal probability densities, $f(x)$ and $f(y)$. That is, by integrating out y we have

$$f(x) = \int_{-\infty}^{\infty} f(x, y) \, dy = \frac{1}{2\sqrt{2\pi}} e^{-x^2/8}$$

This is normal probability density function with zero mean and variance 4 (see Section 6.1). Similarly, by integrating out x we have

$$f(y) = \int_{-\infty}^{\infty} f(x, y) \, dx = \frac{1}{\sqrt{2\pi}} e^{-y^2/2}$$

which is also a normal probability density function with zero mean and variance 1. Since $f(x)f(y) \neq f(x, y)$, the random variables X and Y are not statistically independent. The mean values of X and Y are both zero; hence, we can evaluate the covariance as

$$\text{Cov}[x, y] = \int_{-\infty}^{\infty} \int_{-\infty}^{\infty} xy f(x, y) \, dx \, dy = \sqrt{3}$$

Thus, the correlation coefficient ρ_{xy} becomes $\sqrt{3}/2$. ∎

We now extend the variance and covariance obtained for two random variables to a random vector consisting of n elements. For this, let us write the covariance of any two elements (x_i and x_j) of a random vector **X**, denoted by σ_{ij}, as follows:

$$\text{Cov}[x_i, x_j] = \sigma_{ij} = E\left[(x_i - \mu_i)(x_j - \mu_j)\right] \qquad (3.23)$$

where μ_i, μ_j = mean of random variable X_i, X_j, respectively.

Then, the variance and covariance of a random vector consisting of n elements form a *covariance matrix* denoted by Σ, which can be presented as follows:

$$\Sigma = \begin{pmatrix} \sigma_{11} & \sigma_{12} & \sigma_{13} & \cdots & \sigma_{1n} \\ \sigma_{21} & \sigma_{22} & \sigma_{23} & \cdots & \sigma_{2n} \\ \vdots & & & & \\ \sigma_{n1} & \sigma_{n2} & \sigma_{n3} & \cdots & \sigma_{nn} \end{pmatrix} \quad (3.24)$$

where $\sigma_{ij} = \sigma_{ji}$.

Note that the covariance matrix is symmetric and its diagonal elements, $\sigma_{11}, \sigma_{22}, \sigma_{33}, \ldots, \sigma_{nn}$, are the variances of the random variables X_1, X_2, X_3, and so on, respectively. If the nondiagonal elements of the covariance matrix are zero, then all random variables are uncorrelated.

Following the definition of the correlation coefficient given in Eq. (3.16), the correlation coefficient of any two elements (x_i and x_j) of a random vector **X**, denoted by ρ_{ij}, may be written as

$$\rho_{ij} = \frac{\text{Cov}[x_i, x_j]}{\sqrt{\text{Var}[x_i]\text{Var}[x_j]}} = \frac{\sigma_{ij}}{\sqrt{\sigma_{ii}\sigma_{jj}}} \quad (3.25)$$

Similar to the covariance matrix, the correlation coefficient of a random vector can be presented in the following matrix form:

$$\rho = \begin{pmatrix} 1 & \rho_{12} & \rho_{13} & \cdots & \rho_{1n} \\ \rho_{21} & 1 & \rho_{23} & \cdots & \rho_{2n} \\ \vdots & & & & \\ \rho_{n1} & \rho_{n2} & \rho_{n3} & \cdots & 1 \end{pmatrix} \quad (3.26)$$

3.2.2 Mean and Variance of Sum and Product of Two Random Variables

Let us consider the expected value of the sum and the product of two random variables X and Y assuming that the expected values of each random variable exist. By letting $Z = X + Y$, the expected value of Z, $E[z]$, for discrete type random variables can be obtained as

$$E[z] = \sum_j \sum_i (x_i + y_j) p_{ij} = \sum_i x_i \left(\sum_j p_{ij} \right) + \sum_j y_j \left(\sum_i p_{ij} \right)$$
$$= \sum_i x_i p_i + \sum_j y_j p_j = E[x] + E[y] \quad (3.27)$$

Similarly, for continuous random variables,

$$E[z] = \int_{-\infty}^{\infty} \int_{-\infty}^{\infty} (x+y) f(x, y) \, dx \, dy$$

$$= \int_{-\infty}^{\infty} x \left[\int_{-\infty}^{\infty} f(x, y) \, dy \right] dx + \int_{-\infty}^{\infty} y \left[\int_{-\infty}^{\infty} f(x, y) \, dx \right] dy$$

$$= \int_{-\infty}^{\infty} x f(x) \, dx + \int_{-\infty}^{\infty} y f(y) \, dy = E[x] + E[y] \quad (3.28)$$

It can be seen in Eqs. (3.27) and (3.28), the expected value of the sum of two random variables is equal to the sum of the individual expected values. This holds irrespective of statistical dependence or independence between the two random variables, and this result can be generalized to any finite number of random variables. In general, we have the following theorem:

Theorem 3.2. The expected value of the sum of a finite number of random variables, whose individual expected values exist, is equal to the sum of the individual expected values.

The expected value of the product of two random variables, $Z = XY$, for discrete type random variables can be written, in general, as

$$E[z] = \sum_j \sum_i x_i y_j p_{ij} \quad (3.29)$$

If the two random variables are statistically independent, then we may write

$$E[z] = \sum_j \sum_i x_i y_j P_{ij} = \sum_i x_i P_i \sum_j y_j P_j = E[x] E[y] \quad (3.30)$$

Similarly, for continuous type statistically independent random variables, we have

$$E[z] = \int_{-\infty}^{\infty} \int_{-\infty}^{\infty} xy f(x, y) \, dx \, dy = \int_{-\infty}^{\infty} x f(x) \, dx \int_{-\infty}^{\infty} y f(y) \, dy$$

$$= E[x] E[y] \quad (3.31)$$

MOMENTS OF RANDOM VECTORS

The results derived in Eqs. (3.30) and (3.31) can be generalized for a finite number of random variables. We have the following theorem:

Theorem 3.3. The expected value of the product of a finite number of statistically independent random variables, whose individual expected values exist, is equal to the product of the individual expected values.

We consider next the variance of the sum of two random variables whose individual variances, $\text{Var}[x]$ and $\text{Var}[y]$, exist. By writing the random variable $Z = X + Y$, we can evaluate $\text{Var}[z]$ as

$$\text{Var}[z] = E\big[(x+y)^2\big] - (E[x+y])^2$$

$$= E[x^2] - (E[x])^2 + E[y^2] - (E[y])^2$$

$$+ 2\{E[xy] - E[x]E[y]\}$$

$$= \text{Var}[x] + \text{Var}[y] + 2\,\text{Cov}[x, y] \qquad (3.32)$$

Thus, it can be concluded that the variance of the sum of two random variables is equal to the sum of the individual variances if the covariance is zero; namely if the two random variables are uncorrelated. Note that the two random variables are not necessarily statistically independent as discussed in the previous section. The result derived above can also be generalized to a finite number of uncorrelated random variables. Hence, we have the following theorem:

Theorem 3.4. The variance of the sum of a finite number of uncorrelated random variables is equal to the sum of the individual variances.

For the variance of the product of two random variables, there is no simple relationship with the variance of the individual random variables even though the two random variables are statistically independent. The variance of the product of two statistically independent random variables may be written as follows:

$$\text{Var}[xy] = E[x^2]E[y^2] - (E[x])^2(E[y])^2 \qquad (3.33)$$

3.3 CONDITIONAL MOMENTS

3.3.1 *Conditional Mean and Variance*

The moments discussed in the previous two sections refer to either the probability density function of a single random variable or the joint probability density function of two or more random variables. The same concept can also be applied to the conditional probability density function. The mean and variance in this case are called the conditional mean and conditional variance, respectively, and they are very important in finding statistical properties of random phenomena observed in engineering and physical sciences as will be demonstrated in Example 3.14.

Definition 3.11. The *conditional kth moment* of the random variable Y, given X, denoted by $E[y^k|X = x_i]$, is defined as follows:

$$E\left[y^k | X = x_i\right]$$

$$= \begin{cases} \sum_j y_j^k \cdot p\{y_j | X = x_i\} = \sum_j y_j^k \frac{p_{ij}}{p_i} & \text{for a discrete random variable} \\ \int y^k f(y|x)\, dy & \text{for a continuous random variable} \end{cases}$$

(3.34)

The *conditional mean* denoted as $E[y|x]$ can be obtained by

$$E[y|x] = \begin{cases} \sum_j y_j \cdot p\{y_j | X = x_i\} & \text{for a discrete random variable} \\ \int y f(y|x)\, dy & \text{for a continuous random variable} \end{cases}$$

(3.35)

The *conditional variance*, denoted by $\text{Var}[y|x]$, can be evaluated by using the conditional mean as follows:

$$\text{Var}[y|x] = E\left[\{y - E[y|x]\}^2 | x\right] \quad (3.36)$$

CONDITIONAL MOMENTS

In particular, for a continuous random variable, we can write

$$\text{Var}[y|x] = \int \{y - E[y|x]\}^2 \cdot f(y|x)\, dy \tag{3.37}$$

Example 3.12. The joint probability density function of two random variables X and Y is given by

$$f(x, y) = 4y(x - y)e^{-(x-y)}, \quad 0 \leq y < x < \infty$$

The conditional probability density function $f(x|y)$ can be evaluated as

$$f(x|y) = \frac{f(x, y)}{\int_y^\infty f(x, y)\, dx} = (x - y)e^{-(x-y)}$$

Then, the conditional mean $E[x|y]$ becomes

$$E[x|y] = \int_y^\infty x f(x|y)\, dx = y + 2 \qquad \blacksquare$$

Example 3.13. The following joint probability density function of the random variables X and Y is called the bivariate normal probability density function (see Section 6.1):

$$f(x, y) = \frac{1}{2\pi \sigma_x \sigma_y \sqrt{1 - \rho^2}}$$

$$\times \exp\left\{ -\frac{1}{2(1 - \rho^2)} \left[\left(\frac{x - \mu_x}{\sigma_x} \right)^2 - 2\rho \left(\frac{x - \mu_x}{\sigma_x} \right) \left(\frac{y - \mu_y}{\sigma_y} \right) \right.\right.$$

$$\left.\left. + \left(\frac{y - \mu_y}{\sigma_y} \right)^2 \right] \right\}$$

where μ_x, μ_y are the mean, σ_x^2 and σ_y^2 are the variance of X and Y, respectively, and ρ is the correlation coefficient. Evaluate the conditional mean, $E[y|x]$, and the conditional variance, $\text{Var}[y|x]$. The probability density function of X is a normal distribution written by

$$f(x) = \frac{1}{\sqrt{2\pi}\,\sigma_x} \exp\left[-\left(\frac{x-\mu_x}{\sigma_x}\right)^2\right]$$

Hence, the conditional probability density function $f(y|x)$ can be evalauted by

$$f(y|x) = \frac{f(x,y)}{f(x)} = \frac{1}{\sigma_y\sqrt{2\pi}\sqrt{1-\rho^2}}$$

$$\times \exp\left\{-\frac{1}{2(1-\rho^2)}\right.$$

$$\times \left[\left(\frac{x-\mu_x}{\sigma_x}\right)^2 - 2\rho\left(\frac{x-\mu_x}{\sigma_x}\right)\left(\frac{y-\mu_y}{\sigma_y}\right)\right.$$

$$\left.\left. + \left(\frac{y-\mu_y}{\sigma_y}\right)^2 - (1-\rho^2)\left(\frac{x-\mu_x}{\sigma_x}\right)^2\right]\right\}$$

$$= \frac{1}{\sigma_y\sqrt{2\pi}\sqrt{1-\rho^2}}$$

$$\times \exp\left\{-\frac{1}{2\sigma_y^2(1-\rho^2)}\left[(y-\mu_y)^2 - 2\rho\frac{\sigma_y}{\sigma_x}(y-\mu_y)(x-\mu_x)\right.\right.$$

$$\left.\left. + \rho^2\frac{\sigma_y^2}{\sigma_x^2}(x-\mu_x)^2\right]\right\}$$

$$= \frac{1}{\sigma_y\sqrt{2\pi}\sqrt{1-\rho^2}} \exp\left\{-\frac{1}{2\sigma_y^2(1-\rho^2)}\left[(y-\mu_y) - \rho\frac{\sigma_y}{\sigma_x}(x-\mu_x)\right]^2\right\}$$

This probability density function has the form of a normal probability distribution with mean $\mu_y + \rho(\sigma_y/\sigma_x)(x-\mu_x)$ and variance $\sigma_y^2(1-\rho^2)$. Thus, we have

$$E[y|x] = \mu_y + \rho\frac{\sigma_y}{\sigma_x}(x-\mu_x)$$

$$\text{Var}[y|x] = \sigma_y^2(1-\rho^2)$$

■

CONDITIONAL MOMENTS

3.3.2 *Application of Conditional Mean and Variance*

The following theorem is concerned with the mean and variance of a random variable expressed in terms of the conditional mean and variance, and is extremely useful in practical application.

Theorem 3.5. Consider two random variables X and Y. The mean and variance of the random variable Y can be expressed in terms of the conditional mean and variance as follows:

$$E[y] = E_x(E[y|x])$$

$$\text{Var}[y] = E_x(\text{Var}[y|x]) + \text{Var}_x(E[y|x]) \quad (3.38)$$

where $E_x[\ ]$ and $\text{Var}_x[\ ]$ are the mean and variance with respect to random variable X.

Proof. For brevity, consider a continuous random variable with the probability density function $f(x, y)$. The expected value of Y can be written as

$$E[y] = \int yf(y)\,dy = \iint yf(x, y)\,dx\,dy$$

$$= \int \left\{ \int yf(y|x)\,dy \right\} f(x)\,dx = \int E[y|x]f(x)\,dx$$

$$= E_x(E[y|x])$$

Using the formula derived above, we can write the variance of Y as

$$\text{Var}[y] = E\big[(y - E[y])^2\big] = E_x\big[E(y - E[y])^2|x\big]$$

Here, $E[(y - E[y])^2|x]$ can be expressed as

$$E\big[(y - E[y])^2|x\big] = E\big[\{(y - E[y|x]) + (E[y|x] - E[y])\}^2|x\big]$$

$$= E\big[(y - E[y|x])^2|x\big] + (E[y|x] - E[y])^2$$

$$= \text{Var}[y|x] + \{E[y|x] - E_x(E[y|x])\}^2$$

By taking the expected value of the above formula with respect to X, we have

$$\text{Var}[y] = E_x\left(E\left[(y - E[y])^2 \big| x\right]\right)$$

$$= E_x(\text{Var}[y|x]) + E_x[E[y|x] - E_x(E[y|x])]^2$$

$$= E_x(\text{Var}[y|x]) + \text{Var}_x(E[y|x])$$

Example 3.14. Figure 3.5 shows the time history of nearshore sediment transport moving toward a specified direction due to storms. The number of occurrences of storms in a specified time, denoted by N, as well as the magnitude of sediment transport in each storm, denoted by X, are both random variables with known means and variances. We want to evaluate the mean and variance of the total distance of the movement from the origin in a specified time period.

Let us write the total distance

$$Y = X_1 + X_2 + X_3 + \cdots + X_n$$

Assuming that the magnitudes of transport X in each storm are statistically independent, the mean and variance of Y for a given number of occurrences become

$$E[y|n] = nE[x]$$

$$\text{Var}[y|n] = n\,\text{Var}[x]$$

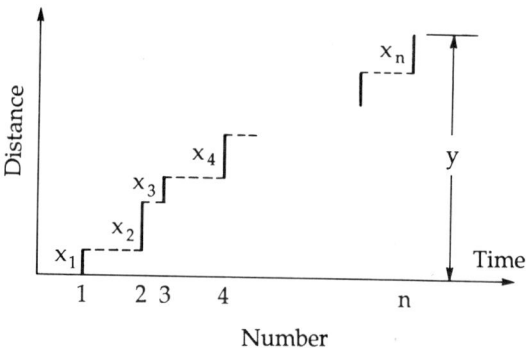

Figure 3.5 Sketch depicting time history of sediment transport X.

CONDITIONAL MOMENTS

Then, from Theorem 3.5, the mean and variance of Y can be obtained by

$$E[y] = E_n(E[y|n]) = E[n]E[x]$$

and

$$\text{Var}[y] = E_n(\text{Var}[y|n]) + \text{Var}_n(E[y|n])$$

Here,

$$\text{Var}_n(E[y|n]) = \text{Var}_n(nE[x])$$

$$= E_n[nE[x] - E(nE[x])]^2$$

$$= E_n\left[n^2(E[x])^2 - (E[n])^2(E[x])^2\right]$$

$$= E\left[n^2 - (E[n])^2\right](E[x])^2$$

$$= \text{Var}[n](E[x])^2$$

Thus, we have

$$\text{Var}[y] = E[n]\text{Var}[x] + \text{Var}[n](E[x])^2 \qquad \blacksquare$$

Another important application of the conditional mean can be made for evaluating the expected value of an arbitrary function of random variables defined in Eq. (3.9) by applying the following theorem:

Theorem 3.6. The expected value of an arbitrary function $u(x, y)$ of the two random variables X and Y can be expressed by

$$E[u(x, y)] = E_x(E[u(x, y)|x]) \qquad (3.39)$$

where $E[u(x, y)|x]$ is the conditional expectation with respect to the random variable X.

Proof. Let us consider some random variables of continuous type with the joint probability density function $f(x, y)$. The conditional expectation can be written as

$$E[u(x, y)|x] = \int_{-\infty}^{\infty} u(x, y) f(y|x) \, dy$$

$$= \int_{-\infty}^{\infty} u(x, y) \frac{f(x, y)}{f_x(x)} \, dy$$

Hence,

$$E_x(E[u(x, y)|x]) = \int_{-\infty}^{\infty} E[u(x, y)|x] f_x(x) \, dx$$

$$= \int_{-\infty}^{\infty} \int_{-\infty}^{\infty} u(x, y) f(x, y) \, dx \, dy$$

$$= E[u(x, y)]$$

If an arbitrary function $u(x, y)$ can be written in the form of $u(x)v(y)$, then we can write Eq. (3.39) as

$$E[u(x)v(y)] = E_x(u(x) E[v(y)|x]) \qquad (3.40)$$

For example, the expected values $E[xy]$ and $E[x^2 y^2]$, and so on, of the two random variables can be expressed as

$$E[xy] = E_x(x E[y|x])$$

$$E[x^2 y^2] = E_x(x^2 E[y^2|x]), \text{ and so on} \qquad (3.41)$$

Example 3.15. Evaluate the expected value $E[xy]$ of the bivariate normal probability distribution given in Example 3.13. The conditional mean $E[y|x]$ is obtained in Example 3.13 as

$$E[y|x] = \mu_y + \rho \frac{\sigma_x}{\sigma_y}(x - \mu_x)$$

CONDITIONAL MOMENTS

Then, from Eq. (3.41), we have

$$E[xy] = E\left[x\left\{\mu_y + \rho\frac{\sigma_x}{\sigma_y}(x - \mu_x)\right\}\right]$$

$$= \mu_y E[x] + \rho\frac{\sigma_y}{\sigma_x}\left\{E[x^2] - \mu_x E[x]\right\}$$

$$= \mu_y E[x] + \rho\sigma_x\sigma_y \qquad \blacksquare$$

Example 3.16. Obtain the expected value $E[x^2y^2]$ of the bivariate normal probability distribution with mean values $\mu_x = \mu_y = 0$. We may apply the formula given in Eq. (3.41), repeated here as

$$E[x^2y^2] = E_x\left[x^2 E[y^2|x]\right]$$

$E[y^2|x]$ can be written as follows from the result obtained in Example 3.13:

$$E[y^2|x] = \text{Var}[y|x] + (E[y|x])^2 = \sigma_y^2(1 - \rho^2) + \left(\rho\frac{\sigma_y}{\sigma_x}x\right)^2$$

Thus, we have

$$E[x^2y^2] = E\left[x^2\left\{\sigma_y^2(1 - \rho^2) + \left(\rho\frac{\sigma_y}{\sigma_x}x\right)^2\right\}\right]$$

$$= \sigma_y^2(1 - \rho^2)E[x^2] + \left(\rho\frac{\sigma_y}{\sigma_x}\right)^2 E[x^4]$$

Since $E[x^2] = \sigma_x^2$ and $E[x^4] = 3\sigma_x^4$, we can write

$$E[x^2y^2] = \sigma_x^2\sigma_y^2(1 - \rho^2) + 3\rho^2\sigma_x^2\sigma_y^2$$

$$= \sigma_x^2\sigma_y^2(1 + 2\rho^2) \qquad \blacksquare$$

EXERCISES

3.1 The sample space of a random variable X is given by (a, b), where $a < b$. Prove that $a \leq E[x] \leq b$ and $\text{Var}[x] \leq \{(b - a)/2\}^2$.

3.2 The random variable X has the following Rayleigh distribution given by

$$f(x) = \frac{2x}{R} e^{-x^2/R}, \qquad 0 \leq x < \infty, \, R > 0$$

Show that the nth moment of the random variable is given by

$$E[x^n] = \begin{cases} \sqrt{\pi/2}\,(1 \cdot 3 \cdot 5 \cdots n)(\sqrt{R})^n & \text{if } n \text{ is odd} \\ 2^{n/2}(n/2)!(\sqrt{R})^n & \text{if } n \text{ is even} \end{cases}$$

3.3 Prove that

$$E\left[\frac{x}{y}\right] = \frac{E[x]}{E[y]}$$

if the random variables X/Y and Y are statistically independent.

3.4 The random variable X has the following normal distribution given by

$$f(x) = \frac{1}{\sigma\sqrt{2\pi}} e^{-x^2/2\sigma^2}, \qquad -\infty < x < \infty$$

Prove that the central moments of the even numbered of the random variables are given by

$$\mu_{2r} = \frac{(\sigma^2)^r}{2^r} \frac{(2r)!}{r!}, \qquad r = 1, 2, 3, \ldots$$

3.5 Derive the standardized probability density function and show the sample space of the random variable given in Problem 3.2.

EXERCISES

3.6 Evaluate the skewness γ_1 of the discrete-type random variable that has a binomial distribution given by

$$p(x) = \binom{n}{x} p^x (1-p)^{n-x}$$

where $x = 0, 1, 2, \ldots, n$, $0 < p < 1$. Show that whether the distribution is skewed to the right or to the left depends on the p value.

3.7 Let two random variables X and Y have the joint probability density function given in Example 3.13. Let us assume that $\mu_x = \mu_y = 0$ and $E[(y - ax)x] = 0$, where a is an arbitrary constant. Prove that $E[y|x] = xE[xy]/E[x^2]$.

3.8 Let the random variable X represent the time from the commencement of its service when a failure of a system occurs. The probability density function of X is given by

$$f(x) = \lambda e^{-\lambda x}, \qquad 0 \leqslant x < \infty, \lambda > 0$$

Show that the expected value of the failure time, given that a failure has not occurred prior to a specified time T, is given by $T + (1/\lambda)$.

CHAPTER 4

Moment Generating Function, Characteristic Function, and Their Application

Moments of a random variable can be evaluated directly following the definitions given in Section 3.1. However, the evaluation of moments is sometimes quite difficult depending on the complexity in the form of the probability density function. The moment generating function defined in Section 4.1 is extremely useful not only for evaluating moments but also for application of probability theory. This is because a function that can generate all the moments of random variable essentially describes the probability law of the random variable. However, the moment generating function does not always exist for some random variables (see Example 4.4). On the other hand, the characteristic function introduced in Section 4.2 always exists, and there is a one-to-one relationship between the probability distribution function and the characteristic function. Subsequently, the characteristic function plays a significant role in deriving a new probability distribution function. For example, the probability density function suitable for describing the statistical properties of non-Gaussian random processes can be derived through the cumulant generating function, which is the logarithm of the characteristic function. The cumulant generating function is discussed in Section 4.3. The probability generating function, which is particularly useful for evaluating moments of discrete random variables, is presented in Section 4.4.

4.1 MOMENT GENERATING FUNCTION

Definition 4.1. The function $m(t) = E[e^{tx}]$ where t is a real number is called the *moment generating function* of the random variable X. That is,

$$m(t) = \begin{cases} \sum_{j \in s} e^{tx_j} p(x_j) & \text{for a discrete-type random variable} \\ \int_s e^{tx} f(x)\, dx & \text{for a continuous-type random variable} \end{cases} \quad (4.1)$$

where s stands for the sample space.

Let us differentiate the moment generating function r times with respect to t. Then, we have

$$\frac{d^r}{dt^r} m(t) = \begin{cases} \sum_{j \in s} x_j^r e^{tx_j} p(x_j) & \text{for a discrete-type random variable} \\ \int_s x^r e^{tx} f(x)\, dx & \text{for a continuous-type random variable} \end{cases}$$

By letting $t = 0$, we have

$$\left[\frac{d^r}{dt^r} m(t)\right]_{t=0} = \begin{cases} \sum_{j \in s} x_j^r p(x_j) \\ \int_s x^r f(x)\, dx \end{cases} = E[x^r] \quad (4.2)$$

This leads to the following theorem.

Theorem 4.1. The rth moment of a random variable, if it exists, can be obtained by differentiating the moment generating function, $m(t) = E[e^{tx}]$, r times with respect to t and by letting $t = 0$.

Example 4.1. We may evaluate the mean and variance of a random variable given in Example 3.1 by applying the moment generating function.

$$m(t) = \sum_{x=0}^{\infty} e^{tx} e^{-\lambda} \frac{\lambda^x}{x!} = e^{-\lambda} \sum_{x=0}^{\infty} \frac{(\lambda e^t)^x}{x!} = e^{-\lambda} e^{\lambda e^t} = e^{\lambda(e^t - 1)}$$

Then,

$$m'(t) = e^{-\lambda} \lambda e^t e^{\lambda e^t}$$

$$m''(t) = e^{-\lambda} \lambda e^t e^{\lambda e^t}(1 + \lambda e^t)$$

By letting $t = 0$, the mean can be obtained as

$$m'(0) = E[x] = \lambda$$

Furthermore, we have

$$m''(0) = E[x^2] = \lambda(1 + \lambda)$$

Hence, $\text{Var}[x] = E[x^2] - (E[x])^2 = \lambda$. ∎

Example 4.2. Evaluate the mean and variance of the probability mass function given in Example 3.4 by using the moment generating function.

$$m(t) = \sum_{x=0}^{n} e^{tx} p(x)$$

$$= \sum_{x=0}^{n} \binom{n}{x} (pe^t)^x (1-p)^{n-x}$$

$$= (pe^t + 1 - p)^n$$

Then, the mean is given by

$$E[x] = m'(0) = np$$

Furthermore,

$$E[x^2] = m''(0) = np + n^2 p^2 - np^2$$

Hence

$$\text{Var}[x] = np(1-p)$$ ∎

Example 4.3. Evaluate the mean and variance of the following probability distribution called the exponential distribution:

$$f(x) = \lambda e^{-\lambda x}, \quad 0 \leq X < \infty$$

MOMENT GENERATING FUNCTION

The moment generating function becomes

$$m(t) = \int_0^\infty \lambda e^{-\lambda x} e^{tx} \, dx = \frac{\lambda}{\lambda - t}$$

Then, we have

$$E[x] = m'(0) = \frac{1}{\lambda}$$

$$E[x^2] = m''(0) = \frac{2}{\lambda^2}$$

and thereby

$$\operatorname{Var}[x] = \frac{1}{\lambda^2}$$ ∎

It should be noted that the moment generating function does not always exist as demonstrated in the following example.

Example 4.4. Evaluate the moment generating function of the Cauchy distribution given by

$$f(x) = \frac{1}{\pi} \frac{1}{1 + x^2}, \quad -\infty < x < \infty$$

It is shown in Example 3.5 that the mean value of the Cauchy distribution does not exist. The moment generating function is given by

$$m(t) = \int_{-\infty}^\infty e^{tx} \frac{1}{\pi} \frac{1}{1 + x^2} \, dx$$

The above integration does not exist. For example, for $t > 0$, we can write

$$m(t) = \frac{1}{\pi} \int_{-\infty}^\infty \frac{e^{tx}}{1 + x^2} \, dx > \frac{1}{\pi} \int_{-\infty}^\infty \frac{tx}{1 + x^2} \, dx = \infty$$

Similarly, we can prove that $m(t)$ becomes indefinite for $t < 0$. ∎

4.2 CHARACTERISTIC FUNCTION

Definition 4.2. The function $\phi(t) = E[e^{itx}]$, where t is a real number and i is an imaginary unit, is called the *characteristic function* of the random variable X. That is,

$$\phi(t) = \begin{cases} \sum_{j \in s} e^{itx_j} p(x_j) & \text{for a discrete-type random variable} \\ \int_s e^{itx} f(x)\, dx & \text{for a continuous-type random variable} \end{cases} \quad (4.3)$$

where s stands for the sample space.

The properties of the characteristic function $\phi(t)$ are as follows:

(i) $\phi(t)$ always exists since $|e^{itx}|$ is a continuous and bounded function for all finite real values of t and x.
(ii) $\phi(0) = 1$
(iii) $|\phi(t)| = |E[e^{itx}]| \leq 1$
(iv) $\phi(-t) = \phi^*(t)$, where $\phi^*(t)$ denotes the complex conjugate to $\phi(t)$. $\quad (4.4)$

Let us differentiate the characteristic function r times with respect to t. Since the absolute moment is finite, we can differentiate the characteristic function under the summation sign as well as under the integral sign. Then, we have

$$\frac{d^r}{dt^r}\phi(t) = \begin{cases} \sum_{j \in s} (ix_j)^r e^{itx_j} p(x_j) \\ \int_s (ix)^r e^{itx} f(x)\, dx \end{cases} = E[(ix)^r e^{itx}]$$

By letting $t = 0$,

$$\frac{d^r}{dt^r}\phi(0) = i^r E[x^r]$$

CHARACTERISTIC FUNCTION

Hence,

$$E[x^r] = \frac{1}{i^r}\phi^{(r)}(0) \tag{4.5}$$

Thus, we have the following theorem:

Theorem 4.2. The rth moment of a random variable, $E[x^r]$, if it exists, can be obtained by $(1/i^r)\phi^{(r)}(0)$, where i is an imaginary unit, and $\phi^{(r)}(0)$ is the rth derivative of the characteristic function with $t = 0$.

Example 4.5. Obtain the mean and variance of the Bernoulli distribution whose probability mass function is given by

$$p(x) = p^x(1-p)^{1-x}, \quad \text{where } x = 0 \text{ and } 1$$

$$\phi(t) = \sum_{x=0}^{1} e^{itx} p(x) = 1 \cdot (1-p) + e^{it}p = 1 + pe^{it} - p$$

$$E[x] = \frac{1}{i}\phi'(0) = p$$

$$E[x^2] = \frac{1}{i^2}\phi''(0) = p$$

and hence

$$\text{Var}[x] = p - p^2 \qquad \blacksquare$$

Example 4.6. Evaluate the mean and variance of a random variable whose probability density function is given by

$$f(x) = \frac{1}{2}e^{-|x|}, \quad -\infty < x < \infty$$

We have

$$\phi(t) = \int_{-\infty}^{\infty} e^{itx} \frac{1}{2} e^{-|x|} \, dx = \frac{1}{1+t^2}$$

Then,

$$E[x] = \frac{1}{i}\phi'(0) = 0$$

$$E[x^2] = \frac{1}{i^2}\phi''(0) = 2$$

and hence

$$\text{Var}[x] = 2 \qquad \blacksquare$$

Example 4.7. Obtain the characteristic function of the Cauchy distribution shown in Example 4.4 for which the moment generating function does not exist. The characteristic function can be written as

$$\phi(t) = \frac{1}{\pi}\int_{-\infty}^{\infty} \frac{e^{itx}}{1 + x^2} dx$$

$\phi(t)$ can be evaluated by applying Cauchy's residue theorem in complex variables. That is, if we consider x as a complex variable, the integrand of $\phi(t)$ is analytic except at points i and $-i$. The residue of the integrand can be evaluated as

$$\text{Res}(i) = \begin{cases} \dfrac{e^{-t}}{2i} & \text{for } t > 0 \\ \dfrac{e^{t}}{2i} & \text{for } t < 0 \end{cases}$$

Then, by the residue theorem, the integration becomes $\pi e^{-|t|}$, and hence we have

$$\phi(t) = e^{-|t|}$$

Although the characteristic function, $\phi(t)$, does exist, it is not differentiable at $t = 0$. Thus, none of the moments of the Cauchy distribution exists. $\blacksquare$

Theorem 4.3. Let $\phi_x(t)$ be the characteristic function of a random variable X and consider a new random variable $Y = aX + b$, where a and b are constants. Then, the characteristic function of the random variable Y is given by $\phi_y(t) = e^{ibt}\phi_x(at)$.

CHARACTERISTIC FUNCTION

Proof. From the definition of the characteristic function, we have

$$\phi_y(t) = E[e^{ity}] = E[e^{it(ax+b)}]$$

$$= e^{ibt}E[e^{iatx}] = e^{ibt}\phi_x(at) \qquad (4.6)$$

Example 4.8. Obtain the characteristic function of the standardized normal distribution shown in Example 3.6. Then, derive the characteristic function of the normal distribution with mean μ and variance σ^2.

The characteristic function of the standardized normal distribution can be evaluated by

$$\phi(t) = \int_{-\infty}^{\infty} \frac{1}{\sqrt{2\pi}} e^{-z^2/2} e^{itz} \, dz$$

$$= e^{-t^2/2} \int_{-\infty}^{\infty} \frac{1}{\sqrt{2\pi}} e^{-\frac{1}{2}(z-it)^2} \, dz$$

$$= e^{-t^2/2}$$

Next, let a random variable X obey the normal probability distribution with mean μ and variance σ^2, and let Z be a standardized random variable. The relationship between X and Z is given by $X = \mu + \sigma Z$. Then, by applying Theorem 4.3, the characteristic function of the random variable X is given by

$$\phi(t) = e^{i\mu t - \frac{1}{2}\sigma^2 t^2} \qquad \blacksquare$$

Theorem 4.4. Let the random variables $X_1, X_2, \ldots, X_n$ be statistically independent. Then the characteristic function of the random variable $Y = X_1 + X_2 + \cdots + X_n$ is the product of the characteristic functions of each random variable. That is,

$$\phi_y(t) = \phi_1(t)\phi_2(t), \ldots, \phi_n(t) \qquad (4.7)$$

Proof

$$\phi_y(t) = E[e^{it(x_1 + x_2 + \cdots + x_n)}]$$

$$= \prod_{j=1}^{n} E[e^{itx_j}] = \prod_{j=1}^{n} \phi_{x_j}(t)$$

Example 4.9. Let two random variables X and Y be statistically independent each having the standardized normal probability distribution shown in Example 3.6. We can show that the sum of these two random variables is also a standardized normal distribution.

From the result obtained in Example 4.8, the characteristic function of each random variable is given by

$$\phi_1(t) = e^{-t_1^2/2}$$

and

$$\phi_2(t) = e^{-t_2^2/2}$$

Hence, by applying Theorem 4.4, the characteristic function of the sum of two independent random variables becomes

$$\phi(t) = \phi_1(t) \cdot \phi_2(t) = e^{-\frac{1}{2}(t_1^2 + t_2^2)}$$

Since the derived characteristic function is in the form of the characteristic function of the standardized normal probability function, it can be stated that the sum of two independent standardized normal probability distribution is also a standardized normal distribution. ∎

By applying Theorem 4.4 together with the property (iv) of the characteristic function given in Eq. (4.4), it can be proved that the characteristic function of the difference of two random variables X_1 and X_2 is given by

$$\phi_{x_1 - x_2}(t) = \phi_{x_1}(t) \cdot \phi_{x_2}^*(t) \tag{4.8}$$

By following the same procedure as shown in Example 4.9, it can be proved from Eq. (4.7) that the difference of two independent random variables each having the standardized normal probability distribution is also the standardized normal distribution.

Next, the formula for evaluating the probability density function from the characteristic function will be shown in the following theorem without proof. For a complete proof of the theorem, the reader is referred to Fisz (1963), among others.

Theorem 4.5. Let $\phi(t)$ be the characteristic function of a random variable X. If X is a discrete-type random variable, its probability mass function, $p(x)$, can be obtained by

$$p(x) = \lim_{T \to \infty} \frac{1}{2T} \int_{-T}^{T} \phi(t) e^{-itx} \, dt \tag{4.9}$$

If X is a continuous-type random variable, its probability density function can be evaluated by

$$f(x) = \frac{1}{2\pi} \int_{-\infty}^{\infty} \phi(t) e^{-itx} \, dt \tag{4.10}$$

where

$$\int_{-\infty}^{\infty} |\phi(t)| \, dt < \infty$$

Example 4.10. The characteristic function of a discrete random variable X is given by

$$\phi(t) = (pe^{it} + q)^n, \qquad \text{where } q = 1 - p$$

The probability mass function $p(x)$ can be obtained by Theorem 4.5 as

$$p(x) = \lim_{T \to \infty} \frac{1}{2T} \int_{-T}^{T} (pe^{it} + q)^n e^{-itx} \, dt$$

By the binomial theorem, we can write

$$(pe^{it} + q)^n = \sum_{r=0}^{n} \binom{n}{r} (pe^{it})^r q^{n-r}$$

Then, we have

$$p(x) = \lim_{T \to \infty} \frac{1}{2T} \sum_{r=0}^{n} \binom{n}{r} p^r q^{n-r} \int_{-T}^{T} e^{i(r-x)t} \, dt$$

where integration yields

$$\int_{-T}^{T} e^{i(r-x)t} \, dt = \begin{cases} \dfrac{2\sin(r-x)T}{(r-x)} & \text{for } r \neq x \\ 2T & \text{for } r = x \end{cases}$$

For $r \neq x$, we have

$$p(x) = \lim_{T \to \infty} \sum_{r=0}^{n} \binom{n}{r} p^r q^{n-r} \frac{\sin(r-x)T}{(r-x)} = 0$$

Hence, the probability mass function $p(x)$ can be obtained for $r = x$ as

$$p(x) = \binom{n}{x} p^x q^{n-x} \qquad \blacksquare$$

Example 4.11. The characteristic function of a random variable X is given by $\phi(t) = \exp\{-t^2/2\}$. The probability density function can be evaluated by Theorem 4.5 as

$$f(x) = \frac{1}{2\pi} \int_{-\infty}^{\infty} e^{-itx} e^{-t^2/2} \, dt$$

$$= \frac{1}{\sqrt{2\pi}} e^{-x^2/2} \frac{1}{\sqrt{2\pi}} \int_{-\infty}^{\infty} e^{-\frac{1}{2}(t+ix)^2} \, dx$$

$$= \frac{1}{\sqrt{2\pi}} e^{-x^2/2}$$

This is the probability density function of the standardized normal distribution. $\blacksquare$

The method for evaluating moments of a random variable by means of the characteristic function can be generalized to a random vector consisting of a finite number of random variables. For brevity, let us consider two random variables X and Y and define the characteristic function and evaluate their moments therefrom.

Definition 4.3. The function defined as

$$\phi(t, u) = E[e^{i(tx+uy)}]$$

$$= \begin{cases} \sum_j \sum_i e^{i(tx_i + uy_j)} p(x_i, y_j) & \text{for a discrete-type random variable} \\ \iint e^{i(tx+uy)} f(x, y) \, dx \, dy & \text{for a continuous-type random variable} \end{cases}$$

$$(4.11)$$

CHARACTERISTIC FUNCTION

where t and u are real numbers. $\phi(t, u)$ is called the *joint characteristic function* of two random variables X and Y.

Theorem 4.6. The moment of order k and l of the random variables X and Y can be obtained from the joint characteristic function, $\phi(t, u)$ as follows:

$$\left[\frac{1}{i^{k+l}}\frac{\partial^{k+l}}{\partial t^k \partial u^l}\phi(t, u)\right]_{\substack{t=0 \\ u=0}} = E[x^k y^l] \tag{4.12}$$

In particular, we have

$$E[x] = \frac{1}{i}\left[\frac{\partial \phi}{\partial t}\right]_{\substack{t=0 \\ u=0}}, \quad E[y] = \frac{1}{i}\left[\frac{\partial \phi}{\partial u}\right]_{\substack{t=0 \\ u=0}}$$

$$E[x^2] = \frac{1}{i^2}\left[\frac{\partial^2 \phi}{\partial t^2}\right]_{\substack{t=0 \\ u=0}}, \quad E[y^2] = \frac{1}{i^2}\left[\frac{\partial^2 \phi}{\partial u^2}\right]_{\substack{t=0 \\ u=0}}$$

$$E[xy] = \frac{1}{i^2}\left[\frac{\partial^2 \phi}{\partial t \partial u}\right]_{\substack{t=0 \\ u=0}} \tag{4.13}$$

Example 4.12. The random variables X and Y have the following joint probability density function (bivariate normal distribution):

$$f(x, y) = \frac{1}{2\pi\sqrt{1-\rho^2}} \exp\left\{-\frac{1}{2(1-\rho^2)}(x^2 - 2\rho xy + y^2)\right\},$$

$$-\infty < x, y < \infty$$

The joint characteristic function is given by

$$\phi(t, u) = \int_{-\infty}^{\infty}\int_{-\infty}^{\infty} \frac{1}{2\pi\sqrt{1-\rho^2}} e^{i(tx+uy)}$$

$$\times \exp\left\{-\frac{1}{2(1-\rho^2)}(x^2 - 2\rho xy + y^2)\right\} dx\, dy$$

$$= e^{-\frac{1}{2}(t^2 + 2\rho tu + u^2)}$$

Thus, it can easily be derived from Eq. (4.13) that

$$E[x] = E[y] = 0$$

$$\text{Var}[x] = \text{Var}[y] = 1$$

and the correlation coefficient $\rho = 1$. ∎

The characteristic function can also be evaluated for the conditional distribution. That is, the *conditional characteristic function* of a random variable X given Y is defined as

$$\phi_x(t|y) = \int_{-\infty}^{\infty} e^{itx} f(x|y) \, dx \qquad (4.14)$$

Since the conditional probability density function $f(x|y)$ can be written as

$$f(x|y) = \frac{f(x, y)}{f(y)} \qquad (4.15)$$

we can express the unconditional characteristic function of X as follows:

$$\phi_x(t) = \int_{-\infty}^{\infty} e^{itx} f(x) \, dx = \int_{-\infty}^{\infty} \phi_x(t|y) f(y) \, dy \qquad (4.16)$$

An example of the application of Eq. (4.16) will be given in Exercise 5.5 in connection with derivation of a nonhomogeneous Poisson distribution.

4.3 CUMULANTS (SEMIINVARIANTS)

As presented in Theorems 4.1 and 4.2, the mean and variance of a random variable can be obtained from the derivatives of either the moment generating function or the characteristic function. There is another function, called the cumulant generating function, for evaluating the mean and variance directly, without carrying out any differentiation. This function does not yield the moments (except the first moment), but it is sometimes very convenient to evaluate the variance of a random variable. More importantly, the cumulant generating function plays a significant role in deriving the probability density function, which is applicable to non-Gaussian random processes as will be shown in Chapter 16.

CUMULANTS (SEMIINVARIANTS)

Let us expand the characteristic function of a random variable X in a power series in the neighborhood of $t = 0$. That is,

$$\phi(t) = E[e^{itx}] = 1 + (it)E[x] + \frac{(it)^2}{2!}E[x^2] + \frac{(it)^3}{3!}E[x^3] + \cdots$$

$$= 1 + \sum_{s=1}^{\infty} (it)^s \frac{m_s}{s!} \tag{4.17}$$

where $m_s = E[x^s]$.

Next, take the logarithm of $\phi(t)$, denoted by $\psi(t)$, and expand it into a power series.

$$\psi(t) = \ln \phi(t) = \ln \left\{ 1 + \sum_{s=1}^{\infty} (it)^s \frac{m_s}{s!} \right\}$$

$$= \left[\sum_{s=1}^{\infty} (it)^s \frac{m_s}{s!} \right] - \frac{1}{2}\left[\sum_{s=1}^{\infty} (it)^s \frac{m_s}{s!} \right]^2 + \frac{1}{3}\left[\sum_{s=1}^{\infty} (it)^s \frac{m_s}{s!} \right]^3$$

$$= m_1 \frac{it}{1!} + (m_2 - m_1^2)\frac{(it)^2}{2!} + (m_3 - 3m_1 m_2 + 2m_1^3)\frac{(it)^3}{3!} + \cdots \tag{4.18}$$

Let us express the function $\psi(t)$ derived in Eq. (4.18) in the following series:

$$\psi(t) = \ln \phi(t) = \sum_{s=1}^{\infty} k_s \frac{(it)^s}{s!} \tag{4.19}$$

From Eqs. (4.18) and (4.19), we have

$k_1 = m_1 = $ mean

$k_2 = m_2 - m_1^2 = $ variance

$k_3 = m_3 - 3m_1 m_2 + 2m_1^3$

$k_4 = m_4 - 3m_2^2 - 4m_1 m_3 + 12m_1^2 m_2 - 6m_1^4$, and so on (4.20)

The function $\psi(t)$ given in (4.19) is called the *cumulant generating function*, and the coefficients k_s are called *cumulants* or *semiinvariants*.

If the cumulants are expressed in terms of the central moments, μ_1, μ_2, and so on, defined in Eq. (3.5), we have the following relationship:

$$k_2 = \mu_2 = \text{variance}$$

$$k_3 = \mu_3$$

$$k_4 = \mu_4 - 3\mu_2^2$$

$$k_5 = \mu_5 - 10\mu_3\mu_2$$

$$k_6 = \mu_6 - 15\mu_4\mu_2 - 10\mu_3^2 + 30\mu_2^3, \text{ and so on} \qquad (4.21)$$

It is also noted that the skewness and kurtosis defined in Eq. (3.8) are expressed by cumulants as follows:

$$\text{Skewness } \gamma_1 = \frac{\mu_3}{(\mu_2)^{3/2}} = \frac{k_3}{(k_2)^{3/2}}$$

$$\text{Kurtosis } \gamma_2 = \frac{\mu_4}{\mu_2^2} = \frac{k_4}{k_2^2} + 3 \qquad (4.22)$$

Example 4.13. Obtain the mean and variance of the Poisson distribution whose probability mass function is given in Example 3.1 through the cumulant generating function.

$$\phi(t) = \sum_{x=0}^{\infty} e^{itx} e^{-\lambda} \frac{\lambda^x}{x!} = e^{-\lambda} \sum_{x=0}^{\infty} \frac{(\lambda e^{it})^x}{x!} = e^{\lambda(e^{it}-1)}$$

Hence, the cumulant generating function becomes

$$\psi(t) = \ln \phi(t) = \lambda(e^{it} - 1)$$

$$= \lambda\left[\sum_{s=0}^{\infty} \frac{(it)^s}{s!} - 1\right] = \lambda \sum_{s=1}^{\infty} \frac{(it)^s}{s!}$$

Thus, we have, $k_1 = \lambda$ (mean) and $k_2 = \lambda$ (variance). ∎

Example 4.14. The random variables X and Y have a linear relationship given by $Y = X + a$. Prove that all cumulants of Y are the same as those of X except k_1. Because this is so, the cumulants are called *semi-invariants*.

Let us write the characteristic functions of X and Y as $\phi_x(t)$ and $\phi_y(t)$, respectively. From Theorem 4.3, we have

$$\phi_y(t) = e^{iat}\phi_x(t)$$

Then, the relationship between the cumulant-generating functions of X and Y becomes

$$\psi_y(t) = iat + \psi_x(t)$$

In expanding $\psi_x(t)$ and $\psi_y(t)$ in a series given in Eq. (4.19), the coefficients are the same except for the first term. Thus, all cumulants except k_1 of the random variables X and Y are the same. ∎

4.4 PROBABILITY GENERATING FUNCTION

Various statistical properties of discrete random variables can often be easily evaluated through the probability generating function defined below:

Definition 4.4. Let X be a discrete random variable and let $p_k = \Pr\{X = k\}$. Then, the function defined by

$$G(s) = \sum_{k=0}^{\infty} p_k s^k, \quad \text{where } -1 \leqslant s \leqslant 1 \qquad (4.23)$$

is called the *probability generating function* of X.

Note that the generating function is essentially equal to the expected value of s^x, where $x = k$, of the random variable X. The series involved in the probability generating function $G(s)$ is absolutely and uniformly convergent in the interval $-1 \leqslant s \leqslant 1$, and hence $G(s)$ is a continuous function. The properties of the probability generating function are as follows:

$$[G'(s)]_{s=1} = E[x]$$
$$[G''(s)]_{s=1} = E[x(x-1)]$$
$$[G'''(s)]_{s=1} = E[x(x-1)(x-2)] \qquad (4.24)$$

Example 4.15. Evaluate the probability generating function of the following probability distribution (Poisson distribution) given in Example 3.1:

$$p(x) = e^{-\lambda}\frac{\lambda^x}{x!}, \qquad x = 0, 1, 2, \ldots$$

The probability generating function becomes

$$G(s) = \sum_{k=0}^{\infty} e^{-\lambda}\frac{\lambda^k}{k!}s^k = e^{-\lambda}\sum_{k=0}^{\infty}\frac{(\lambda s)^k}{k!} = e^{-\lambda(1-s)}$$

Thus, we have $E[x] = \lambda$, $E[x(x-1)] = \lambda^2$, and therefore $\text{Var}[x] = E[x(x-1)] + E[x] - (E[x])^2 = \lambda$ ∎

Example 4.16. Evaluate the probability generating function of the following probability distribution given in Example 3.4 (binomial distribution):

$$p(x) = \binom{n}{x}p^x(1-p)^{n-x}, \qquad x = 0, 1, 2, \ldots, n$$

The probability generating function is given by

$$G(s) = \sum_{k=0}^{n}\binom{n}{k}p^k(1-p)^{n-k}s^k$$

$$= \sum_{k=0}^{n}\binom{n}{k}(ps)^k(1-p)^{n-k}$$

$$= (ps + 1 - p)^n$$

Then, we have $E[x] = np$, $E[x(x-1)] = n(n-1)p^2$, and therefore $\text{Var}[x] = np(1-p)$. ∎

Analogous to Theorem 4.4 regarding the characteristic function of the sum of independent random variables, the probability generating function of independent random variables is equal to the product of the probability generating functions of the random variables. This is stated in the following theorem:

Theorem 4.7. Let the random variables $X_1, X_2, \ldots, X_n$ be statistically independent. Then the probability generating functions of the random variable $Y = X_1 + X_2 + \cdots + X_n$ is given by the product of the probability generating function of the random variables X_i. That is,

$$G_y(s) = G_1(s)G_2(s)\ldots G_n(s) \tag{4.25}$$

Proof

$$G_y(s) = E[s^{x_1+x_2+\cdots+x_n}]$$
$$= E[s^{x_1}]E[s^{x_2}]\ldots E[s^{x_n}]$$
$$= G_1(s)G_2(s)\ldots G_n(s)$$

EXERCISES

4.1 The probability density function of a random variable X is given in the form of

$$f(x) = \frac{k}{2}e^{-x}(1+x^2), \quad -1 \leqslant x < \infty$$

Determine the constant k, and obtained the characteristic function.

4.2 Let the random variable Y be a linear summation of n-independent random variables X_i given by $\Sigma_i a_i X_i$, where $i = 1, 2, \ldots, n$. a_i is constant and X_i takes the values -1 and $+1$ with probability $1/2$. Obtain the characteristic function of Y.

4.3 The rth moment of a continuous-type random variable X is given by

$$m_r = \begin{cases} 0 & \text{if } r \text{ is odd number} \\ r! & \text{if } r \text{ is even number} \end{cases}$$

Derive the probability density function $f(x)$, where $-\infty < X < \infty$, by using the characteristic function.

4.4 Let the random variable Y be a linear summation of n-independent random variables X_i given by $y = \Sigma_i a_i X_i$, where $i = 1, 2, \ldots, n$. Prove that the cumulant generating function of Y is given by $\Sigma_{i=1}^n \psi_i(at)$, where $\psi_i(t)$ is the cumulant generating function of the random variable X_i.

4.5 Let m_r and k_r be the rth moment and the rth cumulant of a random variable. Prove that

$$m_1 = k_1$$

$$m_2 = k_2 + k_1^2$$

$$m_3 = k_3 + 3k_2 k_1 + k_1^3$$

$$m_4 = k_4 + 4k_3 k_1 + 3k_2^2 + 6k_2 k_1^2 + k_1^4$$

4.6 Let $\phi(t_1, t_2)$ be the joint characteristic function of two random variables X and Y, and let $U = a_1 X + b_1 Y + c_1$ and $V = a_2 X + b_2 Y + c_2$. Show that the joint characteristic function of U and V is given by

$$e^{c_1 t_1 + c_2 t_2} \phi(a_1 t_1 + a_2 t_2, b_1 t_1 + b_2 t_2)$$

4.7 Prove that all cumulants of the following discrete probability distribution (Poisson distribution) are the same:

$$p(x) = \frac{\mu^x}{x!} e^{-\mu}, \qquad x = 0, 1, 2, \ldots, n$$

Evaluate the skewness and kurtosis of the distribution.

4.8 Let k_{xs} and k_{ys} be the cumulants of two statistically independent random variables X and Y, respectively. Consider the sum of two random variables $Z = X + Y$. Show that the cumulants of Z, denoted by k_{zs}, are equal to the sum of k_{xs} and k_{ys}.

CHAPTER 5

Discrete Random Variables and Their Distributions

5.1 BINOMIAL AND RELATED DISTRIBUTIONS

5.1.1 Binomial Distribution

Let us consider a random experiment whose outcome is but one of two mutually exclusive events; for example, success or failure or occurrence or nonoccurrence of the event. It is assumed that the probability of success for each trial is p; thus the probability of failure is $q = 1 - p$. Let the random experiment be repeated n-independent times (called the n *Bernoulli trials* with parameter, p), and let us assume that x successes occur in n experiments. Note that there are $\binom{n}{x}$ sequences in which x successes and $(n - x)$ failures take place. Each of these sequences occurs with probability $p^x q^{n-x}$. Hence, the probability of x successes in n trials becomes

$$p(x) = \binom{n}{x} p^x q^{n-x}$$

This leads to the following definition of the binomial distribution:

Definition 5.1. Let p be the probability of success of an event and $q = 1 - p$ be failure of the event in each trial. The discrete random variable X is said to have the *binomial probability distribution* if the probability of obtaining exactly x successes in n Bernoulli trials is given by

$$p(x) = \binom{n}{x} p^x q^{n-x} \tag{5.1}$$

where $x = 0, 1, 2, \ldots, n$, and $0 < p < 1$, $q = 1 - p$.

The binomial distribution is often denoted by $b(n, p)$. In particular, the binomial distribution with $n = 1$ is called the *Bernoulli probability distribution* or the *zero–one probability distribution*.

Example 5.1. The probability of occurrence of flooding of City A when a storm passes in that area is 0.04. The probability of no flooding in 50 independent storms can be evaluated by applying the binomial distribution as

$$Pr\{x = 0\} = \binom{50}{0}(0.04)^0(1 - 0.04)^{50} = 0.130 \qquad \blacksquare$$

Example 5.2. Consider the launching from a submarine of 12 missiles aimed at a target. Assume that the probability of success for each launching is 0.2 and each shot is considered to be not influenced by previous shots. The probability that there are at least two successes in 12 launchings can be evaluated as follows:

$$Pr\{\text{No success in 12 launchings}\} = \binom{12}{0}(0.2)^0(0.8)^{12} = 0.069$$

$$Pr\{\text{One success in 12 launchings}\} = \binom{12}{1}(0.2)^1(0.8)^{11} = 0.206$$

Hence,

$$Pr\{\text{At least two successes in 12 launchings}\}$$
$$= 1 - Pr\{\text{No success}\} - Pr\{\text{One success}\} = 0.725 \qquad \blacksquare$$

The characteristic function of the binomial distribution can be obtained as

$$\phi(t) = \sum_{x=0}^{n} e^{itx} \binom{n}{x} p^x q^{n-x}$$

$$= \sum_{x=0}^{n} \binom{n}{x} (pe^{it})^x q^{n-x}$$

$$= (pe^{it} + q)^n \tag{5.2}$$

Hence, the mean and variance become

$$E[x] = \frac{1}{i}\phi'(0) = np$$

$$E[x^2] = \frac{1}{i^2}\phi''(0) = np + n(n-1)p^2$$

so that

$$\text{Var}[x] = E[x^2] - (E[x])^2 = npq \tag{5.3}$$

Consider x successes in n Bernoulli trials in the binomial distribution. If we are interested in the relative frequency of successes, then let the random variable Y be X/n, where the random variable X has the binomial distribution. The sample space of Y is $(0, 1/n, 2/n, \ldots, 1)$, and the probability that $Y = X/n$ is equivalent to the probability $X = x$ in Eq. (5.1). However, the characteristic function of Y is different from that shown in Eq. (5.2). It can be obtained by applying Theorem 4.3 as

$$\phi(t) = (pe^{it/n} + q)^n \tag{5.4}$$

Therefore, the mean and variance of Y becomes

$$E[y] = E\left[\frac{x}{n}\right] = p$$

and

$$\text{Var}[y] = \text{Var}\left[\frac{x}{n}\right] = \frac{pq}{n} \tag{5.5}$$

Example 5.3. Let the random variables X_1, X_2, and X_3 be statistically independent having the same probability density function $f(x) = e^{-x}$, $0 < X < \infty$. Obtain the probability that two of these three random variables exceed 1.

$$Pr\{\text{Two of the three random variables} > 1\}$$

$$= \binom{3}{2}(1 - Pr\{x < 1\})^2 \cdot Pr\{x < 1\}$$

where

$$Pr\{x < 1\} = \int_0^1 e^{-x}\,dx = 1 - e^{-1}$$

Thus,

$$Pr\{\text{two of the three} > 1\} = \binom{3}{2}(e^{-1})^2(1 - e^{-1}) \qquad \blacksquare$$

Example 5.4. For the random variables X_1, X_2, and X_3 in the previous example, let Y be the smallest of the three. Find the probability density function of Y. Note that two of the three random variables have to be greater than Y. Hence, the probability density function of Y is given by

$$f(y) = \binom{3}{2}e^{-y}\left(\int_y^\infty e^{-x}\,dx\right)^2 = 3e^{-3y} \qquad \blacksquare$$

If the number n is the binomial distribution $b(n, p)$ is large, then the binomial distribution (which is a discrete random variable) is approximately equal to the normal probability distribution (which is a continuous random variable) with the mean value np and variance npq. The proof will be given in Section 6.1.

5.1.2 Negative Binomial Distribution

In the binomial distributioin, we consider x successes in n Bernoulli trials with the probability of success of p. We now consider the case of a total of x successes before the rth failure occurs. This implies x successes and $(r - 1)$ failures in $(x + r - 1)$ trials, followed by a failure in the $(x + r)$th trial. According to the binomial distribution, the probability of x successes in $(x + r - 1)$ trials is given by

$$Pr\left\{\begin{array}{c} x \text{ successes and} \\ r - 1 \text{ failures} \end{array}\right\} = \binom{x + r - 1}{x} p^x q^{r-1}$$

where

$$x = 0, 1, 2, \ldots$$

$$p = \text{probability of success}, \; q = 1 - p$$

Since the $(x + r)$th trial is a failure that takes place with the probability of q, the probability function of x is given by

$$p(x) = \binom{x + r - 1}{x} p^x q^r = (-1)^x \binom{-r}{x} p^x q^r \qquad (5.6)$$

BINOMIAL AND RELATED DISTRIBUTIONS   95

Note that

$$\binom{x+r-1}{x} = \frac{r(r+1)(r+2)\cdots(r+x-1)}{x!}$$

$$= (-1)^x \frac{(-r-x+1)\cdots(-r-2)(-r-1)(-r)}{x!}$$

$$= (-1)^x \frac{(-r)!}{x!(-r-x)!} = (-1)^x \binom{-r}{x} \quad (5.7)$$

This leads to the following definition of the negative binomial distribution.

Definition 5.2. The discrete random variable X is said to be the *negative binomial distribution* if, in the Bernoulli trial with parameter p, the probability of obtaining x successes before the rth failure takes place is given by

$$p(x) = \binom{x+r-1}{x} p^x q^r = (-1)^x \binom{-r}{x} p^x q^r \quad (5.8)$$

where

$$x = 1, 2, \ldots, \infty$$

$$q = 1 - p.$$

The characteristic function of the distribution can be written as

$$\phi(t) = \sum_{x=0}^{\infty} (-1)^x \binom{-r}{x} p^x q^r e^{itx} = q^r \sum_{x=0}^{\infty} \binom{-r}{x} (-pe^{it})^x$$

Here, $|pe^{it}| < 1$; hence, by applying the binomial series, we have

$$\sum_{x=0}^{\infty} \binom{-r}{x} (-pe^{it})^x = (1 - pe^{it})^{-r}$$

Thus, the characteristic function of the negative binomial probability distribution becomes

$$\phi(t) = q^r (1 - pe^{it})^r = \left(\frac{q}{1 - pe^{it}}\right)^r \quad (5.9)$$

It follows that the mean and variance can be obtained as

$$E[x] = rp/q$$

$$\text{Var}[x] = rp/q^2 \qquad (5.10)$$

It will be shown in the next section that if r is large, the negative binomial distribution can be approximated by the Poisson distribution.

5.1.3 *Hypergeometric Distribution*

Let us consider generalization of the binomial distribution. We recall that the binomial distribution considers two mutually exclusive outcomes for which the probabilities of occurrence are p and $q = 1 - p$, respectively. The binomial distribution can be applied to obtain the probability that the number of occurrences is exactly x in N observations. However, if we arbitrarily choose n samples from N and examine the probability of getting exactly x occurrences of the event in these n samples, then the binomial distribution cannot be applied. Instead, the probability distribution of x is given by the following hypergeometric distribution.

Definition 5.3. A sample of size n is drawn without replacement from a population of N in which there are Np materials possessing a certain attribute. The probability of getting exactly x materials having this particular attribute in n samples obeys the *hypergeometric distribution* given by

$$p(x) = \frac{\binom{Np}{x}\binom{Nq}{n-x}}{\binom{N}{n}}$$

where

$$x = 0, 1, 2, \ldots, n,$$

$$q = 1 - p \qquad (5.11)$$

The characteristic function of the hypergeometric distribution is rather difficult to obtain. Hence, the mean and variance of the distribution may be

BINOMIAL AND RELATED DISTRIBUTIONS

evaluated by direct computation as shown in the following. That is,

$$E[x] = \sum_{x=0}^{n} x \frac{\binom{Np}{x}\binom{Nq}{n-x}}{\binom{N}{n}}$$

$$= n\frac{Np}{N} \sum_{x=1}^{n} \frac{\binom{Np-1}{x-1}\binom{Nq}{n-x}}{\binom{N-1}{n-1}}$$

Since the summation from $x = 1$ to n is equal to 1, we have

$$E[x] = np \tag{5.12}$$

Similarly, $E[x(x - 1)]$ can be evaluated as

$$E[x(x-1)] = \sum_{x=0}^{n} x(x-1) \frac{\binom{Np}{x}\binom{Nq}{n-x}}{\binom{N}{n}}$$

$$= n(n-1)\frac{Np(Np-1)}{N(N-1)} \sum_{x=2}^{n} \frac{\binom{Np-2}{x-2}\binom{Nq}{n-x}}{\binom{N-2}{n-2}}$$

$$= n(n-1)\frac{p(Np-1)}{N-1}$$

Thus, the variance becomes

$$\text{Var}[x] = E[x(x-1)] + E[x] - (E[x])^2 = \frac{N-n}{N-1} npq \tag{5.13}$$

Let us consider a special case of the hypergeometric distribution in which N is large. From Eq. (5.11), we have

$$p(x) = \frac{\binom{Np}{x}\binom{Nq}{n-x}}{\binom{N}{n}}$$

$$= \binom{n}{x} \frac{Np(Np-1) \cdots \{Np-(x-1)\} Nq(Nq-1) \cdots \{Nq-(n-x-1)\}}{N(N-1)(N-2) \cdots \{N-(n-1)\}}$$

By dividing both numerator and denominator by N and by letting $N \to \infty$, we have

$$p(x) = \lim_{N \to \infty} \binom{n}{x} \frac{p\left(p - \frac{1}{N}\right) \cdots \left(p - \frac{x-1}{N}\right) q\left(q - \frac{1}{N}\right) \cdots \left(q - \frac{n-x-1}{N}\right)}{1\left(1 - \frac{1}{N}\right)\left(1 - \frac{2}{N}\right) \cdots \left(1 - \frac{n-1}{N}\right)}$$

$$= \binom{n}{x} p^x q^{n-x} \qquad (5.14)$$

Thus, it can be proved that the hypergeometric distribution reduces to the binomial distribution if $N \to \infty$.

5.1.4 *Multinomial Distribution*

As another generalization of the binomial distribution, let us increase the number of outcomes (random variables) in each trial. Two possible outcomes are considered in the binomial distribution. We now consider k mutually exclusive outcomes, $X_1, X_2, \ldots, X_k$, each of which has a probability of occurrence $p_1, p_2, \ldots, p_k$, respectively, where $p_1 + p_2 + \cdots + p_k = 1$. Then, we have the following multinomial distribution:

Definition 5.4. Let $X_1, X_2, \ldots, X_k$ be mutually exclusive outcomes (random variables) in each trial. Let $p_1, p_2, \ldots, p_k$ be the probability of occurrence of each outcome, where $p_1 + p_2 + \cdots + p_k = 1$. Then, the probability that the outcomes occur $x_1, x_2, \ldots, x_k$ times, respectively, in the n-independent trials is given by the following *multinomial distribution*:

$$p\{X_1 = x_1, X_2 = x_2, \ldots, X_k = x_k\} = \frac{n!}{x_1! x_2! \cdots x_k!} p_1^{x_1} p_2^{x_2} \cdots p_k^{x_k}$$

where

$$p_1 + p_2 + \cdots + p_k = 1$$

$$x_1 + x_2 + \cdots + x_k = n \qquad (5.15)$$

Note that if we use an expression similar to that of the binomial distribution, we may write

$$q = 1 - \sum_{j=1}^{k-1} p_j$$

Then, Eq. (5.15) can be written as

$$p\{X_1 = x_1, X_2 = x_2, \ldots, X_k = x_k\}$$

$$= \frac{n!}{x_1! x_2! \cdots x_{k-1}! \left(n - \sum_{j=1}^{k-1} x_j\right)!}$$

$$\times p_1^{x_1} p_2^{x_2} \cdots p_{k-1}^{x_{k-1}} q^{n - \sum_{j=1}^{k-1} x_j} \qquad (5.16)$$

By using the expression given in Eq. (5.16), the characteristic function of the multinomial distribution can be evaluated by

$$\phi(t_1, t_2, \ldots, t_{k-1}) = \sum_{x_1=0}^{n} \sum_{x_2=0}^{n-x_1} \cdots \sum_{x_k=0}^{n-(x_1+\cdots+x_{k-1})}$$

$$\times \exp\{i(t_1 x_1 + t_2 x_2 + \cdots + t_{k-1} x_{k-1})\}$$

$$\times \frac{n!}{x_1! x_2! \cdots x_{k-1}! \left(n - \sum x_j\right)!} p_1^{x_1} p_2^{x_2} \cdots q^{n - \sum x_j}$$

$$= \sum \sum \cdots \sum \frac{n!}{x_1! x_2! \cdots x_{k-1}! \left(n - \sum x_j\right)}$$

$$\times \left(p_1 e^{it_1}\right)^{x_1} \cdots \left(p_{k-1} e^{it_{k-1}}\right)^{x_{k-1}} q^{n - \sum x_j}$$

$$= \left(p_1 e^{it_1} + p_2 e^{it_2} + \cdots + p_{k-1} e^{it_{k-1}} + q\right)^n$$

$$= \left(\sum_{j=1}^{k-1} p_j e^{it_j} + q\right)^n \qquad (5.17)$$

For $k = 2$, Eq. (5.17) reduces to the characteristic function of the binomial distribution given in Eq. (5.2). The mean and variance of each random variable, X_j, are given similar to those of the binomial distribution; namely,

$$E[x_j] = np_j$$

$$\text{Var}[x_j] = np_j(1 - p_j) \qquad (5.18)$$

The random variables $X_1, X_2, \ldots, X_k$ of the multinomial distribution are mutually exclusive, but they are not statistically independent. In order to prove this property, let us consider a simple case of a trinomial distribution, X_1, X_2, and X_3. The joint characteristic function is given by

$$\phi(t_1, t_2) = \left(p_1 e^{it_1} + p_2 e^{it_2} + q\right)^n, \quad \text{where } p_1 + p_2 + q = 1$$

Then, the characteristic function of the marginal distribution of X_1 and X_2 can be obtained as

$$\phi(t_1, 0) = \left(p_1 e^{it_1} + p_2 + q\right)^n$$

$$\phi(0, t_2) = \left(p_1 + p_2 e^{it_2} + q\right)^n$$

Since $\phi(t_1, t_2) \neq \phi(t_1, 0)\phi(0, t_2)$, X_1 and X_2 are not statistically independent.

Example 5.5. Consider the launching of missiles aimed at a target. The probability that a missile hits the target is 0.05, and the probability that a missile falls within 1 mile of the target is 0.15. Evaluate the probability that when 10 missiles are fired, at least one missile will hit the target and at least two missiles will be within one mile of the target.

Let X_1 be the number of missiles hitting the target and X_2 be the number of missiles falling within 1 mile of the target. Then, the desired probability can be evaluated by

$$Pr\{X_1 \geq 1, X_2 \geq 2\} = 1 - [Pr\{X_1 = 0\} + Pr\{X_2 < 2\} - Pr\{X_1 = 0, X_2 < 2\}]$$

where

$$Pr\{X_1 = 0\} = \binom{10}{0}(0.05)^0(0.95)^{10} = 0.599$$

$$Pr\{X_2 < 2\} = \sum_{j=0,1} \binom{10}{j}(0.15)^j(0.85)^{10-j} = 0.544$$

$$Pr\{X_1 = 0, X_2 < 2\} = \sum_{j=0,1} \frac{10!}{0!j!(10-j)!}(0.05)^0(0.15)^j(0.80)^{10-j}$$

$$= 0.308$$

Thus, we have

$$Pr\{X_1 \geq 1, X_2 \geq 2\} = 0.165 \qquad \blacksquare$$

5.2 POISSON AND RELATED DISTRIBUTIONS

5.2.1 Poisson Distribution

Evaluation of the statistical properties of the occurrence of rare events such as the occurrence of accidents, earthquakes, floods, and passage of particles from a radioactive source through a Geiger counter, is an extremely important subject in applied probability in engineering and the physical sciences. Random processes that deal with counting the number of occurrences of rare events are called *counting random processes* and will be discussed in detail in Chapter 17. Among counting processes, the Poisson random process is most frequently observed in natural phenomena, and the basic statistical properties of Poisson random processes can be described by the Poisson probability distribution introduced in this section.

We first derive the Poisson probability distribution from the binomial distribution. Since we consider the occurrence of rare events, it is appropriate to let the probability of occurrence of an event in each trial, p, in the binomial distribution be small and let the number of trials, n, be large. However, the product np, which is equal to the mean value of the binomial distribution is finite. Let us write $np = \mu$. Then, the binomial distribution can be expressed in terms of n and μ as follows:

$$p(x) = \binom{n}{x} p^x (1-p)^{n-x} = \frac{n!}{x!(n-x)!} \left(\frac{\mu}{n}\right)^x \left(1 - \frac{\mu}{n}\right)^{n-x}$$

$$= \frac{\mu^x}{x!} \left(1 - \frac{\mu}{n}\right)^n \frac{n(n-1)(n-2)\cdots(n-x+1)}{n^x} \frac{1}{\left(1 - \frac{\mu}{n}\right)^x}$$

$$= \frac{\mu^x}{x!} \left(1 - \frac{\mu}{n}\right)^n \frac{1\left(1 - \frac{1}{n}\right)\left(1 - \frac{2}{n}\right)\cdots\left(1 - \frac{x-1}{n}\right)}{\left(1 - \frac{\mu}{n}\right)^x} \qquad (5.19)$$

Since

$$\lim_{n \to \infty} \left(1 - \frac{\mu}{n}\right)^n = e^{-\mu}$$

and

$$\lim_{n \to \infty} \frac{1\left(1 - \frac{1}{n}\right)\left(1 - \frac{2}{n}\right) \cdots \left(1 - \frac{x-1}{n}\right)}{\left(1 - \frac{\mu}{n}\right)^x} = 1$$

we have

$$p(x) = \frac{\mu^x}{x!} e^{-\mu}$$

The distribution given above is called the Poisson distribution.

Definition 5.5. A discrete random variable X whose probability mass function is given by the following formula is called the *Poisson distribution*.

$$p(x) = \frac{\mu^x}{x!} e^{-\mu}, \qquad x = 0, 1, 2, \ldots, \infty \tag{5.20}$$

The characteristic function of the Poisson distribution can be obtained directly by applying the formula given in Eq. (4.3). This is shown in Example 4.13. Here, we will show the derivation of the characteristic function from the binomial distribution for $n \to \infty$.

It was obtained in Eq. (5.2) that the characteristic function of the binomial distribution is given by

$$\phi(t) = \left(pe^{it} + q\right)^n, \qquad \text{where } q = 1 - p$$

By letting $p = \mu/n$, we have

$$\phi(t) = \left(pe^{it} + 1 - p\right)^n = \left\{1 + p\left(e^{it} - 1\right)\right\}^n$$

$$= \left\{1 + \frac{\mu}{n}\left(e^{it} - 1\right)\right\}^n$$

Hence, for $n \to \infty$,

$$\phi(t) = \lim_{n \to \infty} \left\{1 + \frac{\mu}{n}(e^{it} - 1)\right\}^n = e^{\mu(e^{it} - 1)} \tag{5.21}$$

The mean and variance of the distribution can be readily obtained from Eq. (5.21). These are

$$E[x] = \frac{1}{i}\phi'(0) = \mu$$

$$E[x^2] = \frac{1}{i^2}\phi''(0) = \mu(\mu + 1)$$

Thus,

$$\text{Var}[x] = E[x^2] - (E[x])^2 = \mu \tag{5.22}$$

The Poisson distribution is often used in connection with the number of occurrences of an event in random processes. In this context, the parameter μ involved in the Poisson distribution may be written as νt, where $\nu > 0$ represents the average number of occurrences of the event per unit time. Then, the probability of exactly x occurrences of the random event in an interval time t is given by

$$p(x) = \frac{(\nu t)^x}{x!} e^{-\nu t}, \qquad x = 0, 1, 2, \ldots \tag{5.23}$$

Example 5.6. Records indicate that the number of passages of a particle from a radioactive source through a Geiger counter is, on the average, 0.4 per minute. What is the probability that there will be at least one passage in the next 2 minutes?

Let X be the number of passages of particles in 2 minutes. Then the probability of no passage in 2 minutes can be evaluated by

$$Pr\{X = 0\} = e^{-(0.4 \times 2)} \frac{(0.4 \times 2)^0}{0!} = 0.449$$

Hence,

$$Pr\{\text{At least one passage}\} = 1 - Pr\{X = 0\} = 0.551 \qquad \blacksquare$$

Example 5.7. Particles emitted from a radioactive source follow the Poisson probability law with an average number ν per unit time. The Geiger counter, however, registers particles with probability p. Hence, the probability that the counter registers exactly x particles in an interval time t can be evaluated by

$$p(x) = \frac{(p\nu t)^x}{x!} e^{-p\nu t}$$ ∎

Next, the relationship between the Poisson distribution and the negative binomial distribution is shown. It can be proved that the negative binomial distribution reduces to the Poisson distribution if the parameter r of the negative binomial distribution is large. In order to prove this, we may consider the characteristic function of the negative binomial distribution given in Eq. (5.9).

By letting $rp/q = \mu$, we have $p = \mu/(\mu + r)$ and $q = r/(\mu + r)$. Then, the characteristic function of the negative binomial distribution becomes

$$\phi(t) = \left(\frac{q}{1 - pe^{it}}\right)^r = \left(\frac{r}{r + \mu - \mu e^{it}}\right)^r$$

$$= \left\{1 - \frac{\mu(e^{it} - 1)}{r}\right\}^{-r} \qquad (5.24)$$

For $r \to \infty$, we have

$$\phi(t) = \lim_{r \to \infty} \left\{1 - \frac{\mu(e^{it} - 1)}{r}\right\}^{-r} = e^{\mu(e^{it} - 1)} \qquad (5.25)$$

This is the characteristic function of the Poisson distribution given in Eq. (5.21). Thus, the negative binomial distribution can be approximated by the Poisson distribution if the parameter r of the negative binomial distribution is large.

Example 5.8. The probability of hitting a target by a missile is 0.42. What is the probability that a total of 10 launchings have been performed before the fifth hit takes place? This problem can also be read as follows: The radioactive particles arrived at a counting system are not always

POISSON AND RELATED DISTRIBUTIONS

registered. The probability of the registration is 0.42 for each arrival. What is the probability that a total of 10 particles have arrived before the fifth registration takes place?

We may apply the negative binomial distribution given in Eq. (5.8). By letting $x =$ number of failures in missile launching before the fifth hit (number of nonregistrations in particle arrival before the fifth registration) $= 6$, $p =$ probability of failure in each launching (probability of nonregistration in each particle arrival) $= 0.58$, and $r = 5$, we have

$$Pr\{X = 6\} = \binom{10}{6}(0.58)^6(0.42)^5 = 0.104$$

We may evaluate the probability by applying the Poisson approximation. For this, by letting $\mu = rp/(1 - p) = 6.905$, we have

$$Pr\{x = 6\} = \frac{(6.905)^6}{6!}e^{-6.905} = 0.151 \qquad \blacksquare$$

5.2.2 Poisson Distribution with Parameter a Random Variable

In the Poisson distribution defined in Eq. (5.20), the parameter μ, which represents the mean value of the distribution, is considered to be a constant. However, in many problems in engineering and physical sciences, the parameter μ may not be a constant; instead, μ is a random variable that has the probability density function $f(\mu)$. This type of Poisson distribution can be applied to a wide variety of practical problems.

In order to obtain the Poisson distribution with parameter μ a random variable, we consider the distribution given in Eq. (5.20) as a conditional probability distribution for a given μ, which may be written as

$$p\{X = x|\mu\} = \frac{\mu^x}{x!}e^{-\mu}, \qquad x = 0, 1, 2, \ldots \qquad (5.26)$$

Here, μ is not a constant but is a random variable. It is often observed in natural sciences that the parameter μ has the following probability density function called the gamma distribution (see Section 6.2):

$$f(\mu) = \frac{1}{\Gamma(m)}\lambda^m \mu^{m-1}e^{-\lambda\mu}, \qquad 0 \leq \mu < \infty \qquad (5.27)$$

Then, the unconditional probability distribution $p\{X = x\}$, can be obtained from Eqs. (5.26) and (5.27) as

$$p\{X = x\} = \int_0^\infty \frac{\mu^x}{x!} e^{-\mu} \cdot \frac{1}{\Gamma(m)} \lambda^m \mu^{m-1} e^{-\lambda\mu} \, d\mu$$

$$= \frac{\lambda^m}{\Gamma(m)} \int_0^\infty \frac{\mu^{x+m-1} e^{-(\lambda+1)\mu}}{x!} \, d\mu$$

$$= \frac{\lambda^m}{\Gamma(m)} \frac{\Gamma(x+m)}{x!(\lambda+1)^{x+m}}$$

$$= \frac{(x+m-1)!}{(m-1)!x!} \left(\frac{1}{\lambda+1}\right)^x \left(\frac{\lambda}{\lambda+1}\right)^m \quad (5.28)$$

By letting $1/(\lambda + 1) = p$ and $\lambda/(\lambda + 1) = q$, we have

$$p\{X = x\} = \binom{x+m-1}{x} p^x q^m$$

$$= (-1)^x \binom{-m}{x} p^x q^m \quad x = 0, 1, 2, \ldots \quad (5.29)$$

This is the negative binomial probability distribution given in Definition 5.2. Thus, we have shown that the Poisson distribution with the parameter μ, which obeys the gamma probability law is a negative binomial distribution. Another method to derive the same result is to apply the conditional characteristic function given in Eq. (4.14). The deviation is left for the reader as an exercise.

5.3 RELATIONSHIP BETWEEN DISTRIBUTIONS OF VARIOUS DISCRETE-TYPE RANDOM VARIABLES

It is shown in the previous two sections that some probability distributions of discrete-type random variables reduce to other probability distributions under certain conditions. The relationship between these probability distribution functions is shown in Figure 5.1 along with conditions required for approximation.

It is noted, as can be seen in the figure, that the binomial distribution and the Poisson distribution, both of which are the distributions of discrete-type random variables, can be approximated by the normal

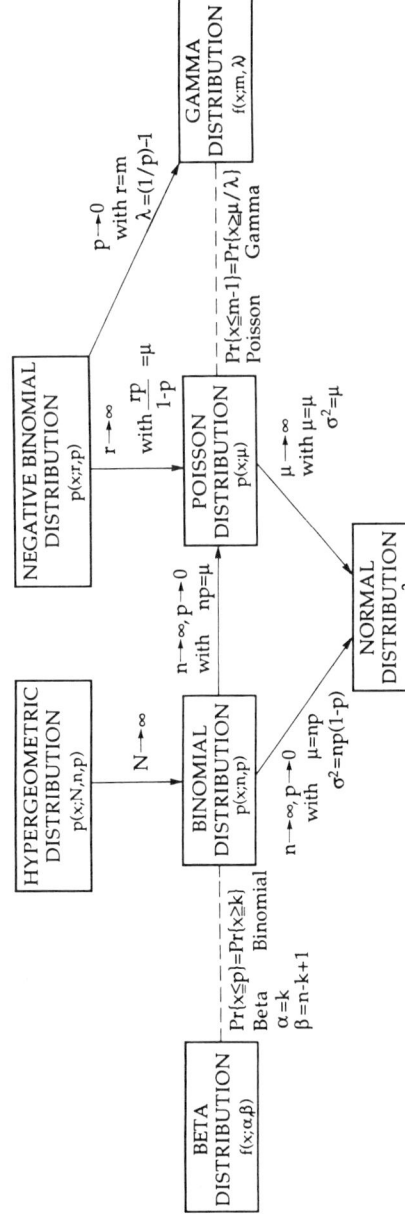

Figure 5.1 Relationship between some discrete-type random variables and some continuous-type random variables.

probability distribution that is a continuous-type random variable. The proof of these approximations will be shown in Section 6.1.

The negative binomial distribution reduces to the gamma probability distribution, which is a continuous random variable if the parameter p is small. Proof of this approximation is left as an exercise in Chapter 6 (see Exercise 6.5).

The following two relationships between cumulative distribution functions of discrete and continuous-type random variables are often extremely useful in applying probability theory to practical problems:

(1) $Pr\{X \leqslant m - 1\}$ in the Poisson distribution with parameter μ
$= Pr\{X \geqslant \mu/\lambda\}$ in the gamma distribution with parameters m and λ,

where

$$\text{Poisson distribution: } p(x) = \frac{\mu^x}{x!}e^{-\mu}, \quad x = 0, 1, 2, \ldots, \infty$$

$$\text{Gamma distribuition: } f(x) = \frac{1}{\Gamma(m)}\lambda^m x^{m-1} e^{-\lambda x}, \quad 0 \leqslant x < \infty \quad (5.30)$$

(2) $Pr\{X \geqslant k\}$ in the binomial distribution with parameter p
$= Pr\{X \leqslant p\}$ in the beta distribution with parameters $\alpha = k$, $\beta = n - k + 1$

where

$$\text{Binomial distribution: } p(x) = \binom{n}{x} p^x q^{n-x}, \quad x = 0, 1, 2, \ldots, n$$

$$\text{Beta distribution: } f(x) = \frac{1}{B(\alpha, \beta)} x^{\alpha - 1}(1 - x)^{\beta - 1}, \quad 0 \leqslant x \leqslant 1$$

$$\text{with } \alpha = k, \beta = n - k + 1 \quad (5.31)$$

Proof of these relationships will be given in sections where the gamma and beta distributions, respectively, are discussed in Chapter 6.

EXERCISES

5.1 Let the probability of occurrence of an event in each trial be p, where $0 < p < 1$. The random variable X represents the number of trials before the first event occurs. Obtain the mean and variance of X.

5.2 Let the probability of occurrence of a particular event in each trial be p, where $0 < p < 1$. Evaluate the probability of occurrence of the kth event in $(x + k)$ trials.

5.3 Derive the characteristic function of the Poisson distribution by applying the formula given in Eq. (4.3).

5.4 Show that the sum of two independent random variables, each obeying the Poisson distribution, is also the Poisson distribution and that the difference of these two random variables is not the Poisson distribution.

5.5 By applying the conditional characteristic function given in Eq. (4.14), prove that the Poisson distribution with the parameter μ, which obeys the gamma probability law, is a negative binomial distribution.

5.6 The random variable X has a binomial distribution with the parameter p, where p is not a constant but instead a random variable given by

$$f(p) = \frac{\Gamma(\alpha + \beta)}{\Gamma(\alpha)\Gamma(\beta)} p^{\alpha-1} q^{\beta-1}, \quad 0 < p < 1, q = 1 - p$$

Prove that the random variable X has the following probability:

$$Pr\{X = x\} = \binom{n}{x} \frac{\Gamma(\alpha + \beta)\Gamma(x + \alpha)\Gamma(n - x + \beta)}{\Gamma(\alpha)\Gamma(\beta)\Gamma(n + \alpha + \beta)},$$

$$x = 0, 1, 2, \ldots, n$$

This distribution is called the negative hypergeometric distribution.

5.7 Consider the random phenomenon that takes the value either $+1$ or -1 by changing sign in a random fashion with time illustrated in

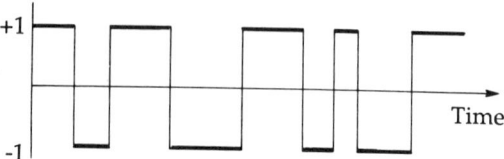

Figure 5.2 Random telegraph signal.

Figure 5.2. Assume that the signal is $+1$ at time $t = 0$, and that the probability of the number of changes of sign in time t, denoted by x, follows the Poisson distribution:

$$Pr\{X = x\} = \frac{(\nu t)^x}{x!}e^{-\nu t}, \qquad x = 0, 1, 2, \ldots, \nu > 0$$

Show that the average number of sign of changes in time t is equal to $e^{-2\nu t}$. This random phenomenon is called the *random telegraph signal*.

CHAPTER 6

Continuous Random Variables and Their Distributions

6.1 NORMAL AND RELATED DISTRIBUTIONS

6.1.1 *Normal Distribution*

Definition 6.1. A random variable X is said to have a *normal distribution* with mean μ and variance σ^2 if its probability density function is given by

$$f(x) = \frac{1}{\sigma\sqrt{2\pi}} e^{-[(x-\mu)^2/2\sigma^2]}, \qquad -\infty < x < \infty \qquad (6.1)$$

The normal distribution is often called the *Gaussian distribution* and it is denoted by $N(\mu, \sigma^2)$. The two parameters, μ and σ^2, involved in the distribution are the mean and variance, respectively, of the distribution. This can be proved from the following characteristic function of the distribution, which was obtained in Section 4.2:

$$\phi(t) = e^{i\mu t - \frac{1}{2}\sigma^2 t^2} \qquad (6.2)$$

By differentiating the characteristic function with respect to t, we have

$$E[x] = \text{mean} = \frac{1}{i}\phi'(0) = \mu$$

and

$$E[x^2] = \frac{1}{i^2}\phi''(0) = \mu^2 + \sigma^2$$

Hence,

$$\text{Variance} = E[x^2] - (E[x])^2 = \sigma^2$$

A normal distribution with $\mu = 0$ and $\sigma^2 = 1$ is called a *standardized normal distribution*, denoted by $N(0,1)$. For convenience, let us denote a random variable having a standardized normal distribution by Z. The probability density function is given by

$$f(z) = \frac{1}{\sqrt{2\pi}} e^{-(z^2/2)}, \quad -\infty < z < \infty \tag{6.3}$$

As shown in Example 3.6, the probability density function, $f(z)$ can be derived from the probability density function $f(x)$ by standardizing the random variable X by

$$Z = \frac{X - \mu}{\sigma} \tag{6.4}$$

The probability density function of the standardized normal distribution $f(z)$ is shown in Figure 6.1. As can be seen in the figure, the probability that the random variable Z falls within the range of $(-3, +3)$ is very large, 0.9973. Then, from the relationship given in Eq. (6.4), we have

$$Pr\{[X - \mu] > 3\sigma\} = 0.0027 \tag{6.5}$$

which implies that the probability that the value of the random variable X deviates from the mean value by more than 3σ is very small, approximately 0.27%. This property of the normal distribution is called the *three-sigma rule*.

NORMAL AND RELATED DISTRIBUTIONS

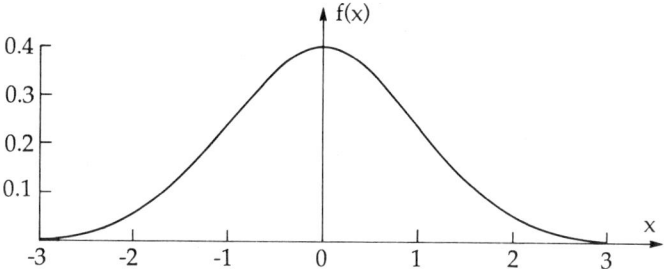

Figure 6.1 Standardized normal probability density function.

The cumulative distribution function of the standardized normal probability distribution is usually denoted by $\Phi(z)$, which is defined as

$$\Phi(z) = \frac{1}{\sqrt{2\pi}} \int_{-\infty}^{z} e^{-z^2/2} \, dz \qquad (6.6)$$

It is noted that the relationship between the cumulative distribution function of the standardized normal distribution, $\Phi(z)$, and the error function often used in mathematical analysis is as follows:

$$\operatorname{erf} x = 2\Phi(\sqrt{2}\,x) - 1$$

where

$$\operatorname{erf} x = \frac{2}{\sqrt{\pi}} \int_{0}^{x} e^{-x^2} \, dx \qquad (6.7)$$

By using the definition given in Eq. (6.6), we can evaluate the probability that the random variable X, which has a normal distribution $N(\mu, \sigma^2)$, falls between a and b as follows:

$$\begin{aligned} Pr\{a < X < b\} &= Pr\left\{ \frac{a-\mu}{\sigma} < Z < \frac{b-\mu}{\sigma} \right\} \\ &= \Phi\left(\frac{b-\mu}{\sigma}\right) - \Phi\left(\frac{a-\mu}{\sigma}\right) \end{aligned} \qquad (6.8)$$

Example 6.1. A random phenomenon is assumed to follow a normal probability distribution with mean $\mu = 50$ and variance $\sigma^2 = 100$. Evaluate (i) $Pr\{x \leq 75\}$, (ii) $Pr\{|x - 75| \leq 10\}$, and (iii) the number a such that $Pr\{x \leq a\} = 0.95$.

Let Z be the standardized random variable, $Z = (X - 50)/10$.

Then (i) $Pr\{x \leq 75\} = Pr\left\{\dfrac{x - 50}{10} \leq \dfrac{75 - 50}{10}\right\}$

$= Pr\{z \leq 2.5\}$

$= \Phi(2.5) = 0.9938$

(ii) $Pr\{|x - 75| \leq 10\} = Pr\{65 \leq x \leq 85\}$

$= Pr\left\{\dfrac{65 - 50}{10} \leq z \leq \dfrac{85 - 50}{10}\right\}$

$= \Phi(3.5) - \Phi(1.5) = 0.0666$

(iii) $Pr\{x \leq a\} = Pr\left\{z \leq \dfrac{a - 50}{10}\right\} = 0.95$.

Hence, $(a - 50)/10 \leq 1.645$, and thereby $a = 66.45$. ∎

It is shown in Figure 5.1 of Section 5.3 that the Poisson probability distribution, which is the distribution of a discrete-type random variable, reduces asymptotically to the normal probability distribution if the parameter of the Poisson distribution is large. The proof of this property is now given as follows:

Consider the Poisson distribution with the parameter μ whose characteristic function is given by Eq. (5.21). That is,

$$\phi_x(t) = e^{\mu(e^{it} - 1)}$$

Let us standardize the distribution by applying the formula given in Eq. (3.7). Since the mean and variance of the Poisson distribution are both μ, the transformation to the standardized distribution is given by $Z = -\sqrt{\mu} + X/\sqrt{\mu}$. Then, by applying Theorem 4.3, the characteristic function of this standardized Poisson distribution becomes

$$\phi_z(t) = \exp\{-i\sqrt{\mu}\,t\} \cdot \exp\{\mu(e^{it/\sqrt{\mu}} - 1)\}$$

$$= \exp\left\{-\dfrac{t^2}{2} + \dfrac{(it)^3}{3!\sqrt{\mu}} + \cdots\right\} \qquad (6.9)$$

NORMAL AND RELATED DISTRIBUTIONS 115

It can be seen from the above equation that the characteristic function converges to that of the standardized normal distribution for large μ. Thus, the Poisson distribution with parameter μ can be approximated by the normal distribution with mean μ and variance μ if μ is large.

Example 6.2. Let X_1 and X_2 be two independent random variables having normal distributions $N(\mu_1, \sigma_1^2)$ and $N(\mu_2, \sigma_2^2)$, respectively. Then, the sum of two random variables $Y = a_1 X_1 + a_2 X_2$, where a_1 and a_2 are constants, has the normal distribution $N(a_1\mu_1 + a_2\mu_2, a_1^2\sigma_1^2 + a_2^2\sigma_2^2)$. In order to prove this property, let us consider the characteristic functions of each random variable. That is,

$$\phi_1(t) = \exp\{i\mu_1 t - \tfrac{1}{2}\sigma_1^2 t_1^2\}$$

$$\phi_2(t) = \exp\{i\mu_2 t - \tfrac{1}{2}\sigma_2^2 t^2\}$$

Then by applying Theorems 4.3 and 4.4, the characteristic function of the random variable Y becomes

$$\phi(t) = \phi_1(t)\phi_2(t) = \exp\{i(a_1\mu_1 + a_2\mu_2)t - \tfrac{1}{2}(a_1^2\sigma_1^2 + a_2^2\sigma_2^2)t^2\}$$

This is the characteristic function of the normal distribution, $N(a_1\mu_1 + a_2\mu_2, a_1^2\sigma_1^2 + a_2^2\sigma_2^2)$, which is to be proved. ∎

One of the most well-known theorems in probability theory is the central limit theorem that enhances the significance of the normal distribution. There are several different presentations of the central limit theorem; however, the following theorem, called the *Lindberg–Levy theorem*, is introduced here since this theorem will be used for probabilistic explanation of the Gaussian random process, which will be discussed in detail in Chapter 9.

Theorem 6.1. Let the random variables $X_1, X_2, X_3, \ldots, X_n$ be statistically independent having the same distribution with mean μ and variance σ^2. Then, the sum of these random variables, $Y = X_1 + X_2 + \cdots + X_n$, is asymptotically normally distributed with mean $n\mu$ and variance $n\sigma^2$ if n is large. This is called the *central limit theorem*.

Proof. Since the random variables $X_1, X_2, \ldots, X_n$ are statistically independent, we have $E[y] = n\mu$ and $\text{Var}[y] = n\sigma^2$. We standardize the random variable Y and let us express it in the following form:

$$Z = \frac{Y - n\mu}{\sqrt{n}\,\sigma} = \frac{1}{\sqrt{n}\,\sigma} \sum_{i=1}^{n} (X_i - \mu) = \frac{1}{\sqrt{n}\,\sigma} \sum_{i=1}^{n} U_i$$

where $U_i = X_i - \mu$.

Let us write the characteristic function of the random variable U as $\phi_u(t)$. Then, the characteristic function of Z can be expressed as

$$\phi_z(t) = \left\{ \phi_u\left(\frac{t}{\sqrt{n}\,\sigma}\right) \right\}^n$$

Here, $\phi_u(\ )$ may be expanded as follows:

$$\phi_u\left(\frac{t}{\sqrt{n}\,\sigma}\right) = \int_{-\infty}^{\infty} \exp\left\{\frac{itu}{\sqrt{n}\,\sigma}\right\} f(u)\, du$$

$$= 1 + \frac{it}{\sqrt{n}\,\sigma} E[u] + \frac{1}{2}\left(\frac{it}{\sqrt{n}\,\sigma}\right)^2 E[u^2] + \cdots$$

$$= 1 - \frac{t^2}{2n} + 0\left(\frac{t^2}{n}\right)$$

Thus, for large n, we have

$$\phi_z(t) = \lim_{n \to \infty} \left\{ 1 - \frac{t^2}{2n} + 0\left(\frac{t^2}{n}\right) \right\}^n = \exp\left\{-\frac{t^2}{2}\right\}$$

This implies that for large n the random variable Z approaches a standardized normal distribution, and thereby the random variable Y is asymptotically normally distributed with mean $n\mu$ and variance $n\sigma^2$.

It should be noted that a significant feature of the central limit theorem is that there is no specific requirement as to the type of probability distribution of the random variable X. As long as the random variables X_i are statistically independent having the same distribution and the first two moments exist, then the sum (and thereby the average) of X_i becomes an asymptotically normal distribution for large number n.

6.1.2 Log-Normal Distribution

Definition 6.2. A random variable X has a *log-normal distribution* if its probability density function is given by

$$f(x) = \frac{1}{\sigma\sqrt{2\pi}\, x} \exp\left[-\frac{(\ln x - \mu)^2}{2\sigma^2}\right], \quad 0 \leq x < \infty \quad (6.10)$$

The log-normal distribution has two parameters, μ and σ^2, and is often denoted by $\Lambda(\mu, \sigma^2)$. It should be noted that the parameters μ and σ^2 are not the mean and variance, respectively, of the random variable X. The logarithm of the random variable X is normally distributed with mean μ and variance σ^2. The mean and variance of the log-normal distribution are given as follows:

$$E[x] = \exp\left\{\mu + \frac{\sigma^2}{2}\right\}$$

$$\text{Var}[x] = \exp\{2\mu + \sigma^2\} \cdot \left(\exp\{\sigma^2\} - 1\right) \quad (6.11)$$

Examples of the log-normal probability density function are shown in Figure 6.2.

The log-normal distribution is frequently used for statistical analysis of random phenomena in natural sciences (see Aitchison and Brown, 1957). The distribution has the following properties, which are extremely useful in practice:

1. If the random variable X has a log-normal distribution, $\Lambda(\mu, \sigma^2)$, then the random variable $1/X$ has a log-normal distribution $\Lambda(-\mu, \sigma^2)$.

2. If the random variable X has a log-normal distribution, $\Lambda(\mu, \sigma^2)$, then the random variable $e^a X^b$, where a and b are constants, has a log-normal distribution $\Lambda(a + b\mu, b^2\sigma^2)$.

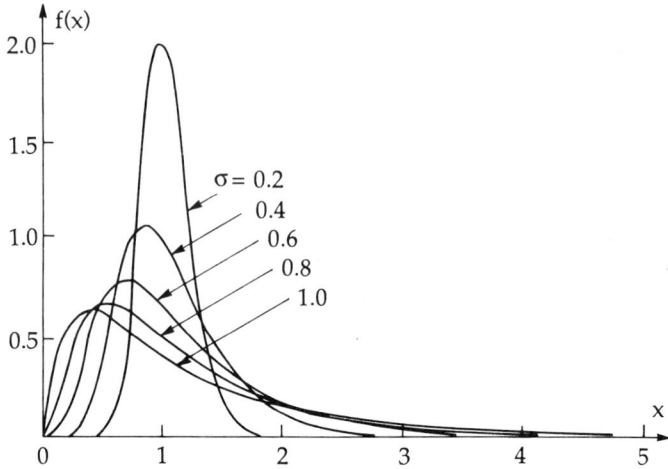

Figure 6.2 Log-normal probability density functions for various σ values at $\mu = 0$.

3. Let X_1 and X_2 be independent random variables each following a log-normal distribution $\Lambda(\mu_1, \sigma_1^2)$ and $\Lambda(\mu_2, \sigma_2^2)$, respectively. Then, the product of the random variables, $X_1 X_2$, has a log-normal distribution $\Lambda(\mu_1 + \mu_2 \cdot \sigma_1^2 + \sigma_2^2)$.

Example 6.3. Proof of Property (2) of the log-normal distribution stated above.

Let $Y = e^a X^b$ be a random variable, and we may write $\ln Y = a + b \ln X$. Since X has the log-normal distribution, $\Lambda(\mu, \sigma)$, $\ln X$ obeys the normal distribution $N(\mu, \sigma^2)$, whose characteristic function can be given by $\exp\{i\mu t - \frac{1}{2}\sigma^2 t^2\}$. By applying Theorem 4.3, we can show that the characteristic function of $\ln Y$ defined above becomes $\exp\{i(a + b\mu)t - \frac{1}{2}b^2\sigma^2 t^2\}$. This is the characteristic function of the normal distribution $N(a + b\mu, b^2\sigma^2)$. Thus, the random variable Y has the log-normal distribution $\Lambda(a + b\mu, b^2\sigma^2)$ ∎

6.1.3 Multivariate Normal Distribution

Definition 6.3. The random variables X and Y have a *bivariate* (or two-dimensional) *normal distribution* if their joint probability density function $f(x, y)$ is given by

$$f(x, y) = \frac{1}{2\pi\sigma_1\sigma_2\sqrt{1-\rho^2}}$$

$$\times \exp\left\{-\frac{1}{2(1-\rho^2)}\left[\left(\frac{x-\mu_1}{\sigma_1}\right)^2 - 2\rho\left(\frac{x-\mu_1}{\sigma_1}\right)\left(\frac{y-\mu_2}{\sigma_2}\right)\right.\right.$$

$$\left.\left.+ \left(\frac{y-\mu_2}{\sigma_2}\right)^2\right]\right\}$$

$$-\infty < x < \infty, -\infty < y < \infty \qquad (6.12)$$

where μ_1, μ_2 are the means, σ_1^2 and σ_2^2 are the variances of X and Y, respectively, while ρ is the correlation coefficient of X and Y.

From integration of the joint probability function, it can easily be proved that the marginal distributions, $f(x)$ and $f(y)$, are the normal distributions $N(\mu_1, \sigma_1^2)$ and $N(\mu_2, \sigma_2^2)$, respectively. That is,

$$f(x) = \int_{-\infty}^{\infty} f(x, y)\, dy = \frac{1}{\sqrt{2\pi}\,\sigma_1} \exp\left[-\frac{(x-\mu_1)^2}{2\sigma_1^2}\right]$$

and

$$f(y) = \int_{-\infty}^{\infty} f(x, y)\, dx = \frac{1}{\sqrt{2\pi}\,\sigma_2} \exp\left[-\frac{(x-\mu_2)^2}{2\sigma_2^2}\right] \qquad (6.13)$$

The conditional distribution of Y given X also becomes a normal distribution. That is,

$$f(y|x) = \frac{f(x, y)}{f(x)} \text{ is } N\left[\mu_2 + \rho\frac{\sigma_2}{\sigma_1}(x-\mu_1),\ (1-\rho^2)\sigma_2^2\right] \qquad (6.14)$$

The derivation of $f(y|x)$ is shown in Example 3.13.

The multivariate normal distribution frequently appears in stochastic analysis of random processes. A typical example may be seen in the analyses presented in Chapters 11 in which the trivariate normal distribution of the displacement, velocity, and acceleration of a random process is considered. The joint probability density function of the multivariate normal distribution is defined as follows:

Definition 6.4. The random variables $X_1, X_2, X_3, \ldots, X_n$ have a *multivariate normal distribution*, denoted by $N(\mu, \Sigma)$, if their joint distribution function is of the form

$$f(\mathbf{x}) = \frac{1}{(2\pi)^{n/2}|\Sigma|^{1/2}} \exp\left\{-\tfrac{1}{2}(\mathbf{x} - \mu)'\Sigma^{-1}(\mathbf{x} - \mu)\right\} \qquad (6.15)$$

where

$$\mathbf{x} = \begin{pmatrix} x_1 \\ x_2 \\ \vdots \\ x_n \end{pmatrix}$$

$$\mu = \text{mean} = \begin{pmatrix} \mu_1 \\ \mu_2 \\ \vdots \\ \mu_n \end{pmatrix}$$

$$\Sigma = \text{covariance matrix} = E[(\mathbf{x} - \mu)(\mathbf{x} - \mu)']$$

$$= \begin{pmatrix} \sigma_1^2 & \sigma_{12} & \cdots & \sigma_{1n} \\ \sigma_{21} & \sigma_2^2 & \cdots & \sigma_{2n} \\ \vdots & & & \\ \sigma_{n1} & \sigma_{n2} & \cdots & \sigma_n^2 \end{pmatrix}$$

Example 6.4. Derive the bivariate normal distribution of two random variables X_1 and X_2 from Eq. (6.15). For this, let the correlation coefficient of X_1 and X_2 be ρ. Then, the covariance matrix becomes

$$\Sigma = \begin{pmatrix} \sigma_1^2 & \rho\sigma_1\sigma_2 \\ \rho\sigma_1\sigma_2 & \sigma_2^2 \end{pmatrix}$$

NORMAL AND RELATED DISTRIBUTIONS 121

and hence, we have

$$|\Sigma| = (1 - \rho^2)\sigma_1^2\sigma_2^2$$

The inverse of the covariance matrix can be obtained as

$$\Sigma^{-1} = \frac{1}{1-\rho^2} \begin{pmatrix} 1/\sigma_1^2 & -\rho/\sigma_1\sigma_2 \\ -\rho/\sigma_1\sigma_2 & 1/\sigma_2^2 \end{pmatrix}$$

By inserting $|\Sigma|$ and Σ^{-1} into Eq. (6.15), we can derive the joint probability density function of X_1 and X_2 given in Eq. (6.12). ∎

The characteristic function of the multivariate normal distribution $N(\mu, \Sigma)$ is given by

$$\phi(t) = e^{it'\mu - \frac{1}{2}t'\Sigma t} \tag{6.16}$$

where $t' = (t_1, t_2, t_3, \ldots, t_n)$. For proof of Eq. (6.16), see Anderson (1966). In particular, for the bivariate normal distribution, the characteristic function becomes

$$\phi(t_1, t_2) = \int_{-\infty}^{\infty}\int_{-\infty}^{\infty} e^{i(t_1 x + t_2 y)} f(x, y)\, dx\, dy$$

$$= \exp\left\{i(\mu_1 t_1 + \mu_2 t_2) - \tfrac{1}{2}(\sigma_1^2 t_1^2 + 2\rho\sigma_1\sigma_2 t_1 t_2 + \sigma_2^2 t_2^2)\right\} \tag{6.17}$$

Example 6.5. Let the random vector $\mathbf{X}$ be distributed following the multivariate normal distribution $N(\mu, \Sigma)$, and let $\mathbf{Y} = \mathbf{CX}$, where $\mathbf{C}$ is a nonsingular matrix. Then $\mathbf{Y}$ is distributed following $N(\mathbf{C}\mu, \mathbf{C\Sigma C}')$.

In order to prove this relationship, we consider the characteristic function of the random vector $\mathbf{Y}$. That is,

$$\phi_y(t) = E[e^{it'y}] = E[e^{it'\mathbf{C}\mathbf{x}}]$$

By letting $t'\mathbf{c} = \tau'$, we have

$$\phi_y(t) = E[e^{i\tau'\mathbf{x}}] = \exp\left\{i\tau'\mu - \tfrac{1}{2}\tau'\Sigma\tau\right\}$$

$$= \exp\left\{it'\mathbf{c}\mu - \tfrac{1}{2}t'\mathbf{c}\Sigma\mathbf{c}'t\right\}$$

122   CONTINUOUS RANDOM VARIABLES AND THEIR DISTRIBUTIONS

This is the characteristic function of the multivariate normal distribution $N(\mathbf{c}\boldsymbol{\mu}, \mathbf{c}\boldsymbol{\Sigma}\mathbf{c}')$. Thus, the random vector $\mathbf{Y}$ is distributed following $N(\mathbf{c}\boldsymbol{\mu}, \mathbf{c}\boldsymbol{\Sigma}\mathbf{c}')$. ∎

6.2 GAMMA AND RELATED DISTRIBUTIONS

6.2.1 Gamma and Exponential Distributions

Definition 6.5. A random variable X has a *gamma distribution* if its probability density function is given by

$$f(x) = \frac{1}{\Gamma(m)} \lambda^m x^{m-1} e^{-\lambda x}, \qquad 0 \leqslant x < \infty, \lambda > 0 \qquad (6.18)$$

where $\Gamma(m) = $ gamma function $= \int_0^\infty x^{m-1} e^{-x}\, dx$.

The characteristic function, mean, and variance of the distribution are as follows:

$$\phi(t) = \left(1 - \frac{it}{\lambda}\right)^{-m}$$

$$E[x] = m/\lambda$$

$$\mathrm{Var}[x] = m/\lambda^2 \qquad (6.19)$$

The gamma distribution has two parameters, m and λ. The probability density functions for various m values for $\lambda = 1$ are shown in Figure 6.3. In particular, the distribution for $m = 1$ is called the exponential distribution defined as follows:

Definition 6.6. A random variable X has an *exponential distribution* if its probability density function is given by

$$f(x) = \lambda e^{-\lambda x}, \qquad 0 \leqslant x < \infty, \lambda > 0 \qquad (6.20)$$

The gamma distribution can be generalized by incorporating the additional parameter c so that it covers many other probability distributions.

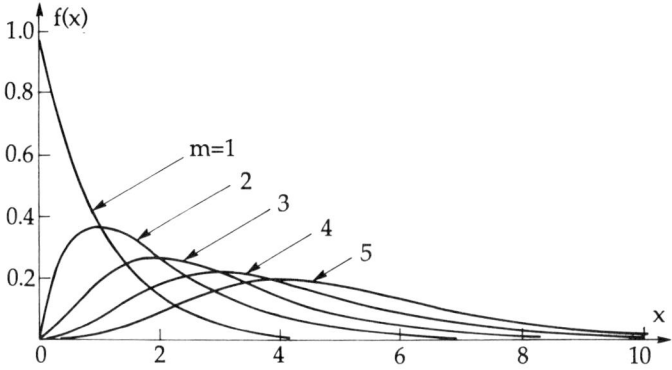

Figure 6.3 Gamma probability density functions for various m values at $\lambda = 1$.

The mathematical expression for this case is given in the following definition.

Definition 6.7. A random variable X has a *generalized gamma distribution* if its probability density function is given by

$$f(x) = \frac{c}{\Gamma(m)} \lambda^{cm} x^{cm-1} e^{-(\lambda x)^c}, \qquad 0 \leq x < \infty \qquad (6.21)$$

The mean and variance of the generalized gamma distribution are given by

$$E[x] = \frac{1}{\lambda} \frac{\Gamma[m + (1/c)]}{\Gamma(m)}$$

$$\text{Var}[x] = \frac{1}{\lambda^2} \left[\frac{\Gamma[m + (2/c)]}{\Gamma(m)} - \left\{ \frac{\Gamma[m + (1/c)]}{\Gamma(m)} \right\}^2 \right] \qquad (6.22)$$

The relationship between the generalized gamma distribution and other probability distributions is presented in Section 6.5.

It is stated in Section 5.3 [Eq. (5.30)] that the cumualtive distribution functions of the Poisson distribution and the gamma distribution have a

124   CONTINUOUS RANDOM VARIABLES AND THEIR DISTRIBUTIONS

functional relationship. That is,

$Pr\{X \leq m - 1\}$ in the Poisson distribution with parameter μ

$= Pr\{X \geq \mu/\lambda\}$ in the gamma distribution with parameters m and λ

Proof of this equality is given below.

For the gamma distribution given in Eq. (6.18), we can evaluate $Pr\{X \geq \mu/\lambda\}$ by integrating by parts as follows:

$$Pr\{X \geq \mu/\lambda\} = \int_{\mu/\lambda}^{\infty} \frac{1}{\Gamma(m)} \lambda^m x^{m-1} e^{-\lambda x} \, dx$$

$$= \frac{\lambda^m}{\Gamma(m)} \left[-\frac{1}{\lambda} [x^{m-1} e^{-\lambda x}]_{\mu/\lambda}^{\infty} + \frac{m-1}{\lambda} \int_{\mu/\lambda}^{\infty} x^{m-2} e^{-\lambda x} \, dx \right]$$

$$= \frac{\lambda^m}{\Gamma(m)} \left[\frac{\mu^{m-1}}{\lambda^m} e^{-\mu} - \frac{m-1}{\lambda^2} [x^{m-2} e^{-\lambda x}]_{\mu/\lambda}^{\infty} \right.$$

$$\left. + \frac{(m-1)(m-2)}{\lambda^2} \int_{\mu/\lambda}^{\infty} x^{m-3} e^{-\lambda x} \, dx \right]$$

$$= \frac{\lambda^m}{\Gamma(m)} \left[\frac{\mu^{m-1}}{\lambda^m} e^{-\mu} + (m-1) \frac{\mu^{m-2}}{\lambda^m} e^{-\mu} \right.$$

$$\left. + (m-1)(m-2) \frac{\mu^{m-3}}{\lambda^m} e^{-\mu} + \cdots \right]$$

$$= \frac{\mu^{m-1}}{(m-1)!} e^{-\mu} + \frac{\mu^{m-2}}{(m-2)!} e^{-\mu} + \frac{\mu^{m-3}}{(m-3)!} e^{-\mu} + \cdots$$

This is equal to $Pr\{X \leq m - 1\}$ in the Poisson distribution.

6.2.2   *Chi-Squared (χ^2) Distribution*

Let us consider the special case of the gamma distribution in which the parameters $m = r/2$ and $\lambda = 1/2$. Then, the distribution is called the χ^2 (chi-square) distribution.

GAMMA AND RELATED DISTRIBUTIONS

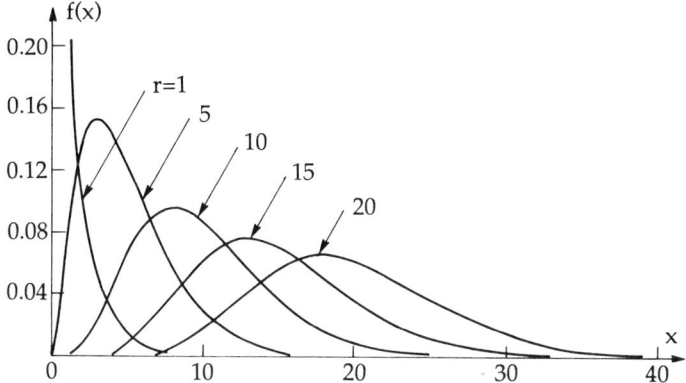

Figure 6.4 Chi-square probability density functions for various r values.

Definition 6.8. A random variable X has a χ^2 (*chi-square*) *distribution* with r degrees of freedom, denoted by $\chi^2_{(r)}$, if its probability density function is given by

$$f(x) = \frac{1}{\Gamma(r/2)} \left(\frac{1}{2}\right)^{r/2} x^{(r/2)-1} e^{-x/2}, \qquad 0 \leq x < \infty \qquad (6.23)$$

Examples of χ^2 distribution are shown in Figure 6.4.

The characteristic function of the χ^2 distribution with r degrees of freedom is given by

$$\phi(t) = (1 - i2t)^{-r/2} \qquad (6.24)$$

The mean and variance of the χ^2 distribution can be obtained from the mean and variance of the gamma distribution by letting $m = r/2$ and $\lambda = 1/2$. Hence, $E[x] = r$ and $\text{Var}[x] = 2r$ for the χ^2 distribution.

Example 6.6. Let the random variable X have the gamma distribution with parameters m and λ. Then, the random variable $Y = 2\lambda X$ has the χ^2 distribution with $2m$ degrees of freedom. The characteristic function of the gamma distribution is given in Eq. (6.19) as

$$\phi_x(t) = \left(1 - \frac{it}{\lambda}\right)^{-m}$$

By applying Theorem 4.3, the characteristic function of the random variable $Y = 2\lambda X$ becomes

$$\phi_y(t) = \phi_x(2\lambda t) = (1 - i2t)^{-m}$$

This is the characteristic function of the χ^2 distribution with $2m$ degrees of freedom. ∎

The χ^2 distribution has important properties that are given in the following two theorems.

Theorem 6.2. The random variables $X_1, X_2, \ldots, X_n$ are statistically independent having a χ^2 distribution with $r_1, r_2, \ldots, r_n$ degrees of freedom, respectively. Then, the sum of random variables

$$Y = \sum_{i=1}^{n} X_i \qquad (6.25)$$

has a χ^2 distribution with $\sum_{i=1}^{n} r_i$ degrees of freedom.

Proof. The characteristic function of the random variable Y can be obtained from Eq. (6.24) as

$$\phi_y(t) = \prod_{j=1}^{n} (1 - 2it)^{-r_j/2} = (1 - 2it)^{\frac{1}{2}\sum_{j=1}^{n} r_j}$$

This is the characteristic function of the χ^2 distribution with $(r_1 + r_2 + \cdots + r_n)$ degrees of freedom.

Theorem 6.3. Let $(X_1, X_2, \ldots, X_n)$ be a random sample of size n from the normal distribution $N(\mu, \sigma^2)$. Then,

$$\sum_{i=1}^{n} \frac{(x_i - \mu)^2}{\sigma^2}$$

has a χ^2 distribution with n degrees of freedom. (6.26)

Proof. Since X has the normal distribution, $N(\mu, \sigma^2)$, the random variable Z defined as $Z = (X - \mu)/\sigma$ has the normal distribution, $N(0, 1)$. Next, let us consider the random variable U defined as $U = Z^2$. The

GAMMA AND RELATED DISTRIBUTIONS

cumulative distribution function of U can be written as

$$F(u) = Pr\{U \leq u\} = Pr\{Z^2 \leq u\} = Pr\{-\sqrt{u} \leq Z \leq \sqrt{u}\}$$

$$= 2\int_0^{\sqrt{u}} \frac{1}{\sqrt{2\pi}} \exp\{-z^2/2\}\, dz$$

By writing $z = \sqrt{u}$, we have

$$F(u) = \int_0^u \frac{1}{\sqrt{2\pi}\sqrt{u}} \exp\{-u/2\}\, du$$

Thus, the probability density function of U becomes

$$f(u) = \frac{1}{\sqrt{2\pi}} \frac{1}{\sqrt{u}} e^{-u/2}$$

which is equal to the χ^2 distribution $\chi^2_{(1)}$. In other words, $U = (x - \mu)^2/\sigma^2$ has the distribution $\chi^2_{(1)}$. Then, from Theorem 6.2, it can be proved that

$$\sum_{i=1}^n \frac{(x_i - \mu)^2}{\sigma^2} = \sum_{i=1}^n U_i$$

has the χ^2 distribution with n degrees of freedom, $\chi^2_{(n)}$.

Example 6.7. The random variable X has the χ^2 distribution with r degrees of freedom. Prove that if r is large (approximately 30 or greater) X is asymptotically equal to the normal distribution with mean r and variance $2r$.

The characteristic function of the random variable X is given in Eq. (6.24) as

$$\phi_x(t) = (1 - i2t)^{-r/2}$$

Since the mean and variance of X are known as $E[x] = r$ and $Var[x] = 2r$, we may standardize the random variable X by letting $Y = (X - r)/\sqrt{2r}$.

Then, from Theorem 4.3, the characteristic function of Y can be written as

$$\phi_y(t) = e^{-i\sqrt{r/2}\,t}\left(1 - i\frac{2t}{\sqrt{2r}}\right)^{-r/2}$$

$$= \left\{e^{i\sqrt{2/r}\,t}\left(1 - i\sqrt{2/r}\,t\right)\right\}^{-r/2}$$

By expanding the exponential part of the above equation and by letting $r \to \infty$, we have

$$\lim_{r \to \infty} \phi_y(t) \sim \lim_{r \to \infty}\left(1 + \frac{t^2}{r}\right)^{-r/2} = e^{-t^2/2}$$

This is the characteristic function of the standardized normal distribution. Hence, X is asymptotically equal to $N(r, 2r)$. ■

6.3 WEIBULL AND RELATED DISTRIBUTIONS

6.3.1 Weibull Distribution

Definition 6.9. A random variable X has a *Weibull distribution* if its probability density function is given in the following form:

$$f(x) = c\lambda^c x^{c-1} e^{-(\lambda x)^c}, \quad 0 \leq x \leq \infty \tag{6.27}$$

The distribution was developed by Weibull on empirical ground (Weibull 1939, 1951) and it has been extensively used in statistical analysis of many practical problems in the natural sciences including extreme value statistics. The cumulative distribution function is given by

$$F(x) = 1 - e^{-(\lambda x)^c} \tag{6.28}$$

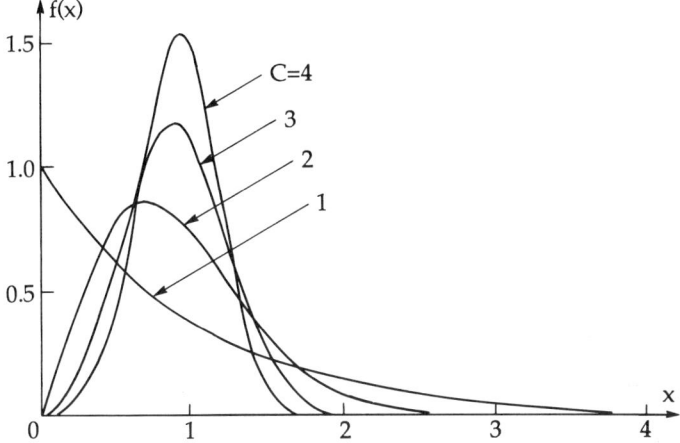

Figure 6.5 Weibull probability density functions for various c values at $\lambda = 1$.

and the mean and variance of the distribution are

$$E[x] = \frac{1}{\lambda}\Gamma\left(1 + \frac{1}{c}\right)$$

$$\text{Var}[x] = \frac{1}{\lambda^2}\left\{\Gamma\left(1 + \frac{2}{c}\right) - \left[\Gamma\left(1 + \frac{1}{c}\right)\right]^2\right\} \quad (6.29)$$

Figure 6.5 shows examples of Weibull probability density functions for various c values for $\lambda = 1$.

Sometimes the following truncated distribution, called the *three-parameter Weibull distribution*, is used for analysis of data:

$$f(x) = c\lambda^c(x - a)^{c-1}e^{-\{\lambda(x-a)\}^c}, \quad a \leqslant x < \infty \quad (6.30)$$

6.3.2 Rayleigh Distribution

Definition 6.10. A random variable X has a *Rayleigh distribution* if its probability density function is given in the following form:

$$f(x) = \frac{2x}{R}e^{-x^2/R}, \quad 0 \leqslant x < \infty \quad (6.31)$$

The mean and variance of the distributions are

$$E[x] = \frac{\sqrt{\pi}}{2}\sqrt{R}$$

$$\text{Var}[x] = \left(1 - \frac{\pi}{4}\right)R \quad (6.32)$$

The Rayleigh distribution is equivalent to $c = 2$, $\lambda = 1/\sqrt{R}$ of the Weibull distribution. However, the distribution has a theoretical background in contrast to the Weibull distribution. The Rayleigh distribution was developed by Lord Rayleigh in 1880 in connection with the resultant intensity of a large number of independent sounds. The following theorem is extremely important for the analysis of amplitudes of a Gaussian random process (see Section 11.1).

Theorem 6.4. Let the random variables X_1 and X_2 both have a normal distribution with the mean zero and variance σ^2, $N(0, \sigma^2)$. Then, the random variable $\sqrt{X_1^2 + X_2^2}$ has a Rayleight distribution with $R = 2\sigma^2$.

In order to prove this theorem, the transformation technique of random variables presented in Section 7.2.2 is required. Hence, the proof is given in Example 7.6.

Figure 6.6 shows the Rayleigh probability density function in which the random variable is expressed in dimensionless form $x/\sqrt{R}$.

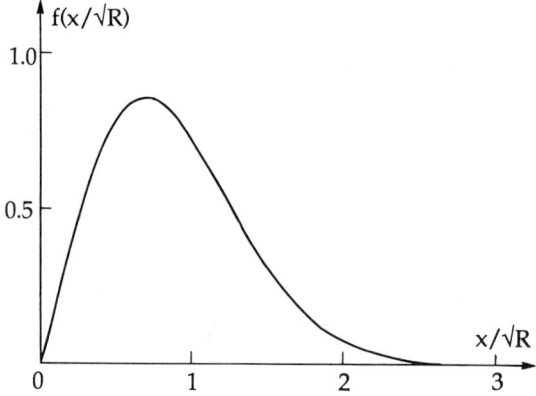

Figure 6.6 Dimensionless Rayleigh probability density function.

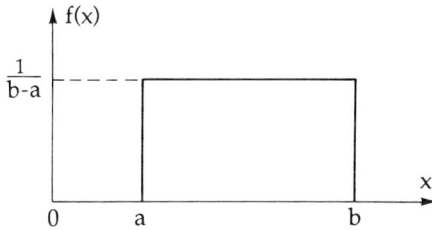

Figure 6.7 Uniform probability density function.

6.4 SOME OTHER DISTRIBUTIONS

Definition 6.11. A random variable X is called a *uniform distribution* if its probability density function is given in the following form:

$$f(x) = \begin{cases} \dfrac{1}{b-a} & \text{for } a \leq x \leq b \\ 0 & \text{otherwise} \end{cases} \qquad (6.33)$$

The shape of the uniform distribution is extremely simple as shown in Figure 6.7. The random variable X may take on any value between a and b with equal probability of being chosen.

Definition 6.12. A random variable X is said to be a *beta distribution* if its probability density function given by

$$f(x) = \frac{1}{B(p,q)} x^{p-1}(1-x)^{q-1}, \qquad 0 \leq x \leq 1 \qquad (6.34)$$

where $B(p,q) = $ beta function $= \int_0^1 x^{p-1}(1-x)^{q-1}\, dx$.

Note that the beta function $B(p,q)$ can be expressed in terms of the gamma function as follows:

$$B(p,q) = \frac{\Gamma(p)\Gamma(q)}{\Gamma(p+q)} \qquad (6.35)$$

If $p = q$, then the shape of the distribution is symmetric about its mean value. Furthermore, if $p = q = 1$, then the distribution reduces to the uniform distribution.

The beta distribution given in Eq. (6.34) is bounded between 0 and 1. If we generalize the limits from $(0, 1)$ to (a, b), then the probability density function can be written as

$$f(x) = \frac{1}{B(p,q)} \frac{(x-a)^{p-1}(b-x)^{q-1}}{(b-a)^{p+q-1}}, \quad a \leq x \leq b \quad (6.36)$$

This distribution may be called the *generalized beta distribution*, and it can be derived from Eq. (6.34) through the transformation techniques of random variables discussed in Section 7.1 (see Exercise 7.3).

It is given in Eq. (5.31) that

$Pr\{X \geq k\}$ in the binomial distribution with parameter p

$= Pr\{X \leq p\}$ in the beta distribution with $\alpha = k$

and $\beta = n - k + 1$

This equality can be proved as follows:

$Pr\{X \leq p\}$ in the beta distribution

$$= \frac{1}{B(k, n-k+1)} \int_0^p x^{k-1}(1-x)^{n-k} dx$$

By integrating by parts, the integration becomes,

$$\int_0^p x^{k-1}(1-x)^{n-k} dx = \frac{1}{k}\left[x^k(1-x)^{n-k}\right]_0^p$$

$$+ \frac{n-k}{k} \int_0^p x^k(1-x)^{n-k-1} dx$$

$$= \frac{1}{k} p^k(1-p)^{n-k} + \frac{n-k}{k(k+1)} p^{k+1}(1-p)^{n-k-1}$$

$$+ \frac{(n-k)(n-k-1)}{k(k+1)} \int_0^p x^{k+1}(1-x)^{n-k-2} dx$$

$$= \frac{1}{k} p^k(1-p)^{n-k} + \frac{n-k}{k(k+1)} p^{k+1}(1-p)^{n-k-1}$$

$$+ \frac{(n-k)(n-k-1)}{k(k+1)(k+2)} p^{k+2}(1-p)^{n-k-2} + \cdots$$

Thus, we have

$$\frac{1}{B(k, n-k+1)} \int_0^p x^{k-1}(1-x)^{n-k} dx$$

$$= \frac{n!}{k!(n-k)!} p^k (1-p)^{n-k}$$

$$+ \frac{n!}{(k+1)!(n-k-1)!} p^{k+1}(1-p)^{n-k-1} + \cdots$$

$$= \sum_{x=k}^n \binom{n}{x} p^x (1-p)^{n-x}$$

This is equal to $\Pr\{x \geq k\}$ in the binomial distribution.

There are many other probability distributions of continuous-type random variables in addition to those introduced here, such as the t distribution and the F distribution. Although these distributions are not included in this chapter, the t distribution will be given in Example 7.9 and the F distribution will be shown in Theorem 12.1.

6.5 RELATIONSHIP BETWEEN DISTRIBUTIONS OF VARIOUS CONTINUOUS-TYPE RANDOM VARIABLES

Many probability distributions presented in the preceding four sections are functionally related. In particular, the linear and asymptotic relationships between probability density functions are summarized as shown in Figure 6.8.

Relationship between probability density functions that are not linear are summarized in the following:

1. If the random variable X has a normal distribution $N(\mu, \sigma^2)$, then the random variable e^X has a log-normal distribution, $\Lambda(\mu, \sigma^2)$.

2. If the random variable X has a normal distribution, $N(0, \sigma^2)$, then the random variable X^2/σ^2 has a χ^2 distribution with one degree of freedom, $\chi^2_{(1)}$.

3. If random variables X and Y both obey a normal distribution, $N(0, \sigma^2)$, then the random variable $\sqrt{X^2 + Y^2}$ follows a Rayleigh distribution with the parameter $R = 2\sigma^2$.

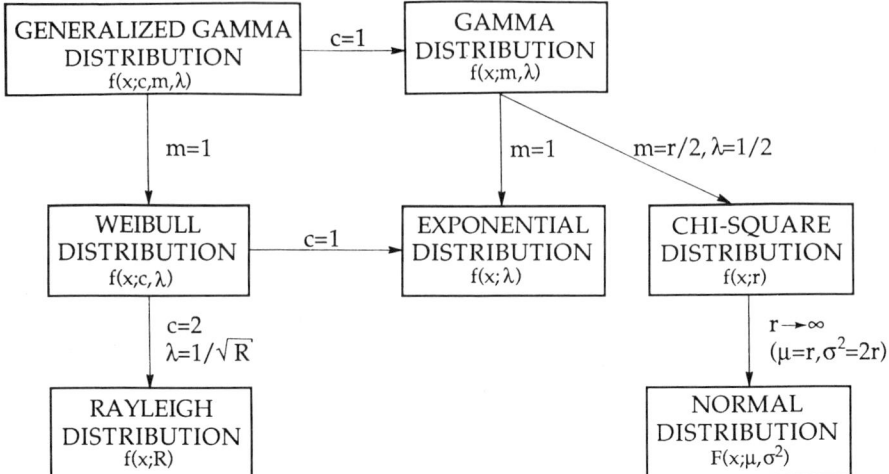

Figure 6.8 Linear and asymptotic relationship between some continuous-type random variables.

4. If the random variable X follows a Rayleigh distribution with the parameter R, then the random variable X^2 has an exponential distribution with the parameter $\lambda = 1/R$.

5. If the random variable X and Y follow χ^2 distributions with p and q degree of freedom, respectively, then the random variable $X/(X+Y)$ obeys a beta distribution with the parameters p and q.

EXERCISES

6.1 The random variable X has a normal probability distribution, $N(\mu, \sigma^2)$. Determine k such that X satisfies the following probability:

$$Pr\{\mu - k\sigma < X < \mu + k\sigma\} = 0.95$$

6.2 The random variable X has a normal probability distribution, $N(\mu, \sigma^2)$. (a) If we are interested in the probability distribution of X, which is greater than a, where a is a constant, prove that the mean value of X is given by

$$E[x] = \mu + \frac{f(a)}{1 - F(a)} \sigma^2$$

(b) If we are interested in the probability distribution of X, where $a < X < b$, prove that the mean value becomes

$$E[x] = \mu + \frac{f(a) - f(b)}{F(a) - F(b)}\sigma^2$$

6.3 The random variable X follows the exponential distribution given in Eq. (6.20). Shows that the conditional probability $Pr\{X \leq x + a | X \geq a\}$ is equal to the cumulative distribution function of X.

6.4 The random variable X obeys the binomial distribution with parameter p. Here, the parameter p is not a constant; instead, it is a random variable having the beta distribution with parameters α and β. Show that the probability $Pr\{X = x\}$ in n trials of the binomial distribution is given by

$$Pr\{X = x\} = \binom{n}{x}\frac{B(\alpha + x, \alpha + \beta - x)}{B(\alpha, \beta)}, \quad x = 0, 1, 2, \ldots, n$$

6.5 Show that the negative binomial distribution given by

$$p(x) = (-1)^x \binom{-r}{x} p^x q^r, \quad \text{where } p + q = 1$$

$$x = 0, 1, 2, \ldots,$$

can be approximated by the gamma probability density function

$$f(x) = \frac{1}{\Gamma(m)} \lambda^m x^{m-1} e^{-\lambda x}, \quad 0 \leq x < \infty$$

if p is much less than unity.

[Hint] Consider characteristic function of the negative binomial distribution, which can be written as $\phi(t) = \{(1 - pe^{it})/q\}^{-r}$, and expand the exponential in a series.

136 CONTINUOUS RANDOM VARIABLES AND THEIR DISTRIBUTIONS

6.6 Show that the characteristic function of the beta distribution given in Eq. (6.34) can be written as

$$\phi(t) = \frac{\Gamma(p+q)}{\Gamma(p)} \sum_{s=0}^{\infty} \frac{\Gamma(p+s)}{\Gamma(p+q+s)\Gamma(s+1)} (it)^s$$

[Hint] Expand e^{itx} of the characteristic function and integrate term by term with respect to x.

6.7 The probability density function of a random variable X is given in the following form:

$$f(x) = kx^{\alpha-1}/(1+x)^{\alpha+\beta}, \quad 0 \leq x < \infty$$

Determine the constant k, and obtain the mean and variance of the distribution.

6.8 The random variables X and Y have a bivariate normal distribution with zero mean, variances σ_1^2 and σ_2^2, and correlation coefficient ρ. Prove that the joint probability density function can be expressed in the following form of a series of the correlation coefficient ρ (Cramér, 1966):

$$f(x, y) = \frac{1}{\sigma_1 \sigma_2} \sum_{j=0}^{\infty} \frac{\rho^j}{j!} \Phi^{(j+1)}(x/\sigma_1) \Phi^{(j+1)}(y/\sigma_2)$$

where $\Phi^{(j)}( ) = j$th derivative of the standardized normal cumulative distribution function.

[Hint] Expand $1/\sqrt{1-\rho^2}$ and $1/(1-\rho^2)$ in the density function into a series of ρ.

6.9 A set of random variables (X_1, X_2, X_3, X_4) obeys the quadravariate normal distribution with zero means. Prove that

$$E[x_1 x_2 x_3 x_4] = E[x_1 x_2] E[x_3 x_4] + E[x_1 x_3] E[x_2 x_4]$$
$$+ E[x_1 x_4] E[x_2 x_3]$$

EXERCISES

[Hint] The joint characteristic function of the quadravariate normal distribution with zero mean can be given by

$$\phi(t_1, t_2, t_3, t_4) = \exp\left\{-\frac{1}{2}\sum_{i=1}^{4}\sum_{j=1}^{4} t_i t_j E[x_i x_j]\right\}$$

Then, evaluate

$$E[x_1 x_2 x_3 x_4] = \frac{\partial^4}{\partial t_1 \, \partial t_2 \, \partial t_3 \, \partial t_4} \phi(0,0,0,0)$$

CHAPTER 7

Transformation of Random Variables

The transformation of random variables dealt with in this chapter is the derivation of the probability distribution of a new random variable(s) from information on the probability distribution of other random variable(s) knowing the functional relationship between them. The transformation techniques are frequently applied to many statistical predictions in practice such as the evaluation of failure in reliability analysis (see Example 7.14) and prediction of amplitudes and periods of random processes (see Chapter 11).

7.1 TRANSFORMATION OF SINGLE RANDOM VARIABLE

Let X be a random variable with a probability density function $f(x)$, and let Y be a new random variable having a single-valued functional relationship with X given by $Y = g(X)$. The cumulative distribution function of Y, denoted here by $F_y(y)$ in order to distinguish it from the random variable X, can be written as

$$F_y(y) = Pr\{Y \leqslant y\} = \Pr\{g(X) \leqslant y\} = \Pr\{X \leqslant h(y)\}$$
$$= F_x[h(y)] \qquad (7.1)$$

TRANSFORMATION OF SINGLE RANDOM VARIABLE

where

$F_x[h(y)]$ = cumulative distribution function of X with $x = h(y)$

$h(y)$ = inverse function of $y = g(x)$

Equation (7.1) is valid for both continuous and discrete random variables if the transformation $Y = g(X)$ does not entail a change of sign in two random variables. Care must be taken when the transformation of random variables involves a change of sign in two random variables such as $Y = -aX$, where $a > 0$, and so on. In this case, Eq. (7.1) becomes

$$F_y(y) = Pr\{Y \leq y\} = Pr\{g(x) \leq y\} = Pr\{-X \leq h(y)\}$$

$$= Pr\{X \geq -h(y)\} = 1 - Pr\{X < -h(y)\}$$

$$= 1 - F_x\{-h(y)\} \quad (7.2)$$

Furthermore, for discrete random variables if $Pr\{X = -h(y)\}$ exists, Eq. (7.2) has to be modified following the definition of the cumulative function for discrete random variables. That is,

$$F_y(y) = Pr\{X \geq -h(y)\} = 1 - [F_x\{-h(y)\} - p_x\{-h(y)\}], \quad (7.3)$$

where $P_x(\ )$ = probability mass function of X.

The probability density function of the random variable Y can be obtained from Eqs. (7.1) through (7.3) as

$$f_y(y) = f_x[h(y)] \left| \frac{d}{dy} h(y) \right| \quad \text{for a continuous random variable}$$

and

$$p_y(y) = p_x[h(y)] \qquad \text{for a discrete random variable} \quad (7.4)$$

$|(d/dy)h(y)|$ in Eq. (7.4) is the Jacobian of the transformation. The absolute value should be taken since the probability density function must be positive. From the above, we can derive the following theorem:

Theorem 7.1. Let X be a continuous random variable with the probability density function $f_x(x)$. Let Y be a new random variable having a single-valued functional relationship with X given by $Y = g(X)$. The probability density function of Y is given by

$$f_y(y) = f_x[h(y)] \left| \frac{d}{dy} h(y) \right|$$

where $h(y)$ = inverse function of $y = g(x)$. For a discrete random variable X with the probability mass function $p_x(x)$, the probability mass function of $p_y(y)$ is given by

$$p_y(y) = p_x[h(y)]$$

Example 7.1. The probability density function of a continuous random variable is given by $f(x) = e^{-x}$, $0 \leqslant x < \infty$ (see Figure 7.1). Consider the transformation of random variable from X to $Y = aX + b$, where a and b are positive constants. We have, $F_x(x) = 1 - e^{-x}$ and $X = (Y - b)/a$. Hence, from Eq. (7.1), the cumulative distribution function of Y becomes

$$F_y(y) = 1 - e^{-(y-b)/a}, \qquad b \leqslant y < \infty$$

The probability density function $f(y)$ can be obtained from Eq. (7.4) as

$$f_y(y) = \frac{1}{a} e^{-(y-b)/a}, \qquad b \leqslant y < \infty \qquad \blacksquare$$

Example 7.2. Consider the transformation $Y = -aX + b$, where a and b are positive constants, for the same probability density function given in Example 7.1. In this case, we have $X = -(Y - b)/a$. Then, from Eq. (7.2), the cumulative distribution function becomes

$$F_y(y) = 1 - \left(1 - e^{(y-b)/a}\right) = e^{(y-b)/a}, \qquad -\infty < y \leqslant b$$

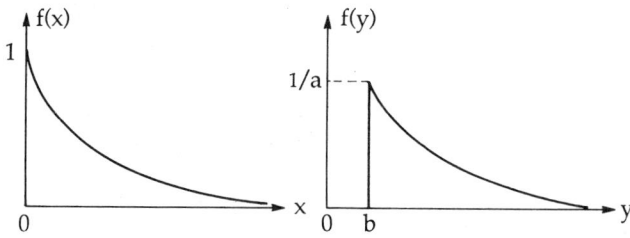

Figure 7.1 Probability density functions $f(x)$ and $f(y)$ in Example 7.1.

TRANSFORMATION OF SINGLE RANDOM VARIABLE

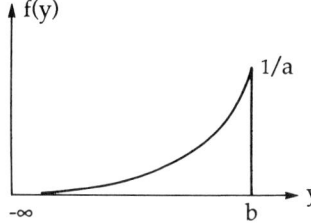

Figure 7.2 Probability density function $f(y)$ in Example 7.2.

and the probability density function can be obtained form Eq. (7.4) as

$$f_y(y) = \frac{1}{a} e^{(y-b)/a}, \quad -\infty < y \leq b$$

Figure 7.2 shows the probability density function $f_y(y)$. ∎

Next, let us consider the transformation of $Y = X^2$. In this case the random variable Y does not take on negative values although the random variable X, corresponding to a specific value of Y, is given by $\pm\sqrt{Y}$. Hence, for a continuous random variable X, we can write the cumulative distribution function of Y as

$$F_y(y) = Pr\{Y \leq y\} = Pr\{X^2 \leq y\} = Pr\{-\sqrt{y} \leq X \leq \sqrt{y}\}$$
$$= F_x(\sqrt{y}) - F_x(-\sqrt{y}), \quad 0 \leq y < \infty \quad (7.5)$$

The probability density function becomes

$$f_y(y) = \frac{d}{dy} F_y(y) = \left|\frac{1}{2\sqrt{y}}\right| \{f_x(\sqrt{y}) + f_x(-\sqrt{y})\} \quad (7.6)$$

If it is known that the sample space of the random variable X is limited to positive values only, then the second term in Eqs. (7.5) and (7.6) should be omitted (see Example 7.3).

For a discrete random variable X, if $Pr\{X = -\sqrt{y}\}$ exists, then the cumulative distribution of Y must be modified as follows:

$$F_y(y) = Pr\{-\sqrt{y} \leq X \leq \sqrt{y}\}$$
$$= F_x(\sqrt{y}) - [F_x(-\sqrt{y}) - Pr\{X = -\sqrt{y}\}] \quad (7.7)$$

Then, the probability mass function of Y is given by

$$p_y(y) = p_x(\sqrt{y}) + p_x(-\sqrt{y}) \tag{7.8}$$

Example 7.3. The hydrodynamic impact pressure, denoted by Y, is a random variable and its magnitude is proportional to the square of the impact velocity, X, that is $Y = kX^2$. Assuming that the impact velocity obeys the Rayleigh probability distribution with the parameter R, the probability density function of the impact pressure can be derived as follows:

The impact pressure X is always positive. Therefore, we have $X = \sqrt{Y/k}$. The probability density function, $f(y)$, then becomes

$$f(y) = \left[\frac{2x}{R}e^{-x^2/R}\right]_{x=\sqrt{y/k}} \frac{1}{2\sqrt{ky}} = \frac{1}{kR}e^{-y/kR}, \qquad 0 \leqslant y < \infty$$

Thus, the random variable Y has an exponential distribution. ∎

Example 7.4. The random variable X has a normal probability distribution with zero mean and unit variance. Consider the transformation of the random variable $Y = X^2$. In this case, we have $X = \pm\sqrt{Y}$. Hence, the probability density function of Y becomes

$$f(y) = \frac{1}{2\sqrt{y}}\{f_x(\sqrt{y}) + f_x(-\sqrt{y})\}$$

$$= \frac{1}{2\sqrt{y}}\left\{\frac{1}{\sqrt{2\pi}}e^{-y/2} + \frac{1}{\sqrt{2\pi}}e^{-y/2}\right\}$$

$$= \frac{1}{\sqrt{2\pi}\sqrt{y}}e^{-y/2}, \qquad 0 \leqslant y < \infty$$

This is the χ^2 distribution with one degree of freedom shown in Figure 7.3. The χ^2 distribution is discussed in detail in Section 6.2. ∎

Example 7.5. The random variable X has a uniform distribution over the interval $(-\pi, \pi)$ as shown in Figure 7.4. Consider the transformation of the random variable $Y = a\cos(X + \theta)$. From $X = \cos^{-1}(Y/a) - \theta$, we

TRANSFORMATION OF SINGLE RANDOM VARIABLE

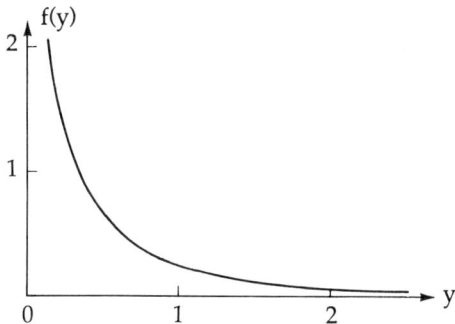

Figure 7.3 Probability density function $f(y)$ in Example 7.4.

have the Jacobian of the transformation as

$$|J| = \left| \frac{1}{\sqrt{a^2 - y^2}} \right|$$

and hence the probability density function $f(y)$ becomes

$$f(y) = \frac{1}{2\pi} \frac{1}{\sqrt{a^2 - y^2}} \times 2 = \frac{1}{\pi} \frac{1}{\sqrt{a^2 - y^2}}, \quad -a < y < a$$

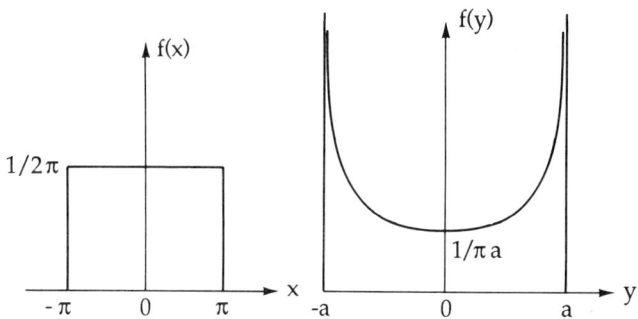

Figure 7.4 Probability density functions $f(x)$ and $f(y)$ in Example 7.5.

The cumulative distribution function is given by

$$F(y) = \int_{-a}^{y} f(y)\, dy = 1 - \frac{1}{\pi}\cos^{-1}\frac{y}{a}$$

∎

7.2 TRANSFORMATION OF SEVERAL RANDOM VARIABLES

7.2.1 *Function of Random Variables*

The concept presented in the previous section can be extended to the transformation of several random variables from one joint probability density function to another. We first consider the transformation of the probability distribution of two random variables.

Let X and Y be two random variables with the joint probability density function $f_{xy}(x, y)$. We now consider two new random variables U and V that have functional relationships with X and Y given by $U = g_1(x, y)$ and $V = g_2(x, y)$, respectively. Here, we assume that the functions g_1 and g_2 have continuous partial derivatives with respect to X and Y. Then, from the definition of the cumulative distribution function, we have

$$F(u, v) = Pr\{U \leq u, V \leq v\} = Pr\{g_1(x, y) \leq u, g_2(x, y) \leq v\}$$

$$= Pr\{X \leq h_1(u, v), Y \leq h_2(u, v)\} = F_{xy}[h_1(u, v), h_2(u, v)] \quad (7.9)$$

where $x = h_1(u, v)$ and $y = h_2(u, v)$ are the inverse functions of $u = g_1(x, y)$ and $v = g_2(x, y)$, and $F_{xy}(\ )$ is the joint probability density function of x and y.

By differentiating $F(u, v)$ with respect to u and v, we can derive the joint probability density function of u and v as

$$f(u, v) = f_{xy}[h_1(u, v), h_2(u, v)] \cdot |J| \quad (7.10)$$

where

$$J = \text{Jacobian of the transformation}$$

$$= \begin{vmatrix} \dfrac{\partial h_1}{\partial u} & \dfrac{\partial h_1}{\partial v} \\ \dfrac{\partial h_2}{\partial u} & \dfrac{\partial h_2}{\partial v} \end{vmatrix}$$

Thus, in summary, we have the following theorem:

Theorem 7.2. Let $f(x, y)$ be the joint probability density function of the random variables X and Y. Consider new random variables $U = g_1(X, Y)$ and $V = g_2(X, Y)$, where the functions g_1 and g_2 have continuous partial derivatives with respect to x and y. Then, the joint probability density function of U and V is given by

$$f(u, v) = f_{xy}[h_1(u, v), h_2(u, v)] \times \begin{vmatrix} \dfrac{\partial h_1}{\partial u} & \dfrac{\partial h_1}{\partial v} \\ \dfrac{\partial h_2}{\partial u} & \dfrac{\partial h_2}{\partial v} \end{vmatrix} \qquad (7.11)$$

where $x = h_1(u, v)$ and $y = h_2(u, v)$ are the inverse functions of $u = g_1(x, y)$ and $v = g_2(x, y)$. The functions $h_1(u, v)$ and $h_2(u, v)$ are differentiable and yield the one-to-one transformation.

Example 7.6. X and Y are independent random variables each obeying a normal distribution $N(0, \sigma^2)$. Hence, the joint probability density function is given by

$$f(x, y) = \frac{1}{2\pi\sigma^2} e^{-(x^2+y^2)/2\sigma^2}$$

Consider the probability density function of the random variable $U = \sqrt{x^2 + y^2}$. In order to find the probability density function of U, it is convenient to use polar coordinates

$$x = u \cos \theta$$
$$y = u \sin \theta \qquad 0 < u < \infty, 0 < \theta < 2\pi$$

and apply Eq. (7.11). The Jacobian for the present problem is $|u|$, and hence the joint probability density function of u and θ becomes

$$f(u, \theta) = \frac{u}{2\pi\sigma^2} e^{-u^2/2\sigma^2}$$

Then, the marginal probability density function $f(u)$ becomes,

$$f(u) = \int_0^{2\pi} f(u, \theta) \, d\theta = \frac{u}{\sigma^2} e^{-u^2/2\sigma^2}$$

This is the Rayleigh probability distribution discussed in Section 6.3, and is

Example 7.7. The random variables X and Y are statistically independent having a gamma distribution with parameters $(m, 1/2)$ and $(n, 1/2)$, respectively. Derive the probability density function of a random variable $U = X/(X + Y)$. The joint probability density function of X and Y can be written as

$$f(x, y) = \frac{1}{\Gamma(m)}\frac{1}{2^m}x^{m-1}e^{-x/2}\frac{1}{\Gamma(n)}\frac{1}{2^n}y^{n-1}e^{-y/2}, \quad \begin{array}{l} 0 \leqslant x < \infty \\ 0 \leqslant y < \infty \end{array}$$

Let us consider another random variable V defined as $V = X + Y$. Then the random variables X and Y can be expressed in terms of U and V as

$$x = uv$$
$$y = (1 - u)v$$

The Jacobian for the transformation of (x, y) to (u, v) can be evaluated as $|J| = v$. In order to determine the sample space for (u, v), we may write

$$u = \frac{x}{x + y} = \frac{1}{1 + (y/x)}$$

and thereby $x \to \infty$ yields $u = 1$, and $y \to \infty$ yields $u = 0$. Thus, we have the sample space $0 \leqslant u \leqslant 1$ and $0 \leqslant v < \infty$. Then, from Eq. (7.11) we have

$$f(u, v) = \frac{1}{\Gamma(m)\Gamma(n)2^{m+n}}u^{m-1}(1 - u)^{n-1}v^{m+n-1}e^{-v/2}$$

The marginal probability density function of u then can be obtained as

$$f(u) = \frac{1}{\Gamma(m)\Gamma(n)2^{m+n}}u^{m-1}(1 - u)^{n-1}\int_0^\infty v^{m+n-1}e^{-v/2}\,dv$$

$$= \frac{1}{\Gamma(m)\Gamma(n)2^{m+n}}u^{m-1}(1 - u)^{n-1}2^{m+n}\Gamma(m + n)$$

$$= \frac{\Gamma(m + n)}{\Gamma(m)\Gamma(n)}u^{m-1}(1 - u)^{n-1}, \quad 0 \leqslant u \leqslant 1$$

This is the beta distribution discussed in Section 6.4. ∎

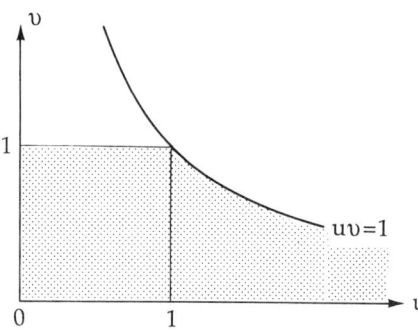

Figure 7.5 Integration domain for Example 7.8.

Example 7.8. X_1 and X_2 comprise a random sample of size two obtained for a distribution having a probability density function $f(x) = 2x$, where $0 \leqslant x \leqslant 1$. We will evaluate $\Pr\{X_1/X_2 < 2\}$ as follows: By letting $x_1/x_2 = u$, and $x_2 = v$, the joint probability density function of u and v becomes

$$f(u, v) = 4uv^3$$

The domain of u and v is shown in Figure 7.5. Taking this sample space into consideration, the marginal probability density function $f(u)$ can be obtained as

$$f(u) = \begin{cases} \int_0^1 f(u, v)\, dv = u & \text{for } 0 \leqslant u \leqslant 1 \\ \int_0^{1/u} f(u, v)\, dv = 1/u^3 & \text{for } 1 \leqslant u \leqslant \infty \end{cases}$$

Thus, we have

$$\Pr\{u < 2\} = 1 - \int_2^\infty (1/u^3)\, du = 7/8 \quad \blacksquare$$

Example 7.9. The random variable X has a normal distribution with zero mean and unit variance, $N(0, 1)$, and the random variable Y has χ^2 distribution with r degrees of freedom. These random variables are statistically independent. Consider a new random variable $U = X/\sqrt{Y/r}$ and derive the probability density function $f(u)$.

The joint probability density function of X and Y can be written

$$f(x, y) = \frac{1}{\sqrt{2\pi}} e^{-x^2/2} \frac{1}{\Gamma(r/2)} \frac{1}{2^{r/2}} y^{(r/2)-1} e^{-y/2}$$

$$= \frac{1}{\sqrt{2\pi}\,\Gamma(r/2)2^{r/2}} y^{(r/2)-1} e^{-\frac{1}{2}(y+x^2)}, \qquad \begin{array}{l} -\infty < x < \infty \\ 0 \leq y < \infty \end{array}$$

We may write the random variable $Y = V$, and consider the transformation from (X, Y) to (U, V). The transformation Jacobian for the present problem is $|\sqrt{v/r}|$. Thus, the joint probability density function of U and V becomes

$$f(u, v) = \frac{1}{\sqrt{2\pi}\,\Gamma(r/2)2^{r/2}} v^{(r/2)-1} \exp\left[-\frac{v}{2}\left(1 + \frac{u^2}{r}\right)\right] \sqrt{\frac{v}{r}} \qquad \begin{array}{l} -\infty < u < \infty \\ 0 \leq u < \infty \end{array}$$

Then, the marginal probability density function of U can be evaluated as

$$f(u) = \frac{1}{\sqrt{2\pi r}\,\Gamma(r/2)2^{r/2}} \int_0^\infty v^{[(r+1)/2]-1} \exp\left[-\frac{1}{2}\left(1 + \frac{u^2}{r}\right)v\right] dv$$

By letting $Z = \frac{1}{2}[1 + (u^2/r)]v$, we have

$$f(u) = \frac{1}{\sqrt{2\pi r}\,\Gamma(r/2)2^{r/2}} \frac{1}{\{\frac{1}{2}[1 + (u^2/r)]\}^{(r+1)/2}} \int_0^\infty z^{[(r+1)/2]-1} e^{-z} dz$$

$$= \frac{1}{\sqrt{2\pi r}\,\Gamma(r/2)2^{r/2}} \frac{2^{(r+1)/2}}{[1 + (u^2/r)]^{(r+1)/2}} \Gamma\left(\frac{r+1}{2}\right)$$

$$= \frac{\Gamma[(r+1)/2]}{\sqrt{\pi r}\,\Gamma(r/2)} \frac{1}{[1 + (u^2/r)]^{(r+1)/2}}, \qquad -\infty < u < \infty$$

This distribution is called the *t distribution* with r degrees of freedom. The distribution plays an important role in statistical inference theory. ∎

Examples 7.6 through 7.9 are all applications of Theorem 7.2 for continuous random variables. It should be noted, however, that Theorem 7.2 can be equally applicable to discrete random variables as shown in the following example.

TRANSFORMATION OF SEVERAL RANDOM VARIABLES   149

Example 7.10. The random variables X and Y are statistically independent and have Poisson distributions with parameters μ and λ, respectively. We can write the joint probability mass function as

$$p(x, y) = \frac{\mu^x}{x!}e^{-\mu} \cdot \frac{\lambda^y}{y!}e^{-\lambda}$$

$$= \frac{\mu^x \lambda^y}{x!y!}e^{-(\mu+\lambda)}, \quad \begin{array}{l} x = 0, 1, 2, \cdots \\ y = 0, 1, 2, \ldots \end{array}$$

In order to find the distribution of the sum of two random variables, denoted by $U = X + Y$, let us write the random variable Y as V, and consider the transformation from (X, Y) to new random variables (U, V). Then, the joint probability mass function of U and V can be written as

$$p(u, v) = \frac{\mu^{(u-v)} \lambda^v}{(u-v)!v!}e^{-(\mu+\lambda)}, \quad \begin{array}{l} u = 0, 1, 2, \ldots \\ v = 0, 1, 2, \ldots \end{array}$$

It is noted here that the random variables U and V have the restriction that $U > V$ since U is the sum of two random variables X and Y. With this in mind, the marginal probability mass function of U can be obtained as

$$p(u) = \sum_{v=0}^{u} p(u, v)$$

$$= e^{-(\mu+\lambda)} \sum_{v=0}^{\mu} \frac{\mu^{(u-v)} \lambda^v}{(u-v)!v!}$$

$$= \frac{e^{-(\mu+\lambda)}}{u!} \sum_{v=0}^{u} \frac{u!}{(u-v)!v!} \mu^{(u-v)} \lambda^v$$

$$= \frac{e^{-(\mu+\lambda)}}{u!}(u + \lambda)^u$$

This is the Poisson probability distribution with parameter $\mu + \lambda$. Thus, we can prove that the sum of two independent Poisson distributions is also a Poisson distribution with the mean value that is equal to the sum of parameters. ∎

The concept given in Theorem 7.2 regarding the transformation of two random variables can be extended to n random variables. The transformation formula applicable to n random variables is given in the following theorem:

Theorem 7.3. Let $f(x_1, x_2, \ldots, x_n)$ be the joint probability density function of the random variables $(X_1, X_2, \ldots, X_n)$. Consider the transformation of random variables to a set of new random variables $(Y_1, Y_2, \ldots, Y_n)$ where $y_1 = g_1(x_1, x_2, \ldots, x_n)$, $y_2 = g_2(x_1, x_2, \ldots, x_n), \ldots, y_n = g_n(x_1, x_2, \ldots, x_n)$. Inversely, we have $x_1 = h_1(y_1, y_2, \ldots, y_n)$, $x_2 = h_2(y_1, y_2, \ldots, y_n), \ldots, x_n = h_n(y_1, y_2, \ldots, y_n)$. Assume that transformations are one-to-one, and the functions h are differentiable. Then, the joint probability density function of $(Y_1, Y_2, \ldots, Y_n)$ is given by

$$f(y_1, y_2, \ldots, y_n) = f(x_1, x_2, \ldots, x_n) \cdot |J| \qquad (7.12)$$

where

$$x_1 = h_1(y_1, y_2, \ldots, y_n)$$
$$x_2 = h_2(y_1, y_2, \ldots, y_n)$$
$$\vdots$$
$$x_n = h_n(y_1, y_2, \ldots, y_n)$$

and

$$|J| = \begin{vmatrix} \dfrac{\partial h_1}{\partial y_1} & \dfrac{\partial h_1}{\partial y_2} & \cdots & \dfrac{\partial h_1}{\partial y_n} \\ \dfrac{\partial h_2}{\partial y_1} & \dfrac{\partial h_2}{\partial y_2} & \cdots & \dfrac{\partial h_2}{\partial y_n} \\ \dfrac{\partial h_n}{\partial y_1} & \dfrac{\partial h_n}{\partial y_2} & \cdots & \dfrac{\partial h_n}{\partial y_n} \end{vmatrix}$$

Example 7.11. A set (X_1, X_2, X_3) is a random sample of size three from a distribution having the probability density function $f(x) = e^{-x}$, $0 < x < \infty$. Consider the transformation $Y_1 = X_1/(X_1 + X_2)$, $Y_2 = (X_1 + X_2)/$

TRANSFORMATION OF SEVERAL RANDOM VARIABLES 151

$(X_1 + X_2 + X_3)$, and $Y_3 = X_1 + X_2 + X_3$, and obtain the marginal probability density function $f(y_1)$, $f(y_2)$, and $f(y_3)$.

The joint probability density function of X_1, X_2, and X_3 is given by

$$f(x_1, x_2, x_3) = e^{-(x_1 + x_2 + x_3)}$$

From the transformation formula, we have the following inverse functions: $X_1 = Y_1 Y_2 Y_3$, $X_2 = Y_2 Y_3(1 - Y_1)$, and $X_3 = Y_3(1 - Y_2)$. Hence, the Jacobian for the transformation becomes $|y_2 y_3^2|$, and the joint probability density function of Y_1, Y_2, and Y_3 becomes

$$f(y_1, y_2, y_3) = f\big(x_1 = y_1 y_2 y_3, x_2 = y_2 y_3(1 - y_1), x_3 = y_3(1 - y_2)\big) y_2 y_3^2$$

$$= y_2 y_3^2 e^{-y_3}$$

where $0 \leq y_1 \leq 1$, $0 \leq y_2 \leq 1$, $0 \leq y_3 \leq \infty$.

The marginal probability density functions can be evaluated as

$$f(y_1) = \int_0^\infty \int_0^1 f(y_1, y_2, y_3)\, dy_2\, dy_3 = 1, \qquad 0 \leq y_1 \leq 1$$

$$f(y_2) = \int_0^\infty \int_0^1 f(y_1, y_2, y_3)\, dy_1\, dy_3 = 2y_2, \qquad 0 \leq y_2 \leq 1$$

$$f(y_3) = \int_0^1 \int_0^1 f(y_1, y_2, y_3)\, dy_1\, dy_2 = \tfrac{1}{2} y_3^2 e^{-y_3}, \qquad 0 \leq y_3 \leq \infty$$

It can be seen from the marginal probability density functions that $f(y_1)f(y_2)f(y_3) = f(y_1, y_2, y_3)$, and hence the random variables Y_1, Y_2, and Y_3 are statistically independent. ∎

7.2.2 Sum, Difference, Product, and Ratio of Two Random Variables

In this section we consider the transformation of the probability distribution associated with four arithmetic operations of two random variables X and Y, application of which is frequently exploited in practical problems.

SUM OF TWO RANDOM VARIABLES, $Z = X + Y$. The cumulative distribution function of the random variable Z may be written in the form of the

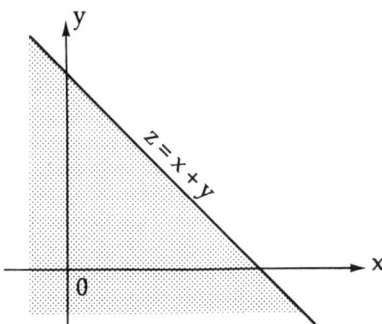

Figure 7.6 Integration domain for Eq. (7.13).

integral of the joint probability density function of X and Y:

$$F(z) = \Pr\{Z \leq z\} = \iint_{x+y \leq z} f_{xy}(x, y)\, dx\, dy \qquad (7.13)$$

where the integration with respect to x and y is carried out below the line $z = x + y$ shown in Figure 7.6. That is,

$$F(z) = \int_{-\infty}^{\infty} \int_{-\infty}^{z-y} f_{xy}(x, y)\, dx\, dy = \int_{-\infty}^{\infty} \int_{-\infty}^{z-x} f_{xy}(x, y)\, dy\, dx \qquad (7.14)$$

We may change the variable of integration either from x to z or y to z. Since the absolute value of the transformation Jacobian is unity, we can write

$$F(z) = \int_{-\infty}^{\infty} \int_{-\infty}^{z} f_{xy}(z - y, y)\, dz\, dy = \int_{-\infty}^{\infty} \int_{-\infty}^{z} f_{xy}(x, z - x)\, dz\, dx \qquad (7.15)$$

Then, by differentiating with respect to z yields

$$f(z) = \int_{-\infty}^{\infty} f_{xy}(z - y, y)\, dy = \int_{-\infty}^{\infty} f_{xy}(x, z - x)\, dx \qquad (7.16)$$

If the two random variables are statistically independent, then the problem density function becomes the following convolution integral:

$$f(z) = \int_{-\infty}^{\infty} f_x(z - y) f_y(y)\, dy = \int_{-\infty}^{\infty} f_x(x) f_y(z - x)\, dx \qquad (7.17)$$

where $f_x(\ )$ and $f_y(\ )$ are the marginal probability density function of X and Y, respectively.

Derivation of the probability density function can also be made by applying Theorem 7.2. For this, we may write $z = x + y$ and $v = x$. Then, the inverse functions are given by $x = h_1(z, v) = v$ and $y = h_2(z, v) = z - v$. Hence, the absolute value of the Jacobian becomes unity. From Eq. (7.11), it follows that

$$f(z, v) = f_{xy}(v, z - v) = f_{xy}(x, z - x) \tag{7.18}$$

The probability density function $f(z)$ can then be obtained as a marginal density function of $f_{xy}(x, z - x)$. That is,

$$f(z) = \int_{-\infty}^{\infty} f_{xy}(x, z - x)\, dx \tag{7.19}$$

This is the same probability density function derived in Eq. (7.16).

Example 7.12. Let (x_1, x_2) be a random sample of size two from a uniform distribution with the interval of $(0, a)$ where $a > 0$, and find the probability density function of the sum of two random variables, $Z = X_1 + X_2$. In applying Eq. (7.16) to this problem, the integration domain must be carefully established from the functional relationship between Z and X_2. As can be seen in Figure 7.7, the integration domain may be divided into

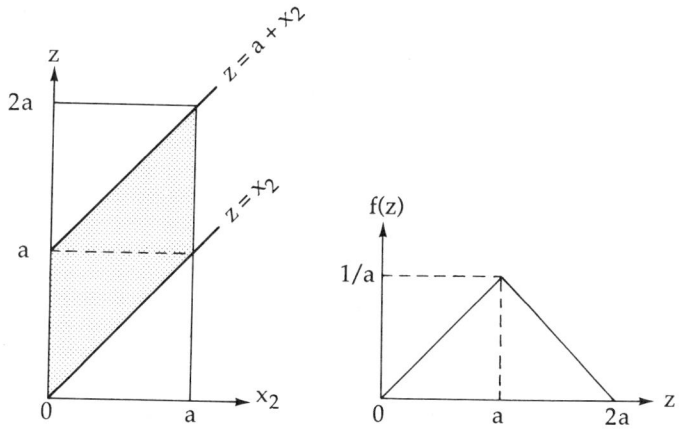

Figure 7.7 Integration domain and probability density function in Example 7.12.

two parts; one for $0 \leqslant z \leqslant a$ and the other for $a \leqslant z \leqslant 2a$. For $0 \leqslant z \leqslant a$,

$$f(z) = \int_0^z f_{x_1 x_2}(z - x_2, x_2) \, dx_2 = z/a^2$$

and for $a \leqslant z \leqslant 2a$,

$$f(z) = \int_{z-a}^a f_{x_1 x_2}(z - x_2, x_2) \, dx_2 = (2a - z)/a^2 \qquad \blacksquare$$

DIFFERENCE OF TWO RANDOM VARIABLES, $Z = X - Y$ The cumulative distribution function of the random variable Z can be expressed by the same formula as given in Eq. (7.13) but with a different integration domain as shown in Figure 7.8. That is

$$F(z) = Pr\{Z \leqslant z\} = \iint_{x-y \leqslant z} f_{xy}(x, y) \, dx \, dy$$

$$= \int_{-\infty}^{\infty} \int_{-\infty}^{z+y} f_{xy}(x, y) \, dx \, dy = \int_{-\infty}^{\infty} \int_{-z+x}^{\infty} f_{xy}(x, y) \, dy \, dx \quad (7.20)$$

The probability density function $f(z)$ can be obtained by the same procedure used in the derivation of Eq. (7.19). That is, by letting $z = x - y$ and $v = x$, we have

$$f(z) = \int_{-\infty}^{\infty} f_{xy}(z, v) \, dv = \int_{-\infty}^{\infty} f_{xy}(x, x - z) \, dx \qquad (7.21a)$$

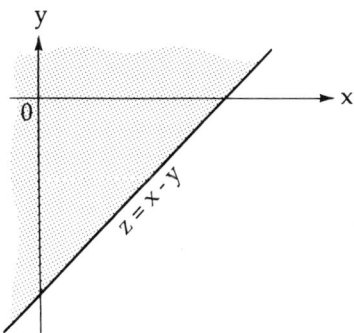

Figure 7.8 Integration domain for Eq. (7.20).

TRANSFORMATION OF SEVERAL RANDOM VARIABLES

Similarly, by letting $z = x - y$ and $v = y$, we have

$$f(z) = \int_{-\infty}^{\infty} f_{xy}(z, v) \, dv = \int_{-\infty}^{\infty} f_{xy}(z + y, y) \, dy \qquad (7.21b)$$

If the random variables are statistically independent, then

$$f(z) = \int_{-\infty}^{\infty} f_x(x) f_y(x - z) \, dx = \int_{-\infty}^{\infty} f_x(z + y) f_y(y) \, dy \qquad (7.22)$$

Example 7.13. The random variables X and Y are statistically independent and each obeys the normal distribution with zero mean and unit variance. The joint probability density function, therefore, is given by

$$f(x, y) = \frac{1}{2\pi} e^{-(x^2 + y^2)/2}$$

Consider the random variable $Z = X - Y$. From Eq. (7.22), we have the density of Z as

$$f(z) = \int_{-\infty}^{\infty} \frac{1}{\sqrt{2\pi}} e^{-x^2/2} \cdot \frac{1}{\sqrt{2\pi}} e^{-(x-z)^2/2} \, dx$$

$$= \frac{1}{2\pi} \int_{-\infty}^{\infty} e^{-(2x^2 - 2xz + z^2)/2} \, dx$$

$$= \frac{1}{2\pi} e^{-z^2/4} \int_{-\infty}^{\infty} e^{-\frac{1}{2}[\sqrt{2}x - (z/\sqrt{2})]^2} \, dx = \frac{1}{2\sqrt{\pi}} e^{-z^2/4}$$

This is the normal probability distribution with zero mean and variance = 2. ∎

Example 7.14. The random variables X and Y represent the loading and structural resistance of a structure. X and Y are statistically independent, each having the probability density function $f_x(x)$ and $f_y(y)$, where $0 \leq x, y < \infty$. Failure of the structure will occur when the loading exceeds the structural resistance; namely, $X > Y$ and thereby $X - Y > 0$ as shown in the shaded area where the two probability density functions overlap in

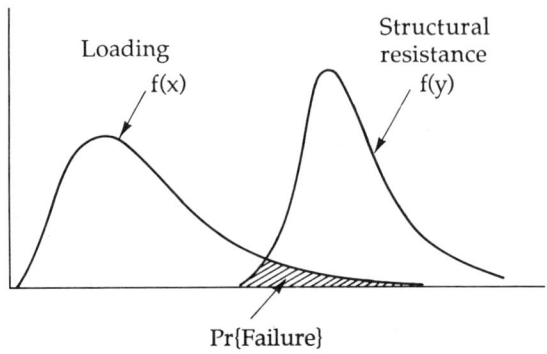

Figure 7.9 Pictorial sketch indicating probability of failure.

Figure 7.9. Then the probability of failure can be evaluated by

$$Pr\{\text{Failure}\} = Pr\{X > Y\} = \iint_{x-y>0} f_x(x) f_y(y)\, dx\, dy$$

$$= \int_0^\infty \int_y^\infty f_x(x) f_y(y)\, dx\, dy = \int_0^\infty \{1 - F_x(y)\} f_y(y)\, dy$$

The formula given above is a fundamental formula in reliability analysis. ∎

PRODUCT OF TWO RANDOM VARIABLES, $Z = XY$ Since the functional relationship $Z = XY$ can be expressed by the two curves shown in Figure 7.10, the integration to obtain the cumulative distribution function of z is carried out independently for the domain $y < 0$ and $y > 0$. That is,

$$F(z) = Pr\{Z \leq z\} = \iint_{xy \leq z} f_{xy}(x, y)\, dx\, dy$$

$$= \int_{-\infty}^0 \int_{z/y}^\infty f_{xy}(x, y)\, dx\, dy + \int_0^\infty \int_{-\infty}^{z/y} f_{xy}(x, y)\, dx\, dy \quad (7.23)$$

Next, let $z = xy$ and $v = y$ and derive the probability density function of z by applying Theorem 7.2. We have

$$f(z) = \int_{-\infty}^0 \frac{1}{|y|} f_{xy}(z/y, y)\, dy + \int_0^\infty \frac{1}{|y|} f_{xy}(z/y, y)\, dy$$

$$= \int_{-\infty}^\infty \frac{1}{|y|} f_{xy}(z/y, y)\, dy \quad (7.24)$$

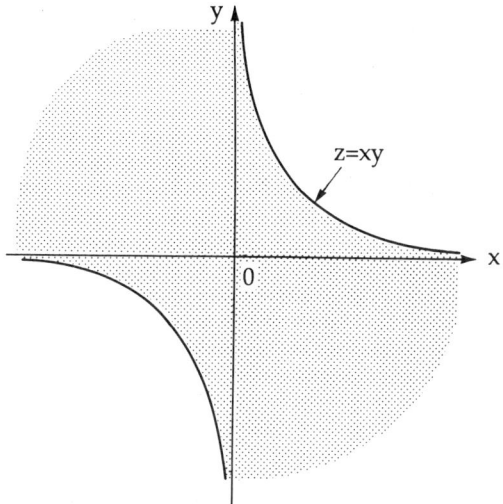

Figure 7.10 Integration domain for Eq. (7.23).

For statistically independent random variables X and Y, we can write

$$f(z) = \int_{-\infty}^{\infty} \frac{1}{|y|} f_x(z/y) f_y(y) \, dy \qquad (7.25)$$

Equations (7.24) and (7.25) can also be written in terms of integration with respect to x. That is,

$$f(z) = \int_{-\infty}^{\infty} \frac{1}{|x|} f_{xy}(x, z/x) \, dx$$

and

$$f(z) = \int_{-\infty}^{\infty} \frac{1}{|x|} f_x(x) f_y(z/x) \, dx$$

if X and Y are statistically independent. (7.26)

Example 7.15. The random variables X and Y are statistically independent, and have uniform distributions between the intervals $(1, a)$ and $(1, b)$, respectively, where $1 \leq a \leq b$. The probability density function of the random variable $Z = XY$ can be obtained by applying Eq. (7.24). For this,

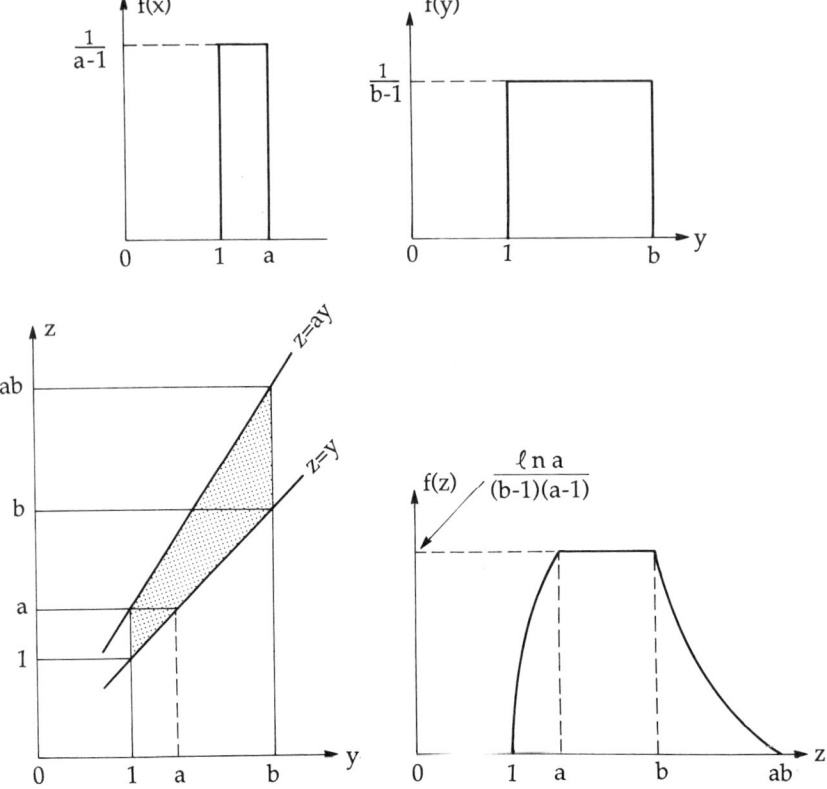

Figure 7.11 Integration domain and probability density functions $f(x)$, $f(y)$, and $f(z)$ in Example 7.15.

it is necessary to consider the relationship between Z and Y and determine the integration domain as shown in Figure 7.11. Then, we can obtain the probability density function as follows:

For $1 \leqslant Z \leqslant a$

$$f(z) = \frac{1}{(a-1)(b-1)} \int_1^z \frac{1}{y} \, dy = \frac{1}{(a-1)(b-1)} \ln Z$$

For $a \leqslant Z \leqslant b$

$$f(z) = \frac{1}{(a-1)(b-1)} \int_{z/a}^z \frac{1}{y} \, dy = \frac{1}{(a-1)(b-1)} \ln a$$

For $b \leq Z \leq ab$

$$f(z) = \frac{1}{(a-1)(b-1)} \int_{z/a}^{b} \frac{1}{y} dy = \frac{1}{(a-1)(b-1)} \ln \frac{ab}{z} \qquad \blacksquare$$

RATIO OF TWO RANDOM VARIABLES, $Z = X/Y$ The domain of integration involved in the cumulative distribution function Z in which $Z \leq X/Y$ is shown in Figure 7.12. That is, the domain is $x \leq zy$ for $y > 0$, and $x \geq zy$ for $y < 0$. Hence, the distribution function becomes

$$F(z) = Pr\{Z \leq z\} = \iint_{x/y \leq z} f_{xy}(x, y) \, dx \, dy$$

$$= \int_0^\infty \int_{-\infty}^{zy} f_{xy}(x, y) \, dx \, dy + \int_{-\infty}^0 \int_{zy}^\infty f_{xy}(x, y) \, dx \, dy \quad (7.27)$$

By letting $z = x/y$ and $v = y$, and by applying Theorem 7.2, the probability density function $f(z)$ can be derived. That is,

$$f(z) = \int_0^\infty |y| f_{xy}(zy, y) \, dy + \int_{-\infty}^0 |y| f_{xy}(zy, y) \, dy$$

$$= \int_{-\infty}^\infty |y| f_{xy}(zy, y) \, dy \qquad (7.28)$$

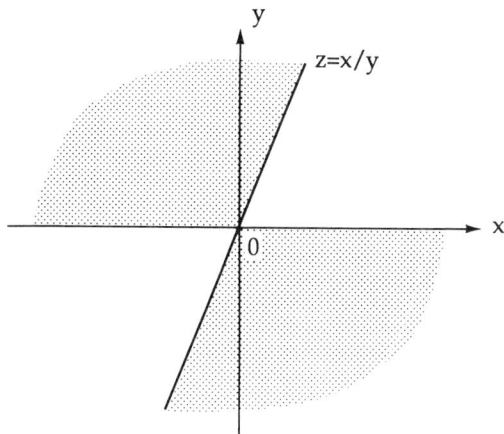

Figure 7.12 Integration domain for Eq. (7.27).

In particular, for two statistically independent random variables X and Y, we can write

$$f(z) = \int_{-\infty}^{\infty} |y| f_x(zy) f_y(y) \, dy \qquad (7.29)$$

As in the case for $Z = XY$, Eq. (7.29) can also be expressed in terms of the integration with respect to X by interchanging the two random variables.

Example 7.16. Let us consider the probability of failure problem given in Example 7.14 from a different viewpoint. In Example 7.14, the solution was obtained by letting the probability of failure, $Pr\{X > Y\}$, be equivalent to $Pr\{(X - Y) > 0\}$. We now consider this probability to be equivalent to $Pr\{X/Y > 1\}$ as shown in Figure 7.13 and apply Eq. (7.27). That is, by letting $z = x/y$ and $y > 0$ for the present problem. Then, we can write

$$Pr\{\text{Failure}\} = Pr\{z > 1\} = 1 - \int_0^\infty \int_{-\infty}^y f_{xy}(x, y) \, dx \, dy$$

$$= \int_0^\infty \int_y^\infty f_x(x) f_y(y) \, dx \, dy = \int_0^\infty \{1 - F_x(y)\} f_y(y) \, dy$$

This is the same result derived in Example 7.14. ∎

Example 7.17. The random variables X and Y are statistically independent, each having a normal distribution $N(0, 1)$. The probability density function of the ratio of these random variables $Z = X/Y$ can be obtained

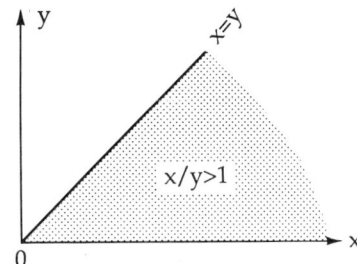

Figure 7.13 Integration domain for Example 7.16.

from Eq. (7.29) as

$$f(z) = \int_{-\infty}^{\infty} |y| \frac{1}{2\pi} e^{-[(z^2+1)y^2]/2} \, dy = \frac{1}{\pi} \int_{0}^{\infty} y e^{-[(z^2+1)y^2]/2} \, dy$$

By letting $(z^2 + 1)y^2/2 = t$, we have

$$f(z) = \frac{1}{\pi} \frac{1}{z^2 + 1} \int_{0}^{\infty} e^{-t} \, dt = \frac{1}{\pi} \frac{1}{z^2 + 1}$$

This probability density function is called the *Cauchy distribution*. ∎

7.3 TRANSFORM THROUGH CHARACTERISTIC FUNCTIONS

The transformation techniques of random variables presented in the previous two sections are commonly employed in finding the probability distribution of a function of random variables. The transformation method introduced in this section is to find the desired probability distribution through application of the characteristic function of the distribution. This method is extremely convenient and simple for deriving the probability density function in some cases, particularly in finding the distribution of a new random variable that is a function of statistically independent random variables. The principle of this method is given below.

Let us consider the functional relationship between two sets of random variables $(X_1, X_2, \ldots, X_n)$ and $(Y_1, Y_2, \ldots, Y_n)$ given in Theorem 7.3. That is,

$$y_1 = g_1(x_1, x_2, \ldots, x_n)$$
$$y_2 = g_2(x_1, x_2, \ldots, x_n), \quad \text{etc.}$$

and inversely, $x_1, x_2, \ldots$ are given by

$$x_1 = h_1(y_1, y_2, \ldots, y_n)$$
$$x_2 = h_2(y_1, y_2, \ldots, y_n), \quad \text{etc.}$$

In order to obtain the probability density function of Y_1 from the knowledge of the joint probability density function $f(x_1, x_2, \ldots, x_n)$, we may consider the following integration:

$$I = \int \cdots \int \int e^{ity_1} \cdot f(x_1, x_2, \ldots, x_n) \, dx_1 \, dx_2 \cdots dx_n \qquad (7.30)$$

Next, the joint probability function $f(x_1, x_2, \ldots, x_n)$ is transformed to the joint probability density function $f(y_1, y_2, \ldots, y_n)$ by applying Eq. (7.12). Then, we have

$$I = \int \cdots \int \int e^{ity_1} \cdot f(y_1, y_2, \ldots, y_n) \cdot dy_1 \, dy_2 \cdots dy_n$$

$$= \int e^{ity_1} \cdot f(y_1) \, dy_1 \qquad (7.31)$$

where $f(y_1, y_2, \ldots, y_n)$ is the joint probability density function and $f(y_1)$ is the marginal probability density function of Y_1. Note that the right side of Eq. (7.31) is, by definition, the characteristic function of Y_1. This implies that Eq. (7.30) represents the characteristic function of Y_1.

Suppose the right side of Eq. (7.31) is equal to a form of the characteristic function of a particular random variable, then from the uniqueness property of the characteristic function, we can find the distribution of Y_1. Even when a particular probability density function cannot be identified from the right side of Eq. (7.31), we may still be able to find the probability density function by integrating the characteristic function following the formula given Theorem 4.5 in Chapter 4. An example of this case is presented in Example 7.21.

Example 7.18. We may apply the method of characteristic function to find the probability density of the sum of statistically independent random variable X and Y having the Poisson distribution discussed in Example 7.10.

The characteristic functions of the Poisson distributions X and Y are given by

$$\phi_x(t) = e^{\mu(e^{it}-1)} \text{ and } \phi_y(t) = e^{\lambda(e^{it}-1)}$$

Since X and Y are statistically independent, the characteristic function of $(X + Y)$ can be written by applying the property of the characteristic function as follows:

$$\phi(t) = \phi_x(t) \cdot \phi_y(t) = e^{(\mu+\lambda)(e^{it}-1)}$$

TRANSFORMATION THROUGH CHARACTERISTIC FUNCTIONS

This is also the characteristic function of the Poisson distribution with the parameter $(\mu + \lambda)$, and hence the random variable $(X + Y)$ obeys the Poisson distribution with parameter $(\mu + \lambda)$. ∎

Example 7.19. $X_1, X_2, \ldots, X_n$ are statistically independent random variables each obeying a normal probability distribution $N(\mu_1, \sigma_1^2)$, $N(\mu_2, \sigma_2^2), \ldots, N(\mu_n, \sigma_n^2)$, respectively. The probability density function of a random variable $Y = k_1 X_1 + k_2 X_2 + \cdots + k_n X_n$, where $k_1, k_2, \ldots, k_n$ are constants, can be derived as follows:

The characteristic function of a random variable X_j can be written as

$$\phi_{X_j}(t) = e^{i\mu_j t - \frac{1}{2}\sigma_j^2 t^2}$$

Then, from the property of the characteristic function, the characteristic function of the random variable $k_j X_j$ becomes

$$\phi_{k_j X_j}(t) = e^{i k_j \mu_j t - \frac{1}{2} k_j^2 \sigma_j^2 t^2}$$

Since the random variables are independent, the characteristic function of $Y = \sum_{j=1}^{n} k_j X_j$ becomes

$$\phi_y(t) = \prod_{j=1}^{n} \phi_{k_j x_j}(t) = e^{i(\Sigma k_j \mu_j)t - \frac{1}{2}(\Sigma k_j^2 \sigma_j^2)t^2}$$

The right side of the above equation is the form of the characteristic function of a normal distribution with the mean $\sum_{j=1}^{n} k_j \mu_j$ and variance $\sum_{j=1}^{n} k_j^2 \sigma_j^2$. Hence, the random variable Y obeys a normal distribution with mean $\sum_{j=1}^{n} k_j \mu_j$ and variance $\sum_{j=1}^{n} k_j^2 \sigma_j^2$. ∎

Example 7.20. We may further extend the results obtained in the previous example to the two-dimensional case. That is, let $(X_1, X_2, \ldots, X_n)$ be a set of random samples of size n from a normal distribution $N(\mu, \sigma^2)$, and derive the joint probability distribution of $Y = \sum_{j=1}^{n} a_j x_j$ and $Z = \sum_{j=1}^{n} b_j x_j$, where a_j and b_j are constants.

The joint characteristic function of Y and Z can be obtained as

$$\phi_{y,z}(t, u) = E[e^{i(ty+uz)}] = E[e^{i\{t(\Sigma a_j x_j) + u(\Sigma b_j x_j)\}}]$$

$$= \int \cdots \int e^{i\{t(\Sigma a_j x_j) + u(\Sigma b_j x_j)\}} \cdot f(x_1, x_2, \ldots, x_n) \, dx_1 \, dx_2 \cdots dx_n$$

Since $x_1, x_2, \ldots, x_n$ are statistically independent, we may write

$$\phi_{y,z}(t, u) = \prod_{j=1}^{n} e^{i(ta_j + ub_j)x_j} f(x_j) \, dx_j$$

From the result obtained in Example 7.19, we have

$$\phi_{y,z}(t, u) = \prod_{j=1}^{n} \exp\left[i(ta_j + ub_j)\mu - \tfrac{1}{2}(ta_j + ub_j)^2 \sigma^2\right]$$

$$= \exp\left[i\left(t\sum a_j\mu + u\sum b_j\mu\right)\right.$$

$$\left. - \tfrac{1}{2}\left(t^2 \sum a_j^2 \sigma^2 + 2tu \sum a_j b_j \sigma^2 + u^2 \sum b_j^2 \sigma^2\right)\right]$$

This is the form of the characteristic function of a bivariate normal distribution with means $\sum_{j=1}^{n} a_j \mu$, $\sum_{j=1}^{n} b_j \mu$, variances $\sum_{j=1}^{n} a_j^2 \sigma^2$, $\sum_{j=1}^{n} b_j^2 \sigma^2$, and covariance $\sum_{j=1}^{n} a_j b_j \sigma^2$. They are statistically independent only if $\sum_{j=1}^{n} a_j b_j = 0$. ∎

Example 7.21. $(X_1, X_2, \ldots, X_n)$ is a set of random samples of size n taken from a uniform distribution over the sample space $(0, a)$. Obtain the probability density function of the random variable $Y = \ln X_1 + \ln X_2 + \cdots + \ln X_n$. For this, we may apply the formula given in Eq. (7.30). First obtain the characteristic function of $\ln X$. That is,

$$\phi_{\ln x}(t) = \int_{-\infty}^{\infty} f(x) e^{it \ln x} \, dx = \int_{-\infty}^{\infty} f(x) x^{it} \, dx$$

Hence, for the present problem, the characteristic function of $\ln X$ becomes

$$\phi_{\ln x}(t) = \int_0^a \frac{1}{a} x^{it} \, dx = \frac{a^{it}}{1 + it}$$

Then, the characteristic function of $Y = \sum_{j=1}^{n} \ln X_j$ is given by

$$\phi_y(t) = \frac{a^{int}}{(1 + it)^n}$$

The distribution of a random variable whose characteristic function has the form given above is unknown. However, the probability density func-

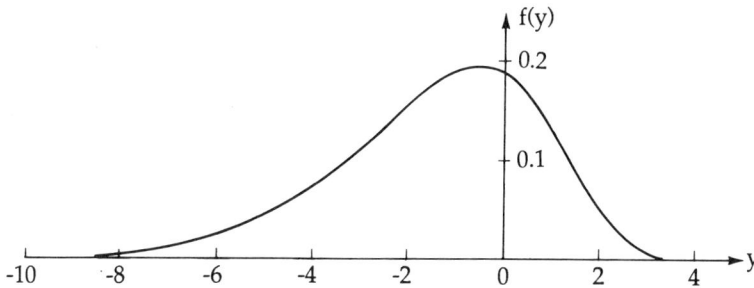

Figure 7.14 Probability density function $f(y)$ in Example 7.21.

tion of Y may be obtained as follows:

$$f(y) = \frac{1}{2\pi}\int_{-\infty}^{\infty} \phi_y(t) e^{-ity}\, dt = \frac{1}{2\pi}\int_{-\infty}^{\infty} \frac{a^{int}}{(1+it)^n} e^{-ity}\, dt$$

$$= \frac{1}{2\pi}\int_{-\infty}^{\infty} \frac{e^{it(n\ln a - y)}}{(1+it)^n}\, dt = \frac{(n\ln a - y)^{n-1} \cdot e^{-(n\ln a - y)}}{\Gamma(n)}$$

Note that the integration is in the form of a Fourier transform and that the sample space of the random variable Y is from $-\infty$ to $n\ln a$. Figure 7.14 shows the probability density function of Y for $a = 2$ and $n = 5$. ∎

EXERCISES

7.1 Let X and Y be statistically independent random variables with the probability density functions $f(x)$ and $f(y)$, respectively. Consider two random variables U and V where U is a function of X alone and V is a function of Y alone. Prove that U and V are also statistically independent.

7.2 X and Y are statistically independent discrete-type random variables, where $X, Y = 0, 1, 2, \ldots, n$. Prove that (a) for the sum of two random variables $Z = X + Y$,

$$Pr\{Z = n\} = \sum_{k=0}^{n} Pr\{X = n - k\} \cdot Pr\{Y = k\}$$

and (b) for the difference of two random variables $Z = X - Y$

$$Pr\{Z = n\} = \sum_{k=0}^{n} Pr\{X = n + k\} \cdot Pr\{Y = k\}$$

7.3 The random variable X has a beta distribution given in Eq. (6.34). Let Y be a new random variable given by $Y = a + (b - a)X$. Derive the distribution of Y that is the generalized beta distribution function given in Eq. (6.36).

7.4 The joint probability density function of two random variables X and Y given by

$$f(x, y) = \lambda_1 \lambda_2 e^{-(\lambda_1 x + \lambda_2 y)}, \qquad 0 \leq x, y < \infty$$

Show that the probability density function of the sum of these random variables, $Z = X + Y$, is given by

$$f(z) = \frac{\lambda_1 \lambda_2}{\lambda_2 - \lambda_1} (e^{-\lambda_1 z} - e^{-\lambda_2 z}), \qquad 0 \leq z < \infty$$

7.5 Two random variables X and Y are independent and have the same distribution, $f(x) = \lambda e^{-\lambda x}$ and $f(y) = \lambda e^{-\lambda y}$, where $0 \leq x, y, < \infty$. Let $U = X + Y$, $V = X/(X + Y)$, and $W = Y/X$. Show that U and V as well as U and W are statistically independent.

7.6 Let the random variable Z be the sum of two independent random variables X and Y. If Z has a gamma distribution with parameter $m_1 + m_2$, λ, and X also has a gamma distribution with parameters m_1, λ, show that the random variable Y is a gamma variate with parameter m_2, λ.

7.7 The random variable X has a uniform distribution $f(x) = 1/a$, $0 \leq x \leq a$. The probability density function of $Y = \ln X_1 + \ln X_2 + \cdots + \ln X_n$, where $X_1, X_2, \ldots, X_n$ are independently observed values from $f(x)$, is given in Example 7.21. By applying the probability density function of Y, obtain the probability density function of the geometric mean of $X_1, X_2, \ldots, X_n$, defined as $G = (X_1 X_2 X_3 \cdots X_n)^{1/n}$.
[Hint] Since $\ln G = Y/n$, the random variable G is equal to $\exp\{Y/n\}$.

7.8 Let X_i be independent random variables from the distribution

$$f(x) = \frac{1}{\Gamma(m)} x^{m-1} e^{-x}, \qquad 0 \leq x < \infty, \lambda > 0$$

Show that the distribution of $\bar{X} = (1/n)\sum_{i=1}^{n} X_i$ asymptotically approaches the normal distribution for large n. Obtain the mean and variance of the normal distribution.

CHAPTER 8

Extreme Value Statistics

Assessment of the probable largest magnitude associated with a random phenomenon is extremely important in applied probability. For example, the magnitude of behavior of a system in engineering and the physical sciences varies in random fashion through excitation induced by the application of external loads. Hence, it is necessary to design the system to be sufficiently strong to withstand the largest load expected to occur during its lifetime. For some systems, the estimation of the smallest value such as the lowest temperature may be as equally significant as estimation of the largest value in the design.

The largest value (or the smallest value) of a random variable that is expected to occur in a certain number of observations or in a certain period of time is called the *extreme value*. A random variable of the continuous type is considered throughout this chapter; however, principles and formulas presented in this chapter are equally applicable to a random variable of the discrete type.

Let the probability density function and the cumulative distribution function of a random variable X be denoted by $f(x)$ and $F(x)$, respectively. These functions are called the *initial probability density function* and the *initial cumulative distribution function*, respectively, in discussing extreme value statistics. The largest value expected to occur in n observations, denoted by Y_n, is also a random variable and follows its own probability law, which is different from that applicable for the random variable X. To avoid possible confusion, let us write the probability density function and the cumulative distribution function of the extreme value as $g(y_n)$ and $G(y_n)$, respectively. Here, the probability functions, $f(x)$, $F(x)$, $g(y_n)$, and $G(y_n)$ have mathematical relationships as will be shown in Section 8.1; therefore, the extreme values can be evaluated precisely from knowledge of the initial probability distribution.

It is a subject of considerable interest in statistics to find the asymptotic behavior of the extreme value cumulative distribution function $G(y_n)$ for a large sample size n. It was shown by Fisher and Tippet (1928), Fréchet (1927), and Gnedenko (1943) that the extreme value distribution $G(y_n)$ converges to three different types of distributions for $n \to \infty$. The distributions were later systematized by Gumbel (1958) who advocated the use of extreme value statistics to practical problems. The asymptotic extreme value distributions are presented in Section 8.2.

Estimation of extreme values from a set of observed data is a significantly important application of the probability theory in practice. Care must be taken in the estimation depending on the available data. This is particularly true if the initial probability distribution is not known. The subject will be discussed in Section 8.3.

8.1 ORDER STATISTICS AND EXTREME VALUES

8.1.1 Order Statistics

Definition 8.1. Let a set of observations $(x_1, x_2, x_3, \ldots, x_n)$ be a random sample of size n from a distribution with probability density function $f(x)$. Let us rearrange the elements of this random sample in ascending order of magnitude such that $y_1 < y_2 < y_3 < \cdots < y_n$. Then $(y_1, y_2, y_3, \ldots, y_n)$ is called the *ordered sample* of size n, and Y_j is called the *jth order statistics*.

It is noted that the random variables $X_1, X_2, \ldots, X_n$ are statistically independent and all have the same probability density function $f(x)$. On the other hand, the random variables $Y_1, Y_2, \ldots, Y_n$ are statistically independent and each has its own probability density function. Prior to discussing the probability density function of these random variables, we first present the following theorem:

Theorem 8.1. Let $Y_1, Y_2, Y_3, \ldots, Y_n$ be a set of order statistics associated with a random sample of size n, $(x_1, x_2, x_3, \ldots, x_n)$, from a distribution having a probability density function $f(x)$. Then, the joint probability density function of $Y_1, Y_2, \ldots, Y_n$, denoted by $g(y_1, y_2, y_3, \ldots, y_n)$, is given by

$$g(y_1, y_2, y_3, \ldots, y_n) = n! f(y_1) f(y_2) f(y_3) \ldots f(y_n) \quad (8.1)$$

where $-\infty < y_1 < y_2 < y_3 < \cdots < y_n$.

Proof. We may consider an arbitrarily specified element Y_i of the ordered sample. It is obvious that each element of the random sample $(x_1, x_2, x_3, \ldots, x_n)$ has an equal possibility of being Y_i and hence n different arrangements are possible for Y_i. Since there are n elements in the ordered sample $(y_1, y_2, y_3, \ldots, y_n)$, a total of $n!$ different arrangements of $x_1, x_2, \ldots$ are possible for the ordered sample. This implies that $n!$ sets of different arrangements of $(x_1, x_2, x_3, \ldots, x_n)$ must be considered for transformation of the joint probability density function of the random sample $(x_1, x_2, x_3, \ldots, x_n)$ to that of the ordered sample $(y_1, y_2, y_3, \ldots, y_n)$. Here, $(x_1, x_2, x_3, \ldots, x_n)$ are statistically independent with the probability density function $f(x)$. Since the absolute value of the Jacobians is unity, the joint probability density function of $(y_1, y_2, y_3, \ldots, y_n)$ can be obtained as follows:

$$g(y_1, y_2, y_3, \ldots, y_n) = n! [f(x_1, x_2, x_3, \ldots, x_n)]_{x_i \to y_i}$$
$$= n! f(y_1) f(y_2) \cdots f(y_n)$$

where $x_i \to y_i$ implies that x_i are changed to y_i.

The probability density function of the largest value, Y_n, can be obtained from Eq. (8.1) in terms of the probability density function $f(x)$ and the cumulative distribution function $F(x)$ as follows:

$$\begin{aligned}
g(y_n) &= \int_{-\infty}^{y_n} \cdots \int_{-\infty}^{y_3} \int_{-\infty}^{y_2} n! f(y_1) f(y_2) \cdots f(y_n) \, dy_1 \, dy_2 \cdots dy_{n-1} \\
&= n! f(y_n) \int_{-\infty}^{y_n} \cdots \int_{-\infty}^{y_3} \left[\int_{-\infty}^{y_2} f(y_1) \, dy_1 \right] \\
&\quad \times f(y_2) \cdots f(y_{n-1}) \, dy_2 \cdots dy_{n-1} \\
&= n! f(y_n) \int_{-\infty}^{y_n} \cdots \int_{-\infty}^{y_4} \left[\int_{-\infty}^{y_3} F(y_2) f(y_2) \, dy_2 \right] \\
&\quad \times f(y_3) \cdots f(y_{n-1}) \, dy_3 \cdots dy_{n-1} \\
&= n! f(y_n) \int_{-\infty}^{y_n} \cdots \int_{-\infty}^{y_5} \left[\int_{-\infty}^{y_4} \frac{1}{2!} \{F(y_3)\}^2 f(y_3) \, dy_3 \right] \\
&\quad \times f(y_4) \cdots f(y_{n-1}) \, dy_4 \cdots dy_{n-1} \\
&= n! f(y_n) \int_{-\infty}^{y_n} \cdots \int_{-\infty}^{y_6} \left[\int_{-\infty}^{y_5} \frac{1}{3!} \{F(y_4)\}^3 f(y_4) \, dy_4 \right] \\
&\quad \times f(y_5) \cdots f(y_{n-1}) \, dy_5 \cdots dy_{n-1} \quad (8.2)
\end{aligned}$$

ORDER STATISTICS AND EXTREME VALUES

By carrying out successive integration, we can derive

$$g(y_n) = n!f(y_n)\frac{\{F(y_n)\}^{n-1}}{(n-1)!}$$

$$= n \cdot f(y_n)\{F(y_n)\}^{n-1}, \quad -\infty < y_n < \infty \quad (8.3)$$

The cumulative distribution function of Y_n becomes,

$$G(y_n) = \int_{-\infty}^{y_n} g(y_n)\, dy_n = \{F(y_n)\}^n \quad (8.4)$$

The probability density function of the smallest value Y_1 in n observations can be derived in the same fashion as that for the largest value but the integration procedure is reversed. That is,

$$g(y_1) = \int_{y_1}^{\infty} \cdots \int_{y_{n-2}}^{\infty}\int_{y_{n-1}}^{\infty} n!f(y_n)f(y_{n-1})\cdots f(y_2)f(y_1)\, dy_n\, dy_{n-1}\cdots dy_2$$

$$= n!f(y_1)\int_{y_1}^{\infty}\cdots\int_{y_{n-2}}^{\infty}\left[\int_{y_{n-1}}^{\infty} f(y_n)\, dy_n\right]$$
$$\times f(y_{n-1})\cdots f(y_2)\, dy_{n-1}\cdots dy_2$$

$$= n!f(y_1)\int_{y_1}^{\infty}\cdots\int_{y_{n-3}}^{\infty}\left[\int_{y_{n-2}}^{\infty} \{1 - F(y_{n-1})\}f(y_{n-1})\, dy_{n-1}\right]$$
$$\times f(y_{n-2})\cdots f(y_2)\, dy_{n-1}\cdots dy_2$$

$$= n!f(y_1)\int_{y_1}^{\infty}\cdots\int_{y_{n-4}}^{\infty}\left[\int_{y_{n-3}}^{\infty} \frac{1}{2!}\{1 - F(y_{n-2})\}^2 \cdot f(y_{n-2})\, dy_{n-2}\right]$$
$$\times f(y_{n-3})\cdots f(y_2)\, dy_{n-3}\cdots dy_2 \quad (8.5)$$

By carrying out successive integration, we have

$$g(y_1) = n\{1 - F(y_1)\}^{n-1}f(y_1), \quad -\infty < y_1 < \infty \quad (8.6)$$

The cumulative distribution function of Y_1 becomes

$$G(y_1) = \{1 - F(y_1)\}^n \quad (8.7)$$

The probability density function of the jth order statistics, $g(y_j)$, can be derived in terms of $f(x)$ and $F(x)$ through a similar integration procedure used for $g(y_n)$ and $g(y_1)$. The density function is given in the following without derivation:

$$g(y_j) = \frac{n!}{(j-1)!(n-j)!}\{F(y_j)\}^{j-1}\{1 - F(y_j)\}^{n-j}f(y_j),$$

$$-\infty < y_j < \infty \quad (8.8)$$

8.1.2 Evaluation of Extreme Values

Various statistical properties of extreme values can now be evaluated from the probability density functions given in Eqs. (8.3) and (8.6). The following discussion pertains to the largest value in n observations, Y_n, but the concept is equally applicable to the smallest value Y_1.

As can be seen in Eq. (8.3), the probability density function of Y_n is a function of n. The shape of the probability density function $g(y_n)$ is, in general, much more sharply concentrated about its modal value than the initial probability density function, and the concentration increases with increase in number n. This is illustrated in the following example.

Example 8.1. Evaluate the probability density function $g(y_n)$ of the Rayleigh probability distribution (in dimensionless form) for various n values.

The initial probability density function is given by

$$f(x) = 2xe^{-x^2}$$

Then, we can obtain $g(y_n)$ from Eq. (8.3) as

$$g(y_n) = 2ny_n e^{-y_n^2}\left(1 - e^{-y_n^2}\right)^{n-1}$$

The probability density functions $g(y_n)$ for various n values are shown in Figure 8.1. ∎

An explanatory sketch illustrating the relationship between the probability density functions $f(x)$ and $g(y_n)$ is given in Figure 8.2. The modal value of $g(y_n)$, denoted by $\overline{Y}_n$, is called the *probable extreme value* or *characteristic largest value*. It is the extreme value most likely to occur in n observa-

ORDER STATISTICS AND EXTREME VALUES

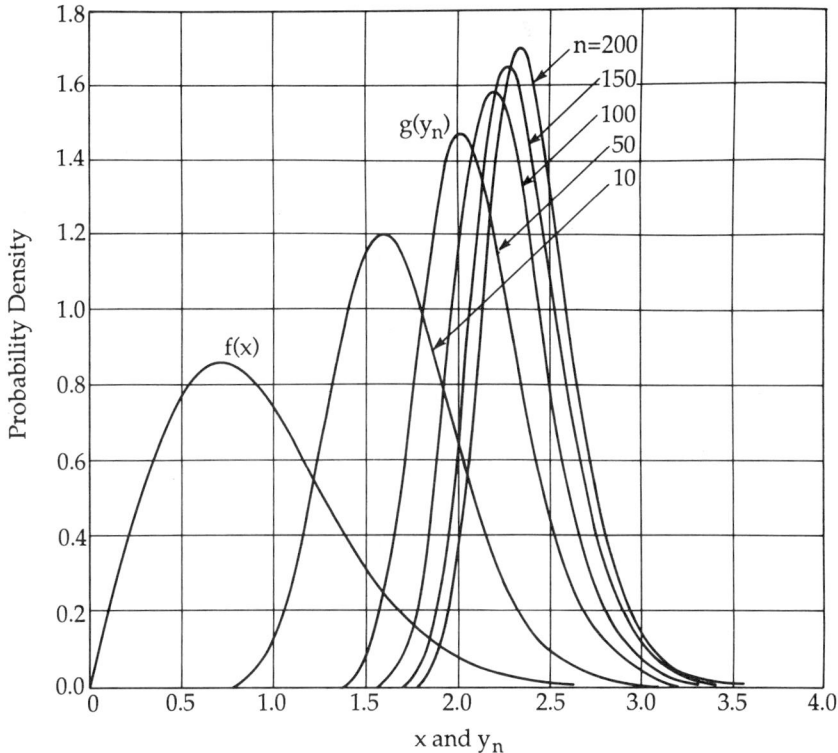

Figure 8.1 Probability density function of extreme value $g(y_n)$ of the Rayleigh distribution for various n values.

tions and it can be evaluated as the solution of the following equation:

$$\frac{d}{dy_n} g(y_n) = 0 \tag{8.9}$$

which yields

$$f'(y_n) F(y_n) + (n-1)\{f(y_n)\}^2 = 0 \tag{8.10}$$

By denoting y_n that satisfies the above equation as $\bar{y}_n$, we have

$$1 = -\frac{(n-1)\{f(\bar{y}_n)\}^2}{f'(\bar{y}_n) \cdot F(\bar{y}_n)} \tag{8.11}$$

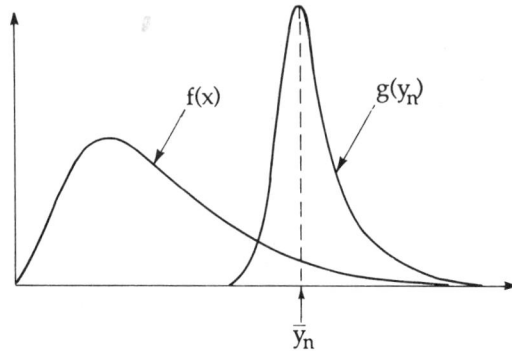

Figure 8.2 Initial probability density function, $f(x)$, extreme value probability density function, $g(y_n)$, and probable extreme value, $\bar{y}_n$.

Then, by dividing Eq. (8.11) by $1 - F(\bar{y}_n)$, we have

$$\frac{1}{1 - F(\bar{y}_n)} = -\frac{n-1}{F(\bar{y}_n)} \frac{f(\bar{y}_n)}{1 - F(\bar{y}_n)} \frac{f(\bar{y}_n)}{f'(\bar{y}_n)} \qquad (8.12)$$

If we assume that the initial distribution satisfies the L'Hospital rule in the form of

$$\frac{f(\bar{y}_n)}{1 - F(\bar{y}_n)} = -\frac{f'(\bar{y}_n)}{f(\bar{y}_n)} \qquad \text{for large } \bar{y}_n \qquad (8.13)$$

then Eqs. (8.12) and (8.13) yield

$$\frac{1}{1 - F(\bar{y}_n)} = \frac{n-1}{F(\bar{y}_n)} \sim n \qquad \text{for large } n \text{ and } \bar{y}_n \qquad (8.14)$$

The left-hand side of Eq. (8.14) is defined as the return period in Chapter 2. Thus, Eq. (8.14) implies that the probable extreme value expected to occur in n observations, $\bar{Y}_n$ (where n is large), can be evaluated from the initial cumulative distribution function as the x value for which the return period is equal to n or the probability of exceeding this x value is $1/n$.

As an example of the probability distribution that satisfies the L'Hospital rule, let us consider the initial cumulative distribution function $F(x)$ that

ORDER STATISTICS AND EXTREME VALUES

can be expressed in the form

$$F(x) = 1 - e^{-q(x)} \tag{8.15}$$

where $q(x)$ is a positive real-valued function that satisfies the conditions required for $F(x)$ to be a cumulative distribution function. $F(x)$ given in Eq. (8.15) covers a variety of distribution functions frequently used in practical problems. From Eq. (8.15), we have

$$\frac{f(x)}{1 - F(x)} = q'(x) \tag{8.16}$$

and

$$\frac{f'(x)}{f(x)} = -q'(x) + q''(x)/q'(x)$$

Since the second term of $f'(x)/f(x)$ is small in comparison with the first term, the probability function satisfies the condition given in Eq. (8.13). It is noted that the left side of Eq. (8.16) is defined as the hazard function in Example 2.20.

Example 8.2. Evaluate the probable extreme value for the Rayleigh probability distribution given by

$$f(x) = \frac{2x}{R} e^{-x^2/R}$$

The cumulative distribution function is given by

$$F(x) = 1 - e^{-x^2/R}$$

Hence, from Eq. (8.14), we have

$$\exp\{-y_n^2/R\} = 1/n$$

and thereby its solution is given by

$$\bar{y}_n = \sqrt{\ln n}\, \sqrt{R} \qquad \blacksquare$$

It was stated earlier that the probable extreme value, $\bar{y}_n$, is interpreted as that being most likely to occur since it is the value for which the probability density function $g(y_n)$ peaks. It is noted, however, that the extreme value Y_n is a random variable and it may exceed the probable extreme value, $\bar{y}_n$. In fact, the chance of occurrence of an extreme value greater than $\bar{y}_n$ is rather high. Let us evaluate the probability that the extreme value will exceed $\bar{y}_n$. That is, for large n,

$$\lim_{n \to \infty} \Pr\{\text{Extreme value} > \bar{y}_n\} = 1 - G(\bar{y}_n)$$

$$= \lim_{n \to \infty} \left[1 - \{F(\bar{y}_n)\}^n\right]$$

$$= \lim_{n \to \infty} \left[1 - \left(1 - \frac{1}{n}\right)^n\right]$$

$$= 1 - e^{-1} = 0.632 \qquad (8.17)$$

As shown in Eq. (8.17), there is a 63.2% chance that the extreme value will exceed $\bar{y}_n$. Since this probability is extremely high, the probable extreme value $\bar{y}_n$ should not be used for design in engineering problems. It is highly desirable from a design consideration to estimate the extreme value for which the probability of being exceeded is very small. One way to achieve this concept is to choose a very small value α that may be called the *risk parameter*, and evaluate the extreme value $\hat{y}_n$ for which the following relationship holds:

$$\int_0^{\hat{y}_n} g(y_n)\, dy_n = \{F(\hat{y}_n)\}^n = 1 - \alpha \qquad (8.18)$$

Considering that α is small and n is large, we have

$$F(\hat{y}_n) = (1 - \alpha)^{1/n} \sim 1 - \frac{\alpha}{n} + 0(\alpha^2) \qquad (8.19)$$

Thus, the design extreme value with risk parameter α can be evaluated as the solution of the following equation:

$$1 - F(\hat{y}_n) = \frac{\alpha}{n} \qquad (8.20)$$

Example 8.3. Evaluate the extreme value in n observations with the risk parameter α for the following Weibull probability distribution:

$$f(x) = c\lambda^c x^{c-1} e^{-(\lambda x)^c} \qquad 0 \leq x < \infty$$

The cumulative distribution function is given by

$$F(x) = 1 - \exp\{-(\lambda x)^c\}$$

By applying Eq. (8.20), we have

$$\exp\{-(\lambda y_n)^c\} = \alpha/n$$

Hence, the extreme value $\hat{y}_n$ can be obtained as

$$\hat{y}_n = \frac{1}{\lambda}\left(\ln\frac{n}{\alpha}\right)^{1/c}$$ ∎

Figure 8.3 is an explanatory sketch indicating the probabilities of exceeding the probable extreme and design extreme values. It may be well to summarize here these extreme values: Let Y_n be the largest value of a random variable X in n observations. Then, we define

1. Probable extreme value (characteristic largest value), $\bar{y}_n$, is the value of Y_n for which the probability density function peaks. We have,

$$Pr\{\text{Extreme value } Y_n > \bar{y}_n\} = 1 - e^{-1} = 0.632$$

$$Pr\{\text{Random variable } X > \bar{y}_n\} = 1/n$$

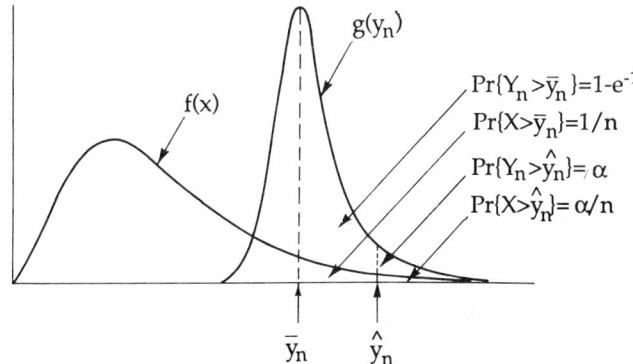

Figure 8.3 Probabilities of exceeding the probable extreme value, $\bar{y}_n$, and design extreme value, $\hat{y}_n$.

2. Design extreme value, $\hat{y}_n$, is the value of Y_n for which the probability of exceeding this value is a specified small value α, called the risk parameter. We have

$$Pr\{\text{Extreme value } Y_n > \hat{y}_n\} = \alpha$$

$$Pr\{\text{Random variable } X > \hat{y}_n\} = \alpha/n$$

The extreme value is developed based on order statistics; hence, it is evaluated as a function of the number of observations, n. For a random process, the magnitude of which varies in random fashion with time, however, it may be more meaningful to express the extreme value in terms of time rather than as a function of number of occurrences. This can be done by using the estimated average number of occurrences of the event per unit time. Evaluation of the extreme value in terms of time will be discussed in Chapter 11.

8.2 ASYMPTOTIC DISTRIBUTIONS OF EXTREME VALUES

In this section, the asymptotic behavior of the distribution function of extreme values and its limiting form for large n will be discussed.

Let us write the cumulative distribution function of the largest value Y_n given in Eq. (8.4) in the following form:

$$G(y_n) = \{F(y_n)\}^n = e^{n \ln F(y_n)} \tag{8.21}$$

It can be seen in the above equation that $\ln F(y_n) \to 0$ for $n \to \infty$. Hence, the value of $G(y_n)$ depends entirely on the asymptotic behavior of the initial cumulative distribution function $F(x)$ toward the extreme value. In general, a continuous random variable may categorized into three different types: (1) the distribution is unlimited toward the extreme value and all moments of the distribution exist; (2) the distribution is unlimited toward the extreme value but only a finite number of moments exist; and (3) the distribution is limited at the upper and/or lower bounds. We will see that there exists a different type of asymptotic extreme value distribution for each of the initial probability density functions classified above.

In the development of the asymptotic distribution functions of the extreme values, Fréchet, Fisher, and Tippet considered a total of m sets of random samples, each comprising a sample of size n. Here, the cumulative distribution function of the largest value in n observations is $G(y_n)$. Then, we may write the cumulative distribution function of the largest value in mn observations as $\{G(y_n)\}^m$, and this is equal to $G(y_n)$ with a linear transformation of the random variable Y_n. Hence, we have the following relationship:

$$\{G(y_n)\}^m = G(a_m y_n + b_m) \qquad (8.22)$$

where a_m, b_m are constants.

The first solution of the above equation was derived by Fréchet (1927), which later became known as the Type II asymptotic extreme value distribution. The same solution along with two additional solutions were obtained under different conditions by Fisher and Tippett (1928). Later, it was proved by Gnedenko (1943) that there exist only three solutions of Eq. (8.22). These three types of asymptotic distributions were studied in detail and were systematized by Gumbel (1958) as Type I, II, and III asymptotic extreme value distributions.

The derivation of these asymptotic distributions is quite complicated and is beyond the scope of this book. Here, only a simplified derivation of the Type I asymptotic distribution, which is most frequently used in analysis of observed data, is outlined in Section 8.2.1. The reader who is interested in the detailed derivation of the three types of asymptotic distributions is referred to Gumbel (1958).

For extreme values, we consider the largest value in n observations of a random phenomenon as well as the smallest value in the observation, in general. Since there are three asymptotic extreme value distributions, a total of six extreme value random variables will be discussed in this section. In order to identify these random variables, the largest value in n observations for each type of distribution (Type I, II, and III) are denoted by ^1y_n, ^2y_n, and ^3y_n, respectively, throughout this chapter, while the smallest value in n observations for each type of distribution is denoted by $^1y_1$, $^2y_1$, and $^3y_1$, respectively.

8.2.1 Type I Asymptotic Distribution

In the derivation of the asymptotic Type I extreme value distribution, let us assume that the initial distribution can be expressed in the form given in Eq. (8.15). We may present the equation again here.

$$F(x) = 1 - \exp\{-q(x)\} \qquad (8.23)$$

It was proved, under the condition of Eq. (8.23), that the probability distribution satisfies the L'Hospital rule and thereby the probable extreme value (characteristic largest value), $\overline{Y}_n$, satisfies the relationship given in Eq. (8.14). That is, the probability that the random variable X exceeds the probable extreme value becomes

$$\Pr\{X > \bar{y}_n\} = 1 - F(\bar{y}_n) = \exp\{-q(\bar{y}_n)\} = 1/n \qquad (8.24)$$

From the above two equations, we may write $F(x)$ as

$$F(x) = 1 - \frac{1}{n}e^{-\{q(x)-q(\bar{y}_n)\}} \qquad (8.25)$$

Hence, the cumulative distribution function of the extreme value for large n becomes

$$G(y_n) = \lim_{n \to \infty}\left[1 - \frac{1}{n}e^{-\{q(y_n)-q(\bar{y}_n)\}}\right]^n$$

$$= \exp\{-e^{-\{q(y_n)-q(\bar{y}_n)\}}\} \qquad (8.26)$$

Next, let us write $q(y_n) - q(\bar{y}_n) = A(y_n)$, and expand $A(y_n)$ in the neighborhood of $\bar{y}_n$ by applying the Taylor expansion series as follows:

$$A(y_n) = A(\bar{y}_n) + A'(\bar{y}_n)(y_n - \bar{y}_n) + \tfrac{1}{2}A''(\bar{y}_n)(y_n - \bar{y}_n)^2 + \cdots \qquad (8.27)$$

Since $A(\bar{y}_n) = 0$, we can write approximately $A(y_n) = A'(\bar{y}_n)(y_n - \bar{y}_n)$; therefore we have

$$q(y_n) - q(\bar{y}_n) = q'(\bar{y}_n)(y_n - \bar{y}_n) \qquad (8.28)$$

Thus, from Eqs. (8.26) and (8.28), we can write

$$G(y_n) = \exp\{-e^{-q'(\bar{y}_n)(y_n-\bar{y}_n)}\} \qquad (8.29)$$

Note that from Eqs. (8.16) and (8.24), $q'(x)$ for $x = \bar{y}_n$ becomes

$$q'(\bar{y}_n) = nf(\bar{y}_n) \qquad (8.30)$$

By writing y_n in the above equation as ^1y_n and by letting

$$q'(\bar{y}_n) = nf(\bar{y}_n) = \alpha_n \quad \text{and} \quad \bar{y}_n = u_n \qquad (8.31)$$

ASYMPTOTIC DISTRIBUTIONS OF EXTREME VALUES

the cumulative distribution function $G({}^1y_n)$ can be expressed as

$$G({}^1y_n) = \exp\{-e^{-\alpha_n({}^1y_n - u_n)}\}, \quad -\infty < {}^1y_n < \infty \quad (8.32)$$

The cumulative distribution function given in Eq. (8.32) is called the *Type I asymptotic extreme value distribution*, and extreme values of many probability distributions whose cumulative distribution function is of the exponential type as shown in Eq. (8.15) asymptotically converge to the Type I distribution for large n. A more precise definition of the *exponential-type distribution* is given by von Mises as the distribution that satisfies the following condition:

$$\lim_{x \to \infty} \frac{d}{dx}\left[\frac{1 - F(x)}{f(x)}\right] = 0 \quad (8.33)$$

The exponential type distribution is unlimited toward the extreme value ${}^1y_n$ and all moments exist. Examples of probability distributions that belong to this category are the exponential, normal, log-normal, chi-square, and gamma distributions, and so on, among others.

In order to evaluate the mean and variance of the Type I distribution, let us apply the cumulant-generating function introduced in Eq. (4.19). For this, transform the random variable ${}^1Y_n$ by letting

$$z = \alpha_n({}^1y_n - u_n) \quad (8.34)$$

and derive the cumulative distribution function of Z as

$$G(z) = \exp\{-e^{-z}\} \quad (8.35)$$

Figure 8.4 shows the dimensionless probability density function of the Type I asymptotic extreme value distribution given in Eq. (8.35).

The characteristic function of Z can be written as

$$\phi_z(t) = \int_{-\infty}^{\infty} e^{itz} \cdot g(z)\, dz$$

$$= \int_{-\infty}^{\infty} \exp\{-(1 - it)z\} \cdot \exp\{-e^{-z}\}\, dz \quad (8.36)$$

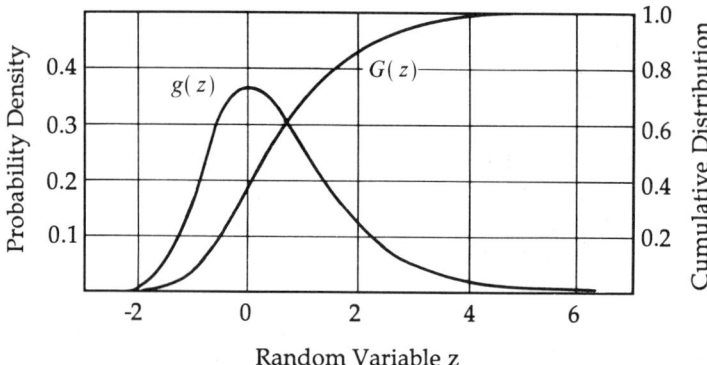

Figure 8.4 Dimensionless probability density function $g(z)$, and cumulative distribution function, $G(z)$, of Type I asymptotic extreme value distribution.

Then, by letting $e^{-z} = s$, we have

$$\phi_z(t) = \int_0^\infty s^{it} e^{-s} \, ds = \Gamma(1 - it) \qquad (8.37)$$

Next, by taking the logarithm of $\phi_z(t)$, the cumulant-generating function becomes

$$\psi_z(t) = \ln \phi_z(t) = \ln \Gamma(1 - it)$$

$$= \gamma(it) + \sum_{\nu=2}^{\infty} \frac{1}{\nu} \left(\sum_{\lambda=1}^{\infty} \lambda^{-\nu} \right) (it)^\nu \qquad (8.38)$$

where γ = Euler's constant, 0.577. Here, the first term and second terms of Eq. (8.38) yield the mean and variance, respectively, of the random variable Z. These are,

$$E[z] = \gamma$$

and

$$\text{Var}[z] = \sum_{\lambda=1}^{\infty} \lambda^{-2} \sim \frac{\pi^2}{6} \qquad (8.39)$$

Next, from the transformation formula given in Eq. (8.34), the mean and variance of the random variable ^1Y_n can be obtained as

$$E[^1y_n] = u_n + \frac{\gamma}{\alpha_n}$$

$$\text{Var}[^1y_n] = \frac{1}{\alpha_n^2}\frac{\pi^2}{6} \qquad (8.40)$$

From the above equation, the parameters α_n and u_n associated with the extreme value ^1y_n can be expressed in terms of the mean and variance as

$$\alpha_n = \frac{\pi/\sqrt{6}}{\sqrt{\text{Var}[^1y_n]}}$$

$$u_n = E[^1y_n] - \frac{\gamma}{\alpha_n} = E[^1y_n] - \gamma\frac{\sqrt{6}}{\pi}\sqrt{\text{Var}[^1y_n]} \qquad (8.41)$$

In practice, the mean and variance, $E[^1y_n]$ and $\text{Var}[^1y_n]$, respectively, can be evaluated from data; and hence the cumulative distribution function of the extreme value given in Eq. (8.32) can be evaluated with the aid of Eq. (8.41).

There are several other methods for estimating the parameters α_n and u_n of the distribution in addition to the formula given in Eq. (8.41). These include the graphical method and the maximum likelihood method (see Gumbel 1958, among others).

Example 8.4. Figure 8.5(*a*) shows the histogram of the monthly highest temperature in degrees Fahrenheit constructed from data consisting of 240 observations in 20 years at City A. Figure 8.5(*b*) shows the cumulative distribution plotted on extreme value paper. Included also in the figure is the Type I asymptotic distribution, the parameters of which are evaluated by Eq. (8.41). As can be seen in the figure, the asymptotic Type I distribution reasonably represents the extreme values. By applying the method for predicting the probable extreme value presented in the previous section to a Type I distribution, the highest temperature expected in 20 years is 100.2°F as compared with the observed highest of 96.8°F. The highest temperature expected in 50 years is estimated to be 109.0°F. ∎

The probability density function of the Type I asymptotic extreme value distribution can be obtained from Eq. (8.32) as

$$g(^1y_n) = \alpha_n \exp\{-\alpha_n(^1y_n - u_n)\} \cdot \exp\{-e^{-\alpha_n(^1y_n - u_n)}\} \qquad (8.42)$$

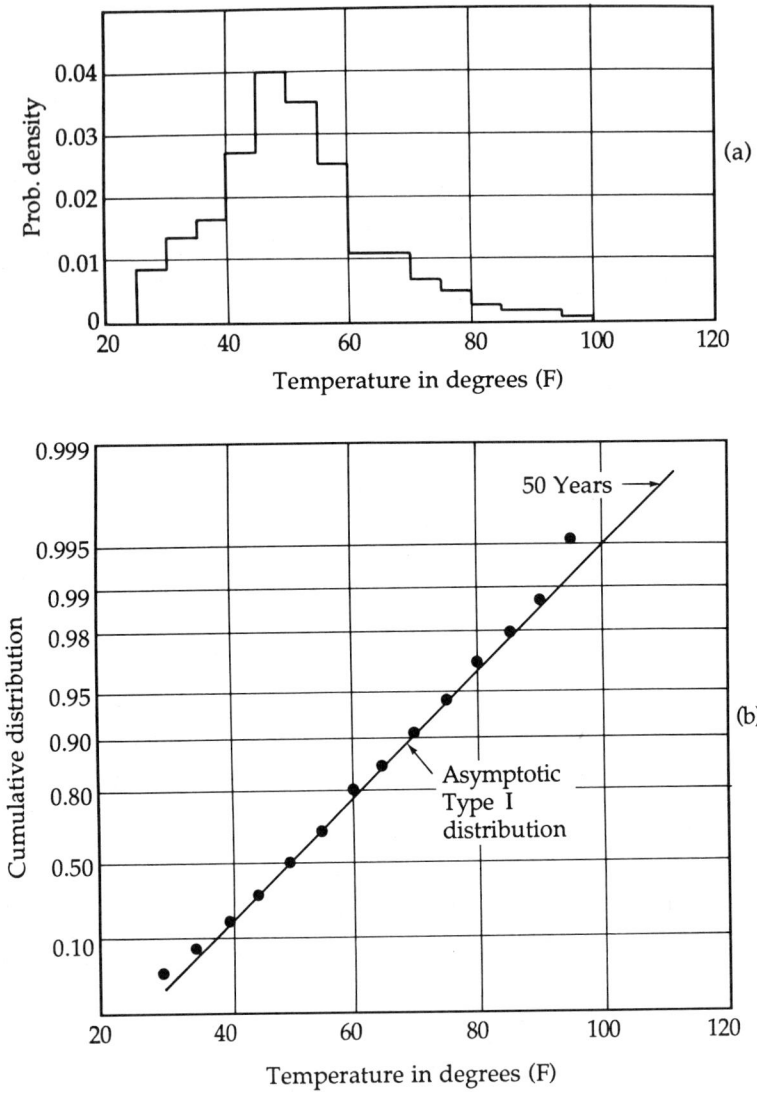

Figure 8.5 Histogram of monthly highest temperature and cumulative distribution plotted on extreme value paper.

Thus far, the sample space of the asymptotic Type I distribution is considered to be from $-\infty$ to ∞. However, it is often encountered that the sample space of the random variable ^1Y_n is limited to $(0, \infty)$. In this case, the probability functions given in Eqs. (8.32) and (8.42) must be truncated at $y_n = 0$. The truncated cumulative distribution function and probability density function, denoted by $G_*(y_n)$ and $g_*(y_n)$, respectively, become as follows [see Eqs. (2.35) and (2.36)]:

$$G_*(^1y_n) = \frac{1}{1 - \exp\{-e^{\alpha_n u_n}\}} \left[\exp\{-e^{-\alpha_n(^1y_n - u_n)}\} - \exp\{-e^{\alpha_n u_n}\}\right]$$

$$g_*(^1y_n) = \frac{1}{1 - \exp\{-e^{\alpha_n u_n}\}} \left[\alpha_n \exp\{-\alpha_n(^1y_n - u_n)\}\exp\{-e^{-\alpha_n(^1y_n - u_n)}\}\right]$$

$$0 \leq {}^1y_n < \infty \tag{8.43}$$

The derivation of the truncated cumulative distribution is left as an exercise for the reader.

The Type I asymptotic extreme value distribution presented so far pertains to the largest value in n observations. The concept of the distribution function developed for the largest value ^1Y_n can be equally applicable to the smallest value in n observations, denoted by $^1Y_1$.

Although the sample space of the random variable $^1Y_1$ is from $-\infty$ to $+\infty$ (which is the same as for the largest value ^1Y_n), the cumulative distribution function becomes unity at $^1y_1 = -\infty$. This implies that the cumulative distribution function of $^1Y_1$ can be obtained by changing the sign of the distribution for ^1Y_n. Hence, analogous to Eq. (8.35), we may write the cumulative distribution function for the smallest value as

$$G(z) = \exp\{-e^z\} \tag{8.44}$$

Here, the random variable z can be expressed in the same form as defined in Eq. (8.34). That is,

$$z = \alpha_1(^1Y_1 - u_1) \tag{8.45}$$

Thus, the cumulative distribution function of the Type I asymptotic distribution applicable for the smallest value becomes

$$G(^1y_1) = \exp\{-e^{\alpha_1(^1y_1 - u_1)}\}, \quad -\infty < {}^1y_1 < \infty \tag{8.46}$$

It is noted again that for the cumulative distribution function defined above we have $G(\infty) = 0$ and $G(-\infty) = 1$.

The probability density function of ${}^1Y_1$, denoted by $g({}^1Y_1)$ is given by

$$g({}^1y_1) = \alpha_1 \exp\{\alpha({}^1y_1 - u_1)\} \cdot \exp\{-e^{\alpha_1({}^1y_1 - u_1)}\} \quad (8.47)$$

The mean and variance of the distribution are given as follows:

$$E[{}^1y_1] = u_1 - \frac{\gamma}{\alpha_1}$$

$$\mathrm{Var}[{}^1y_1] = \frac{1}{\alpha_1^2}\frac{\pi^2}{6} \quad (8.48)$$

The derivations of $E[{}^1y_1]$ and $\mathrm{Var}[{}^1y_1]$ are left as an exercise for the reader.

From Eq. (8.48), the parameters α and u can be expressed in terms of the mean and variance as

$$\alpha_1 = \frac{\pi/\sqrt{6}}{\sqrt{\mathrm{Var}[{}^1y_1]}}$$

$$u_1 = E[{}^1y_1] + \gamma\frac{\sqrt{6}}{\pi}\sqrt{\mathrm{Var}[{}^1y_1]} \quad (8.49)$$

As in the case for the largest value, the parameters α_1 and u_1 can be evaluated from the mean and variance obtained from the data.

8.2.2 Type II Asymptotic Distribution

The Type II asymptotic extreme value distribution is associated with the initial distribution, which is unlimited toward the extreme value and has only a finite number of moments. Distributions carrying this property are called *Cauchy-type distributions*, and the Cauchy distribution shown in Example 4.4 is a typical example.

The Cauchy-type distribution must satisfy the following condition:

$$\lim_{x \to \infty} \{1 - F(x)\} \cdot x^k = a \quad (8.50)$$

where $k > 0$ and $a > 0$, and no moments of order equal to or greater than k exist.

If we write the probable extreme value of the Cauchy-type distribution as v_n, then from Eq. (8.14) v_n must satisfy the following equation for large n:

$$1 - F(v_n) = 1/n \qquad (8.51)$$

Hence, from Eqs. (8.50) and (8.51), the constant a can be obtained as

$$a = (1/n) v_n^k \qquad (8.52)$$

This results in Eq. (8.50) yielding for large x

$$F(x) = 1 - \frac{1}{n}\left(\frac{v_n}{x}\right)^k \qquad (8.53)$$

Based on the initial distribution given in Eq. (8.53), the cumulative distribution function of the extreme value, denoted by $G(^2y_n)$, can be written for large n as

$$G(^2y_n) = \lim_{n \to \infty} \left\{1 - \frac{1}{n}\left(\frac{v_n}{^2y_n}\right)^k\right\}^n = \exp\left\{-\left(\frac{v_n}{^2y_n}\right)^k\right\},$$

$$0 \leq {}^2y_n < \infty, \, v_n > 0 \quad (8.54)$$

The asymptotic distribution derived in Eq. (8.54) is called the *Type II asymptotic extreme value distribution*. The probability density function is given by

$$g(^2y_n) = \frac{k v_n^k}{{}^2y_n^{k+1}} \exp\left\{-\left(\frac{v_n}{^2y_n}\right)^k\right\} \qquad (8.55)$$

It is noted that the Type II asymptotic distribution can be derived from the Type I asymptotic distribution through a transformation of the random variables. That is, by letting ${}^1Y_n = \ln {}^2Y_n$, and $u_n = \ln v_n$, and $\alpha_n = k$ in Eq. (8.32), we can write the cumulative distribution function as

$$G(^2y_n) = \exp\left\{-e^{-k(\ln {}^2y_n - \ln v_n)}\right\}$$

$$= \exp\left\{-e^{\ln(v_n/{}^2y_n)^k}\right\} = \exp\left\{-\left(\frac{v_n}{{}^2y_n}\right)^k\right\}, \quad 0 \leq {}^2y_n < \infty \quad (8.56)$$

This is the cumulative distribution function of the Type II asymptotic distribution defined in Eq. (8.54).

By using the functional relationship given in the derivation of Eq. (8.56), the mean and variance of the Type II asymptotic distribution can be evaluated through the characteristic function of the standardized Type I extreme value random variable Z shown in Eq. (8.34). That is, we may write

$$Z = \alpha_n(^1Y_n - u) = k(\ln {}^2Y_n - \ln v_n) = k \ln({}^2Y_n/v_n) \quad (8.57)$$

Hence, we have

$$^2Y_n = v_n \exp\{Z/k\} \quad (8.58)$$

The mean value of the random variable ^2Y_n can be written as

$$E[^2y_n] = v_n E[e^{z/k}] = v_n E[e^{i(1/ik)z}] \quad (8.59)$$

where $E[e^{i(1/ik)z}]$ is in the form of the characteristic function of Z given in Eq. (8.36) with $t = 1/ik$. Hence, from Eq. (8.37) we have $E[e^{z/k}] = \Gamma(1 - 1/k)$, and thereby

$$E[^2y_n] = v_n \Gamma\left(1 - \frac{1}{k}\right) \quad (8.60)$$

Similarly, the second moment, $E[^2y_n^2]$ can be obtained from the characteristic function of Z with $t = 2/ik$. That is,

$$E[^2y_n^2] = v_n^2 \Gamma\left(1 - \frac{2}{k}\right) \quad (8.61)$$

Thus, in summary, the mean and variance of the Type II extreme variate are given by

$$E[^2y_n] = v_n \Gamma\left(1 - \frac{1}{k}\right)$$

$$\text{Var}[^2y_n] = v_n^2 \left\{\Gamma\left(1 - \frac{2}{k}\right) - \Gamma^2\left(1 - \frac{1}{k}\right)\right\} \quad (8.62)$$

The following relationship can be derived from Eqs. (8.60) and (8.61):

$$\frac{E[^2y_n^2]}{\left(E[^2y_n]\right)^2} = \frac{\Gamma[1-(2/k)]}{\Gamma^2[1-(1/k)]} \qquad (8.63)$$

Since the left side of the above equation is known from the sample data, the parameter k can be estimated therefrom. Subsequently, the parameter v_n can be estimated from Eq. (8.60).

The cumulative distribution function of the Type II asymptotic distribution applicable for the smallest value $^2Y_1$ can be derived by the same procedure used in the derivation of Eq. (8.56) for the largest value. However, care must be taken that the sample space of the random variable $^2Y_1$ is from 0 to $-\infty$, while that of $^1Y_1$ is from $+\infty$ to $-\infty$. Therefore transformation of the random variable from $^1Y_1$ to $^2Y_1$ by $^1Y_1 = -\ln {}^2Y_1$ and let $u_1 = -\ln v_1$ and $\alpha_1 = k$. Then, from Eq. (8.46) we can derive the following cumulative distribution function of Type II asymptotic distribution applicable for the smallest value:

$$G(^2y_1) = \exp\left\{-\left(\frac{v_1}{{}^2y_1}\right)^k\right\}, \qquad -\infty < {}^2y_1 \leq 0 \qquad (8.64)$$

It is noted that for the cumulative distribution function defined above, we have $G(0) = 0$ and $G(-\infty) = 1$.

The probability density function of $^2Y_1$ can be derived from Eq. (8.64) as

$$g(^2y_1) = \frac{kv_1^k}{{}^2y_1^{k+1}} \exp\left\{-\left(\frac{v_1}{{}^2y_1}\right)^k\right\} \qquad (8.65)$$

Note that $^2Y_1 < 0$ in the above equation; hence, the absolute value should be taken as the probability density function.

The mean and variance of the Type II extreme value $^2Y_1$ are given by

$$E[^2y_1] = -v_1\Gamma\left(1 - \frac{1}{k}\right)$$

$$\text{Var}[^2y_1] = v_1^2\left\{\Gamma\left(1 - \frac{2}{k}\right) - \Gamma^2\left(1 - \frac{1}{k}\right)\right\} \qquad (8.66)$$

8.2.3 Type III Asymptotic Distribution

The Type III asymptotic extreme value distribution is associated with initial distributions that are bounded by either an upper or a lower limit value, called limited-type distributions. The direct derivation of the Type III asymptotic distribution is extremely complicated, hence a simple derivation from the Type I asymptotic distribution is given here. That is, consider the following transformation of random variables from ^1Y_n to ^3Y_n:

$$^1Y_n - u_n = -\ln\left(\frac{w_n - {}^3Y_n}{w_n - v_n}\right) \qquad (8.67)$$

By writing $\alpha_n = k$ in Eq. (8.32), the transformation yields

$$G({}^3y_n) = \exp\left\{-\exp\left\{-k\left[-\ln\left(\frac{w_n - {}^3y_n}{w_n - v_n}\right)\right]\right\}\right\} = \exp\left\{-\left(\frac{w_n - {}^3y_n}{w_n - v_n}\right)^k\right\}$$

$$-\infty < {}^3y_n < w_n \qquad (8.68)$$

This is the *Type III asymptotic extreme value distribution* with an upper limit value w_n.

It is noted that the Type III asymptotic distribution reduces to the Type I asymptotic distribution if the upper limit value w_n is very large such that v_n/w_n is much smaller than unity. Under this condition, the Type III cumulative distribution function can be written as

$$G({}^3y_n) = \exp\left\{-\left(\frac{w_n - {}^3y_n}{w_n - v_n}\right)^k\right\} \sim \exp\left\{-\left(1 - \frac{{}^3y_n - v_n}{w_n}\right)^k\right\}$$

$$= \exp\left\{-\left[1 - k\left(\frac{{}^3y_n - v_n}{w_n}\right) + \frac{k(k-1)}{2!}\left(\frac{{}^3y_n - v_n}{w_n}\right)^2 \cdots\right]\right\}$$

$$\sim \exp\left\{-\left[1 - \frac{k}{w_n}({}^3y_n - v_n) + \frac{1}{2!}\left(\frac{k}{w_n}\right)^2({}^3y_n - v_n)^2 \cdots\right]\right\}$$

$$= \exp\left\{-e^{-(k/w_n)({}^3y_n - v_n)}\right\} \qquad (8.69)$$

By letting $k/w_n = \alpha_n$ and $v_n = u_n$, Eq. (8.69) becomes the Type I asymptotic distribution given in Eq. (8.32).

ASYMPTOTIC DISTRIBUTIONS OF EXTREME VALUES

The probability density function, denoted by $g(^3y_n)$, can be obtained by differentiating $G(^3y_n)$ with respect to ^3y_n. That is,

$$g(^3y_n) = \frac{k}{w_n - v_n}\left(\frac{w_n - {}^3y_n}{w_n - v_n}\right)^{k-1} \exp\left\{-\left(\frac{w_n - {}^3y_n}{w_n - v_n}\right)^k\right\} \quad (8.70)$$

The mean and variance of the Type III asymptotic distribution can be evaluated through the following method:

Let us write

$$\left(\frac{w_n - {}^3y_n}{w_n - v_n}\right)^k = z \quad (8.71)$$

and evaluate the moment $E[z^{m/k}]$ of the random variable Z. By applying the change of random variable technique, the moment can be obtained as

$$E[z^{m/k}] = \int_0^\infty z^{m/k} e^{-z}\, dz = \Gamma\left(1 + \frac{m}{k}\right) \quad (8.72)$$

From Eqs. (8.71) and (8.72), we have

$$E\left[\left(w_n - {}^3y_n\right)^m\right] = (w_n - v_n)^m \Gamma\left(1 + \frac{m}{k}\right) \quad (8.73)$$

By letting $m = 1$,

$$E\left[\left(w_n - {}^3y_n\right)\right] = (w_n - v_n)\Gamma\left(1 + \frac{1}{k}\right) \quad (8.74)$$

and hence, the mean $E[{}^3y_n]$ becomes

$$E[{}^3y_n] = w_n - (w_n - v_n)\Gamma\left(1 + \frac{1}{k}\right) \quad (8.75)$$

Next, by letting $m = 2$,

$$E\left[\left(w_n - {}^3y_n\right)^2\right] = (w_n - v_n)^2 \Gamma\left(1 + \frac{2}{k}\right) \quad (8.76)$$

Then, from Eqs. (8.74) and (8.76), $\text{Var}[^3y_n]$ can be obtained as

$$\text{Var}[^3y_n] = \text{Var}[w_n - {}^3y_n] = E\left[(w_n - {}^3y_n)^2\right] - \left(E[w_n - {}^3y_n]\right)^2$$

$$= (w_n - v_n)^2 \left\{ \Gamma\left(1 + \frac{2}{k}\right) - \Gamma^2\left(1 + \frac{1}{k}\right)\right\} \quad (8.77)$$

By using the mean and variance obtained above, let us evaluate the skewness defined in Eq. (3.8) of the extreme value ^3Y_n. It becomes

$$\text{Skewness } \gamma = \frac{E\left[\left({}^3y_n - E[^3y_n]\right)^3\right]}{\left(\text{Var}[^3y_n]\right)^{3/2}}$$

$$= \left[\Gamma\left(1 + \frac{3}{k}\right) - 3\Gamma\left(1 + \frac{2}{k}\right)\Gamma\left(1 + \frac{1}{k}\right) + 2\Gamma^3\left(1 + \frac{1}{k}\right)\right]$$

$$\times \{B(k)\}^3 \quad (8.78)$$

where

$$B(k) = 1 \Big/ \left\{ \Gamma\left(1 + \frac{2}{k}\right) - \Gamma^2\left(1 + \frac{1}{k}\right)\right\}^{1/2}$$

It can be seen in Eq. (8.78) that the skewness of the Type III asymptotic distribution is expressed solely as a function of the parameter k. Hence, the parameter k of the distribution can be evaluated from Eq. (8.78) by calculating the skewness of the observed data.

The parameters v_n and w_n can be determined subsequently from the following relationships:

$$v_n = E[^3y_n] - \left\{1 - \Gamma\left(1 + \frac{1}{k}\right)\right\} B(k) \left(\text{Var}[^3y_n]\right)^{1/2} \quad (8.79)$$

and

$$w_n = E[^3y_n] + \Gamma\left(1 + \frac{1}{k}\right) B(k) \left(\text{Var}[^3y_n]\right)^{1/2} \quad (8.80)$$

There are other methods to estimate the parameters of the Type III asymptotic distribution such as the maximum likelihood method. Readers

ASYMPTOTIC DISTRIBUTIONS OF EXTREME VALUES

who are interested in these estimation methods are referred to Gumbel (1958).

The lower limit value of the Type III asymptotic distribution $G({}^3y_n)$ is unlimited, $-\infty$. Therefore, for a random variable that has an upper bound and whose sample space is nonnegative, the distribution must be truncated at $y_n = 0$. In this case, the cumulative distribution function and the probability density function become as follows:

$$G_*({}^3y_n) = \frac{1}{1 - \exp\left\{-\left(\frac{w_n}{w_n - v_n}\right)^k\right\}}$$

$$\times \left[\exp\left\{-\left(\frac{w_n - {}^3y_n}{w_n - v_n}\right)^k\right\} - \exp\left\{-\left(\frac{w_n}{w_n - u_n}\right)^k\right\}\right]$$

$$g_*({}^3y_n) = \frac{1}{1 - \exp\left\{-\left(\frac{w_n}{w_n - v_n}\right)^k\right\}} \left(\frac{k}{w_n - v_n}\right)\left(\frac{w_n - {}^3y_n}{w_n - v_n}\right)^{k-1}$$

$$\times \exp\left\{-\left(\frac{w_n - {}^3y_n}{w_n - v_n}\right)^k\right\} \tag{8.81}$$

The derivation of the truncated cumulative distribution function is left as an exercise for the reader.

Next, the Type III asymptotic distribution for the smallest value, denoted by ${}^3Y_1$, is outlined. The cumulative distribution function can be derived from the Type I asymptotic distribution in the same manner as for the derivation of the largest value. However, the sample space of ${}^3Y_1$ is given by $w_1 < {}^3Y_1 < \infty$, where w_1 corresponds to $-\infty$ for the random variable ${}^1Y_1$. Hence, let us write the functional relationship between the two random variables ${}^1Y_1$ and ${}^3Y_1$ as

$${}^1Y_1 - u_1 = -\ln\left(\frac{v_1 - w_1}{{}^3Y_1 - w_1}\right) \tag{8.82}$$

Then, by writing $\alpha_1 = k$ in Eq. (8.46) we can derive the following cumulative distribution function for Y_1:

$$G({}^3y_1) = \exp\left\{-\exp\left[-k\ln\left(\frac{v_1 - w_1}{{}^3y_1 - w_1}\right)\right]\right\}$$

$$= \exp\left\{-\left(\frac{{}^3y_1 - w_1}{v_1 - w_1}\right)^k\right\}, \quad w_1 < {}^3y_1 < \infty \quad (8.83)$$

where $g(\infty) = 0$ and $G(w_1) = 1$.

The mean and variance of the smallest value can be obtained by the same procedure as that used for the derivation of these quantities for the largest value. Hence, only the results are given as follows:

$$E[{}^3y_1] = w_1 + (v_1 - w_1)\Gamma\left(1 + \frac{1}{k}\right) \quad (8.84)$$

and

$$\text{Var}[{}^3y_1] = (v_1 - w_1)^2\left\{\Gamma\left(1 + \frac{2}{k}\right) - \Gamma^2\left(1 + \frac{1}{k}\right)\right\} \quad (8.85)$$

8.3 ESTIMATION OF EXTREME VALUES FROM OBSERVED DATA

Methods for estimating extreme values from observed data based on the extreme value theory presented in the preceding sections are discussed in this section. It is noted that data employed as examples of the analysis methods pertain to certain specific areas of engineering; however, the estimation methods are equally applicable to any area of the general sciences.

Estimation methods of extreme values may be categorized into the following two classes depending on the data.

1. The initial probability distribution is known.

If the initial probability distribution of the data is theoretically known, then the extreme values can be estimated exactly through the extreme value probability function presented in Section 8.1. As an example, let us consider a set of data $(x_1, x_2, \ldots, x_n)$ obtained by reading the peaks and

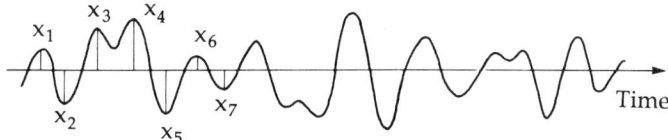

Figure 8.6 Peaks and troughs of a stochastic process with continuous state and time.

troughs of the random phenomenon shown in Figure 8.6. Suppose the random phenomenon is known to be a normal random process, then the probability distribution of its peaks and troughs, x, is theoretically known as will be discussed in Chapter 11. Therefore, the data should be represented by a known initial probability density function, and then the extreme values can be evaluated by applying order statistics.

2. The initial probability distribution is unknown.

This is the usual situation for most of the data from which we want to estimate the extreme values. In this case, there is no theoretical basis for selecting any particular probability distribution to characterize the observed data, except for the case where the data consist of the largest (or smallest) values of a random phenomenon as will be discussed later.

As an example, Table 8.1 shows data representing sea severity tabulated at 0.5 m intervals. The data consist of a total of 5412 observations accumulated over 3 years. The data are typical examples for which there is no theoretical basis for selecting any particular distribution. Hence, the data may be plotted on available probability paper, such as log-normal probability paper (see Figure 8.7) and Weibull probability paper (Figure 8.8).

As can be seen in these figures, the data appear to follow the log-normal distribution for a cumulative distribution up to 0.99, and the data also appear to be represented by the Weibull probability distribution except for small x values. Care must be taken in using Weibull probability paper, because the paper is usually constructed by taking the logarithm of the cumulative distribution function twice. Therefore, a small difference between the data and cumulative distribution function drawn on the Weibull probability paper may result in a substantial difference between the histogram and theoretical probability density function.

If we want to estimate the extreme value expected in 50 years based on the plots shown in Figures 8.7 and 8.8, the estimation can be made by Eq. (8.14) in which the number n in 50 years becomes 90,200. This implies that

Table 8.1

Data on Sea Severity for a 3-Year Period Obtained from Measurements in the North Sea (Bouws, 1978)

Sea Severity (Random Variable X)	Number of Observations
0-0.5 m	1,280
0.5-1.0	1,549
1.0-1.5	1,088
1.5-2.0	628
2.0-2.5	402
2.5-3.0	192
3.0-3.5	115
3.5-4.0	63
4.0-4.5	38
4.5-5.0	18
5.0-5.5	21
5.5-6.0	7
6.0-6.5	8
6.5-7.0	2
7.0-7.5	1
Total	5,412 in 3 Years

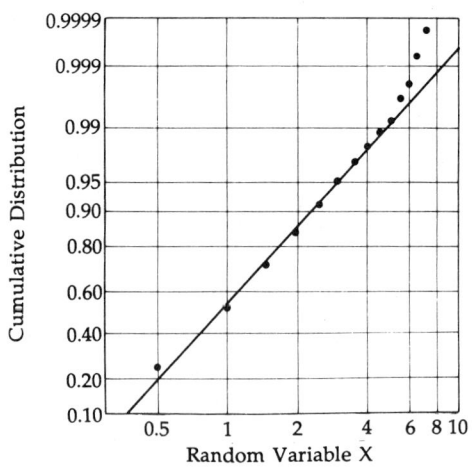

Figure 8.7 Cumulative distribution function of sea severity plotted on log-normal probability paper.

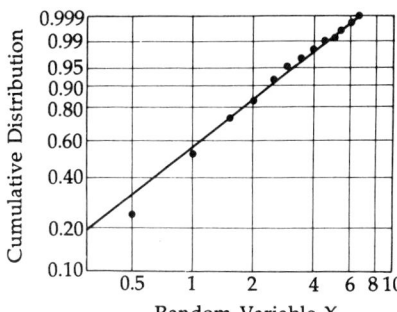

Figure 8.8 Cumulative distribution function of sea severity plotted on Weibull probability paper.

the extreme value is evaluated by extending the line to a value of the cumulative distribution function of $F(x) = 1 - (1/n) = 0.9999889$, which may leave some reservation as to the accuracy of the estimation.

Another way to plot the data is to take the logarithm of the return period as shown in Figure 8.9. By extending the line established by connecting the data points to the value equivalent to $\ln n$, the extreme value expected in 50 years can be estimated. For information, the extreme value expected in 10 years is also included in Figure 8.9.

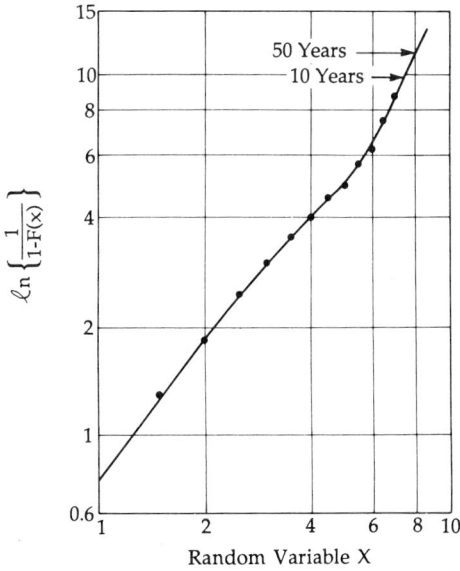

Figure 8.9 Estimation of extreme sea severity from the extension of data points.

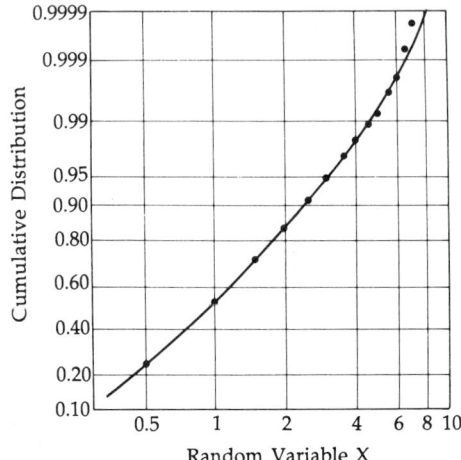

Figure 8.10 Cumulative distribution function based on Eq. (8.86) plotted on log-normal probability paper.

It should be noted that, in general, data toward the extreme value are always sparse. For example, less than 1% of the total number of observations are in the range of the larger x values as can be seen in Table 8.1. Hence, the question always remains as to how reliable the prediction technique is if we estimate the extreme values by extending a line plotted on probability paper taking into account the large (or small) x values, which are extremely unreliable data.

For the estimation of extreme values, it is highly desirable to represent the data precisely by a certain probability distribution over the entire range of the cumulative distribution. It may not be necessary to represent the data points at very high cumulative distribution values such as 0.999 or higher because of the reason discussed in the previous paragraph.

One way to represent data points precisely is to express the cumulative distribution function in the form given in Eq. (8.15), and to express $q(x)$ in the following form (Ochi and Whalen, 1980):

$$q(x) = ax^m \exp\{-px^k\} \tag{8.86}$$

The parameters involved in $q(x)$ are determined numerically by a nonlinear minimization procedure. The extreme value can then be estimated based on this probability function. The cumulative distribution function obtained by applying this method to the data given in Table 8.1 is shown in Figure 8.10. As can be seen in the figure, the data are well represented by the cumulative distribution function over the entire range except very high x values.

When the data consist of the largest (or smallest) values of a random phenomenon observed during a certain time period, such as those observed hourly, daily, or weekly, there is a theoretical basis to choose a particular probability distribution for estimating extreme values. That is, the probability distribution applicable for the random variable under this situation must be one of three asymptotic extreme value distributions discussed in Section 8.2. An example of the estimation of extreme values belonging to this category was given in Example 8.4.

EXERCISES

8.1 The cumulative distribution function of the Type I asymptotic extreme value distribution is given in Eq. (8.32) for a sample space $-\infty < {}^1Y_n < \infty$. Prove that the cumulative distribution function for the sample space $0 < {}^1Y_n < \infty$ is given by

$$G_*({}^1y_n) = \frac{1}{1 - \exp\{-e^{\alpha_n u_n}\}} \left[\exp\{-e^{-\alpha_n({}^1y_n - u_n)}\} - \exp\{-e^{\alpha_n u_n}\} \right]$$

8.2 Prove that the mean and variance of the asymptotic Type I distribution applicable for the smallest value, ${}^1Y_1$, are given by

$$E[{}^1Y_1] = u_1 - \frac{\gamma}{\alpha_1}$$

$$\mathrm{Var}[{}^1y_1] = \frac{1}{\alpha_1^2} \frac{\pi^2}{6}$$

8.3 The Type III asymptotic extreme value distribution is bounded by an upper limit value w_n, but the lower bound is unlimited, $-\infty$. Prove that if the sample space of the random variable is nonnegative, $0 \leqslant {}^3Y_n \leqslant w_n$, then the cumulative distribution function is given by

$$G_*({}^3y_n) = \frac{1}{1 - \exp\left\{-\left(\frac{w_n}{w_n - v_n}\right)^k\right\}}$$

$$\times \left[\exp\left\{-\left(\frac{w_n - {}^3y_n}{w_n - v_n}\right)^k\right\} - \exp\left\{-\left(\frac{w_n}{w_n - u_n}\right)^k\right\} \right]$$

8.4 The Type II asymptotic extreme value distribution given in Eq. (8.54) is applicable for the extreme value ^2Y_n whose sample space is $0 \leqslant {}^2Y_n < \infty$. Show that the cumulative distribution function applicable for $\varepsilon_n \leqslant {}^2Y_n < \infty$ is given by

$$G({}^2y_n) = \exp\left\{-\left(\frac{v_n - \varepsilon_n}{{}^2y_n - \varepsilon_n}\right)^k\right\}$$

where $v_n - \varepsilon_n > 0$ and $k > 0$.

Similarly, the cumulative distribution function applicable for $-\infty < Y_1 \leqslant \varepsilon_1$ is given by

$$G({}^2y_1) = \exp\left\{-\left(\frac{\varepsilon_1 - {}^2y_1}{\varepsilon_1 - v_1}\right)^k\right\}$$

where $\varepsilon_1 - v_1 > 0$ and $k > 0$.

CHAPTER 9

Stochastic Processes

9.1 INTRODUCTION

In the preceding chapters the probability theory relevant to problems in engineering and the physical sciences was introduced. It was shown that the characteristics of a random phenomenon can be described through the probability distribution of a random variable representing the phenomenon. In probability theory, however, statistical properties of random phenomena (events) that occur in random fashion with respect to time are not considered.

Consideration of the time element cannot be ignored in many random phenomena observed in engineering and the physical sciences. Just a few examples are (1) particle movement in Brownian motion, (2) emissions from a radioactive source, (3) fluctuating current in an electric circuit, (4) wave profile in the ocean, (5) response of an airplane to a wind gust and motion of a ship in a seaway, and (6) vibration of a building caused by an earthquake. In order to evaluate the statistical characteristics of these random phenomena, it is necessary to consider the concept of a family of random variables that is a function of time. Random phenomena whose characteristics can be determined by this concept are called stochastic or random processes. A stochastic process is defined as follows:

Definition 9.1. A family of random variables $x(t)$ where t is a parameter belonging to an index set T is called a *stochastic process* (or a *random process*), and is denoted by $\{x(t), t \in T\}$.

The parameter t is most commonly interpreted as time, and hereafter t is considered as time. A stochastic process $\{x(t), t \in T\}$ is, in the strict sense, a function of two arguments $\{x(t, \omega); t \in T, \omega \in \Omega\}$, where Ω is the sample space. For fixed t, $x(\omega)$ is a family of random variables, called an *ensemble*, for a fixed ω, $x(t)$ is a function of time that may be called a *sample function*.

In order to elaborate on the definition of a stochastic process given above, let us consider a set of n records indicating the time history of a random phenomenon shown in Figure 9.1. As an example of a random phenomenon, we consider the wave profile in the ocean that is recorded by n recorders dispersed in a certain area. Since the waves are coming from various directions in random fashion, every recorder would register a different time history of the wave profile. The magnitude of deviation from the mean-line, $x(t)$, is a random variable representing the wave profile. At a specific time t_1, we have a set (family) of random variables consisting of n elements, $\{^1x(t_1), {^2}x(t_2), \ldots, {^n}x(t_1)\}$. At another time t_2, we have another set of random variables $\{^1x(t_2), {^2}x(t_2), \ldots, {^n}x(t_2)\}$. The collection of n recorders simultaneously observed at a specified time is called an *ensemble*.

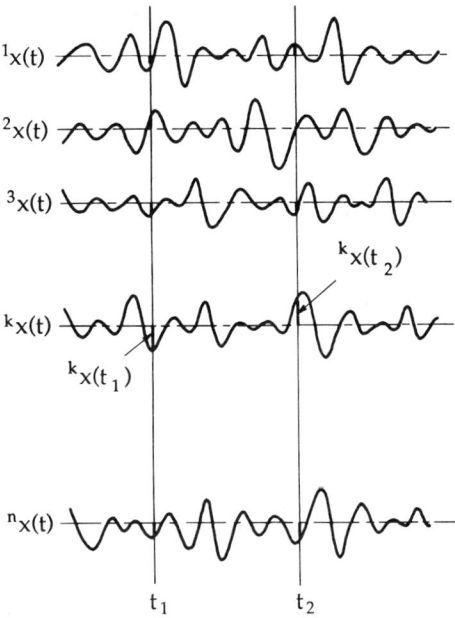

Figure 9.1 Definition of ensemble of a stochastic process.

CLASSIFICATION OF STOCHASTIC PROCESSES

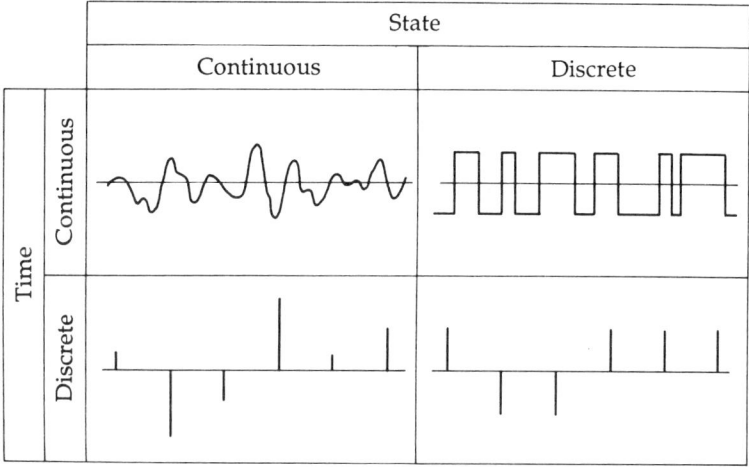

Figure 9.2 Pictorial sketch illustrating stochastic processes of continuous and discrete state and time.

In principle, statistical properties of a random phenomenon $x(t)$ must be obtained with respect to the ensemble, a set of many simultaneously observed data. However, the statistical properties may be obtained from analysis of a single record if we assume that the random phenomenon satisfies the ergodic property that will be discussed later.

The magnitude of the random process $x(t)$ shown in Figure 9.1 can be any quantity that may be called a *continuous state process*. The process $x(t)$ is also continuous with respect to time, and hence it may be called a *continuous time process*. In general, however, both time and state may be discrete. Therefore, there exist four different types of stochastic processes as illustrated in Figure 9.2 where only a sample element representing each stochastic process is shown.

9.2 CLASSIFICATION OF STOCHASTIC PROCESSES

9.2.1 *Stationary Process*

Statistical properties of a random process are evaluated based on an ensemble, and hence they may or may not be the same as time progresses. If the statistical properties of a process are invariant under translation of time, then a process is called a stationary process and is defined below:

Definition 9.2. The joint distribution of n-dimensional random vectors, $\{^1x(t), {}^2x(t), \ldots, {}^nx(t)\}$ and $\{^1x(t+\tau), {}^2x(t+\tau), \ldots, {}^nx(t+\tau)\}$ is the same for all τ, then the stochastic process $x(t)$ is said to be a *stationary* (or *steady-state*) *stochastic process*.

If a stochastic process does not satisfy the condition given in Definition 9.2, the process is called an *evolutionary stochastic process*. In order to clarify the difference between stationary and evolutionary stochastic processes, let us consider the number of emissions from a radioactive source or the number of floodings due to a storm. The emissions as well as the floodings occur randomly in time. If we consider the number of occurrences during a period from the beginning of the observation until time t, denoted by $N(t)$, then $N(t)$ is an evolutionary stochastic process since it depends on time t. On the other hand, if we consider the number of occurrences during a specified time interval τ, then the number $\{N(t+\tau) - N(t)\}$ may not depend on time t and thereby it may be a stationary random process. The latter is called an increment stochastic process and will be discussed in more detail in the next section.

The condition for a stationary stochastic process requires that the joint distribution be invariant irrespective of time, which is somewhat severe in practice. A stochastic process that satisfies this condition is often called a *strictly stationary stochastic process*. A more relaxed condition for the stationarity of a stochastic process that is commonly acknowledged in the analysis of random data is that called weakly stationary. In order to define a weakly stationary stochastic process, it is necessary to introduce the auto-covariance function of a stochastic process.

Analogous to the definition of the covariance of two random variables given in Eq. (3.14), the *autocovariance function* of a random process $x(t)$, denoted by $C_{xx}(t_1, t_2)$, is defined with respect to two random vectors $\mathbf{x}(t_1)$ and $\mathbf{x}(t_2)$ as follows:

$$C_{xx}(t_1, t_2) = \text{Cov}[\mathbf{x}(t_1), \mathbf{x}(t_2)]$$

$$= E[(\mathbf{x}(t_1) - E[\mathbf{x}(t_1)])(\mathbf{x}(t_2) - E[\mathbf{x}(t_2)])]$$

$$= \frac{1}{n}\sum_{k=1}^{n} \{^kx(t_1) - \bar{x}(t_1)\}\{^kx(t_2) - \bar{x}(t_2)\} \qquad (9.1)$$

where

$$\bar{x}(t_1) = E[\mathbf{x}(t_1)] = \frac{1}{n} \sum_{k=1}^{n} {}^k x(t_1)$$

$$\bar{x}(t_2) = E[\mathbf{x}(t_2)] = \frac{1}{n} \sum_{k=1}^{n} {}^k x(t_2)$$

and $E[\mathbf{x}(t)]$ is called the *mean value function*.

As can be seen in Eq. (9.1), the autocovariance function $C_{xx}(t_1, t_2)$ depends on time t_1 and t_2. However, if $C_{xx}(t_1, t_2)$ is a function of only the time difference $(t_2 - t_1)$, irrespective of when t_1 is chosen, then we may write the autocovariance function as

$$\text{Cov}[\mathbf{x}(t), \mathbf{x}(t + \tau)] = R(\tau) \qquad (9.2)$$

and this leads to the definition of a weakly stationary stochastic process.

Definition 9.3. A stochastic process said to be *weakly stationary* (or *covariance stationary*) if its mean value function $E[\mathbf{x}(t)]$ is constant independent of t and its autocovariance function $\text{Cov}[\mathbf{x}(t), \mathbf{x}(t + \tau)]$ depends on τ for all t.

9.2.2 Ergodic Process

Stochastic processes were defined in the previous section in terms of an ensemble of records. The ergodic theorem developed by statisticians deals with the conditions under which the time average of a single record (sample mean) is equivalent to the ensemble average. The theorem was first developed for a strictly stationary stochastic process. Parzen derived the necessary and sufficient conditions for a sample mean to be ergodic and showed that the strictly stationary condition is not necessarily required (Parzen 1958, 1962). The ergodic theorem is a subject of great interest in mathematical statistics. Details of the theorem, however, appear to be beyond the scope of our prime interest; therefore, we simply state the definition of the ergodic process as follows:

Definition 9.4. A stochastic process $x(t)$ is said to be an *ergodic process* if the time average of a single record is approximately equal to the ensemble average. That is,

$$E[\mathbf{x}(t)] = \begin{cases} \dfrac{1}{n} \sum_{n=1}^{n} x(t_i) & \text{for a discrete-time process} \\ \dfrac{1}{T} \int_{0}^{T} x(t)\, dt & \text{for a continuous-time process} \end{cases} \quad (9.3)$$

The ergodic property of a stochastic process is commonly assumed to be true in the analysis of data observed in engineering and the physical sciences, and thereby statistical properties of a random process may be evaluated from analysis of a single record. It is taken for granted that the ergodic property holds for stochastic processes discussed in this text; therefore, following discussions pertain to a single record $x(t)$ instead of an ensemble of data $\mathbf{x}(t)$.

Example 9.1. The stochastic process $x(t)$ is given by $x(t) = A_0 \cos(\omega_0 t + \varepsilon)$, whereas A_0 and ω_0 are constants and ε is random variable uniformly distributed between $-\pi$ and π. The mean value function and covariance function can be obtained as

$$E[x(t)] = E[A_0 \cos(\omega_0 t + \varepsilon)] = 0$$

$$\begin{aligned}
\text{Cov}[x(t), x(t+\tau)] &= E[x(t)x(t+\tau)] - E[x(t)]E[x(t+\tau)] \\
&= A_0^2 E[\cos(\omega_0 t + \varepsilon) \cdot \cos\{\omega_0(t+\tau) + \varepsilon\}] \\
&= \frac{A_0^2}{2} E[\cos(2\omega_0 t + \omega_0 \tau + 2\varepsilon) + \cos \omega_0 \tau] \\
&= \frac{A_0^2}{2} \cos \omega_0 \tau \qquad \blacksquare
\end{aligned}$$

Example 9.2. The stochastic process $x(t)$ is given by

$$x(t) = \sum_{i=1}^{n} (A_i \cos \omega_i t + B_i \sin \omega_i t)$$

where ω_i are known, and A_i, B_i are all statistically independent random variables with $E[A_i] = E[B_i] = 0$ and $\text{Var}[A_i] = \text{Var}[B_i] = \sigma_i^2$ (known). It can be proved that this stochastic process is weakly stationary as follows:

$$E[x(t)] = \sum_i \{E[A_i]\cos \omega_i t + E[B_i]\sin \omega_i t\} = 0$$

and

$$\text{Cov}[x(t_1), x(t_2)] = E\left[\sum_i \sum_j (A_i \cos \omega_i t_1 + B_i \sin \omega_i t_1)\right.$$
$$\left. \times (A_j \cos \omega_j t_2 + B_j \sin \omega_j t_2)\right]$$
$$= \sum_i \sum_j E[A_i A_j] \cos \omega_i t_1 \cos \omega_j t_2$$
$$+ \sum_i \sum_j E[B_i B_j] \sin \omega_i t_2 \sin \omega_j t_1$$
$$= \sum_i \left(E[A_i^2]\cos \omega_i t_1 \cos \omega_i t_2 + E[B_i^2]\sin \omega_i t_2 \sin \omega_i t_1\right)$$
$$= \sum_i \sigma_i^2 \cos \omega_i (t_2 - t_1)$$

Since the mean value function is constant and the autocovariance function is a function of $(t_2 - t_1)$, the stochastic process is weakly stationary. ∎

9.2.3 Independent Increment Process

Definition 9.5. A stochastic process $x(t)$ is said to be an *independent increment process* if $x(t_{i+1}) - x(t_i)$, where $i = 0, 1, 2, \ldots$, is statistically independent (and thereby statistically uncorrelated).

Some stochastic processes are stationary as well as independent increments that possess both properties given in Definitions 9.2 and 9.5. Therefore, if a stochastic process is a stationary independent increment, then the joint probability density function of $\{x(t_2 + \tau) - x(t_1 + \tau)\}$ and $\{x(t_2) - x(t_1)\}$ are identical irrespective of t_1, t_2, and $\tau > 0$.

Example 9.3. Let a stochastic process $x(t)$ be a stationary independent increment process. Then, the variance of the independent increment $x(t_2) - x(t_1)$, where $t_1 < t_2$, is proportional to $t_2 - t_1$.

Because of the independent increment property, we can write

$$\text{Var}[x(t_2)] = \text{Var}[\overline{x(t_2) - x(t_1)} + \overline{x(t_1) - x(0)}]$$
$$= \text{Var}[x(t_2) - x(t_1)] + \text{Var}[x(t_1) - x(0)]$$

Since the increment is a stationary process,

$$\text{Var}[x(t_2) - x(t_1)] = \text{Var}[x(t_2 - t_1)]$$
$$\text{Var}[x(t_1) - x(0)] = \text{Var}[x(t_1)]$$

By inserting these relationships to the formula of $\text{Var}[x(t_2)]$, we can derive

$$\text{Var}[x(t_2 - t_1)] = \text{Var}[x(t_2)] - \text{Var}[x(t_1)]$$

If we write $\text{Var}[x(t)]$ as $v(t)$, where $v(t)$ is a real-valued, nonnegative function, then we have the following relationship:

$$v(t_2 - t_1) = v(t_2) - v(t_1)$$

The solution that satisfies the above equation is given by $v(t) = \sigma^2 t$, where σ^2 is a positive constant. Thus, we can prove

$$\text{Var}[x(t_2) - x(t_1)] = \sigma^2(t_2 - t_1), \quad t_1 < t_2 \qquad \blacksquare$$

Example 9.4. Let a stochastic process $x(t)$ be an independent increment process. Then, the autocovariance function, $\text{Cov}[x(t_1), x(t_2)]$ where $t_1 < t_2$, is equal to $\text{Var}[x(t_1)]$.

We may write the autocovariance function as

$$\text{Cov}[x(t_1), x(t_2)] = \text{Cov}[x(t_1), \overline{x(t_2) - x(t_1)} + x(t_1)]$$
$$= \text{Cov}[x(t_1), x(t_2 - t_1)] + \text{Cov}[x(t_1), x(t_1)]$$

The first term becomes zero because of the independent increment property, and the second term is the variance. Thus, the autocovariance, $\text{Cov}[x(t_1), x(t_2)]$ is equal to $\text{Var}[x(t_1)]$. $\blacksquare$

Example 9.5. Let $N(t)$ be a stochastic process with independent increments. Consider the stochastic process $x(t)$ defined as $x(t) =$

CLASSIFICATION OF STOCHASTIC PROCESSES

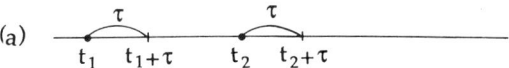

Figure 9.3 Illustration of $(t_1 + \tau)$ being less (a) or greater than (b) t_2.

$N(t + \tau) - N(t)$ where $\tau > 0$, and evaluate the autocovariance function, $\text{Cov}[x(t_1), x(t_2)]$ where $t_1 < t_2$.

Two cases must be considered for this problem depending on the magnitude of τ as shown in Figure 9.3.

(a) If $t_1 + \tau < t_2$

$$\text{Cov}[x(t_1), x(t_2)] = \text{Cov}[N(t_1 + \tau) - N(t_1), N(t_2 + \tau) - N(t_2)]$$

Since $N(t_1 + \tau) - N(t_1)$ and $N(t_2 + \tau) - N(t_2)$ are independent, we have $\text{Cov}[x(t_1), x(t_2)] = 0$.

(b) If $t_1 + \tau > t_2$

$$\text{Cov}[x(t_1), x(t_2)] = \text{Cov}[N(t_1 + \tau) - N(t_1), N(t_2 + \tau) - N(t_2)]$$

$$= \text{Cov}\big[\overline{N(t_1 + \tau) - N(t_2)} + \overline{N(t_2) - N(t_1)},$$

$$N(t_2 + \tau) - N(t_2)\big]$$

$$= \text{Cov}[N(t_1 + \tau) - N(t_2), N(t_2 + \tau) - N(t_2)]$$

$$+ \text{Cov}[N(t_2) - N(t_1), N(t_2 + \tau) - N(t_2)]$$

Here, the second term becomes zero because of the statistical independence. The first term can be further written as

$$\text{Cov}\big[N(t_1 + \tau) - N(t_2), \overline{N(t_2 + \tau) - N(t_1 + \tau)} + \overline{N(t_1 + \tau) - N(t_2)}\big]$$

$$= \text{Cov}[N(t_1 + \tau) - N(t_2), N(t_2 + \tau) - N(t_1 + \tau)]$$

$$+ \text{Cov}[N(t_1 + \tau) - N(t_2), N(t_1 + \tau) - N(t_2)]$$

The first term is zero and the second term is equal to $\text{Var}[N(t_1 + \tau) - N(t_2)]$. Thus, we have

$$\text{Cov}[x(t_1), x(t_2)] = \begin{cases} \text{Var}[N(t_1 + \tau) - N(t_2)] & \text{for } 0 < t_2 - t_1 < \tau \\ 0 & \text{otherwise} \end{cases}$$

■

9.2.4 Markov Process

The Markov process plays an extremely important role in applied stochastic process theory and a detailed discussion will be given in Chapter 13. Only the definition and some preliminary information on Markov properties are given here.

Definition 9.6. A stochastic process $x(t)$ is said to be a *Markov process* if it satisfies the following conditional probability:

$$Pr\{x(t_n) \leq x_n | x(t_1) = x_1, x(t_2) = x_2, \ldots, x(t_{n-1}) = x_{n-1}\}$$
$$= Pr\{x(t_n) = x_n | x(t_{n-1}) = x_{n-1}\}, \quad \text{where } t_1 < t_2 < \cdots t_{n-1} < t_n$$

(9.4)

Equation (9.4) implies that the conditional probability of a random process $x(t) = x_n$ at time t_n, given that its values at some earlier times are known, depends only on the immediate past value at time t_{n-1} and is independent of the history of the process prior to t_{n-1}.

The state and time of a Markov process can be discrete as well as continuous. In particular, a Markov process with a discrete state is called the *Markov chain*, and a Markov process with a continuous state is called a *diffusion process*.

We may write the relationship given in Eq. (9.4) in terms of the probability density function. That is,

$$f\{x(t_n)|x(t_1), x(t_2), \ldots, x(t_{n-1})\} = f\{x(t_n)|x(t_{n-1})\} \quad (9.5)$$

By applying Eq. (9.5), the joint probability density function can be written as follows:

$$f\{x(t_1), x(t_2), \ldots, x(t_n)\}$$
$$= f\{x(t_n)|x(t_1), x(t_2), \ldots, x(t_{n-1})\}$$
$$\times f\{x(t_1), x(t_2), \ldots, x(t_{n-1})\}$$
$$= f\{x(t_n)|x(t_{n-1})\} \cdot f\{x(t_1), x(t_2), \ldots, x(t_{n-1})\} \quad (9.6)$$

CLASSIFICATION OF STOCHASTIC PROCESSES 211

Similarly, we may write

$$f\{x(t_1), x(t_2), \ldots, x(t_{n-1})\}$$
$$= f\{x(t_{n-1})|x(t_{n-2})\} \cdot f\{x(t_1), x(t_2), \ldots, x(t_{n-2})\} \quad (9.7)$$

By repeating the same procedure, Eq. (9.6) yields

$$f\{x(t_1), x(t_2), \ldots, x(t_n)\} = f\{x(t_1)\} \prod_{r=2}^{n} f\{x(t_r)|x(t_{r-1})\} \quad (9.8)$$

Thus, we have shown that the joint probability density function of a Markov process can be expressed by the marginal probability density $f\{x(t_1)\}$ and a set of conditional probability density functions $f\{x(t_r)|x(t_{r-1})\}$, which is called the *transition probability density*.

A Markov process is said to be *homogeneous in time* if the transition probability density is invariant with time τ. That is,

$$f\{x(t_r + \tau)|x(t_{r-1} + \tau)\} = f\{x(t_r)|x(t_{r-1})\} \quad (9.9)$$

9.2.5 Counting Process

An important phase of applied stochastic process theory in engineering and the physical sciences is the prediction of statistical properties associated with the occurrence of a random phenomenon (event). Prediction as to how frequently the random phenomenon takes place and what is the time interval between successive occurrences of the event, and so on, provides information vital for understanding the stochastic behavior of the event. Stochastic processes that deal with the frequency of occurrence of random events are called counting processes as defined below:

Definition 9.7. An integer-valued continuous-time stochastic process $N(t)$ is called a *counting process* of the series of events if $N(t)$ represents the total number of occurrences of the event in the time interval $t = 0$ to t.

Figure 9.4 shows a pictorial sketch of a counting process. The time intervals between successive occurrences of the events $T_1 = t_1$, $T_2 = t_2 - t_1$, $T_3 = t_3 - t_2, \ldots$ are defined as the *interarrival times*. If the interarrival times are independent, identically distributed random variables, then the process is called a *renewal process* (or *renewal counting process*). In particular, if the interarrival times obey an exponential distribution, the stochastic process is called a *Poisson process*. The counting process will be discussed in detail in Chapter 17.

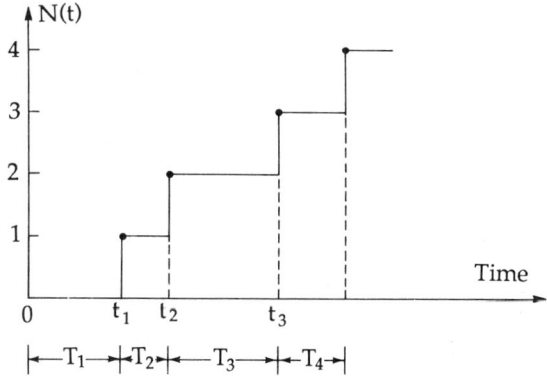

Figure 9.4 Pictorial sketch indicating counting process $N(t)$.

9.2.6 Narrow-Band Process

In the analysis of continuous-state and continuous-time stochastic processes, it is often assumed that the stochastic process is narrow-banded, which implies that the amplitude and phase of the process vary slowly and randomly with time while the frequency retains a constant value as shown in Figure 9.5. The narrow-band stochastic process is defined as follows:

Definition 9.8. A continuous-state and continuous-time stationary stochastic process $x(t)$ is called a *narrow-band process* if $x(t)$ can be expressed by

$$x(t) = A(t)\cos\{\omega_0 t + \varepsilon(t)\} \qquad (9.10)$$

where ω_0 = constant. The amplitude $A(t)$ and the phase $\varepsilon(t)$ are random

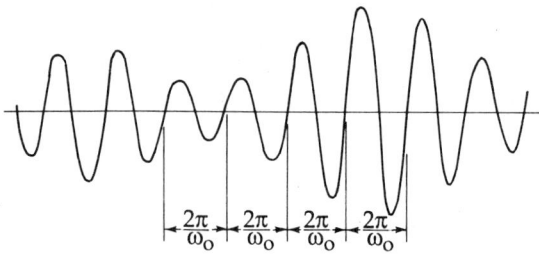

Figure 9.5 Example of narrow-band process.

STOCHASTIC PROCESSES FOR ANALYSIS OF PHYSICAL PHENOMENA 213

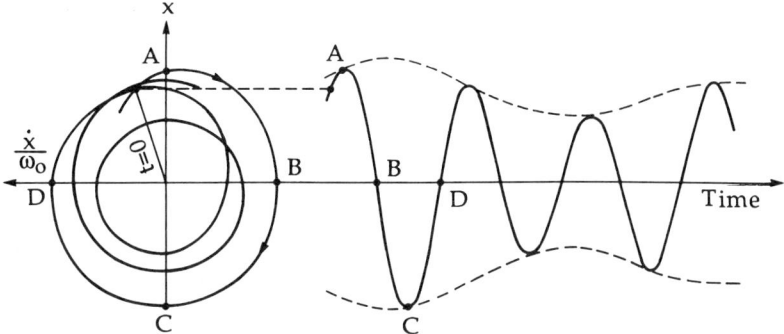

Figure 9.6 Phase-plane diagram of displacement, $x(t)$, and velocity/frequency, $\dot{x}(t)/\omega_0$, for a narrow-band process.

variables whose sample spaces are $0 \leqslant A(t) < \infty$ and $0 \leqslant \varepsilon(t) \leqslant 2\pi$, respectively.

The terminology "narrow-band" originated from the fact that the spectral density function of the process is sharply concentrated in the neighborhood of the frequency ω_0. For a better understanding of the narrow-band stochastic process, let us draw a phase-plane diagram of displacement, $x(t)$, versus velocity/frequency, $\dot{x}(t)/\omega_0$. As can be seen in Figure 9.6, the phase-plane curve of the process has a spiral shape moving slowly inward and outward in a random fashion. If envelopes are drawn by connecting the peaks (and troughs), they form a symmetric pair of smooth curves, and there exists a single peak (or trough) for every half cycle. Prediction of

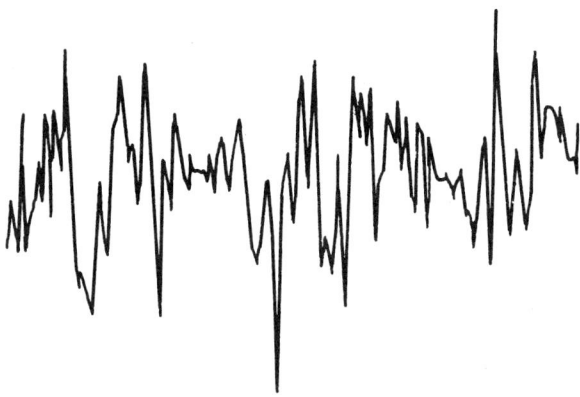

Figure 9.7 Example of wide-band random process.

statistical properties of narrow-band stochastic processes is presented in Chapter 11.

A stochastic process that does not satisfy the condition given in Eq. (9.10) is called a *non-narrow-band* or *wide-band process*. The process consists of several different frequencies as shown in the example in Figure 9.7.

9.3 SOME STOCHASTIC PROCESSES FOR ANALYSIS OF PHYSICAL PHENOMENA

In this section definitions of several stochastic processes often used as the basis of analysis of random phenomena observed in the fields of engineering and general physics are outlined for convenience. For a detailed discussion of each process the reader is referred to the appropriate chapters.

9.3.1 *Normal (Gaussian) Process*

Definition 9.9. A stochastic process $x(t)$ is said to be a *normal* (or *Gaussian*) *process* if for any given time t the random variable $x(t)$ is normally distributed.

It is noted that the definition given above is based on the assumption that the stochastic process satisfies the ergodic property. In general, the definition should read "A stochastic process $x(t)$ is said to be a normal (or Gaussian) process if for any $t_1, t_2, \ldots, t_n$, the random variables $x(t_1), x(t_2), \ldots, x(t_n)$ are jointly normally distributed." The joint normal distribution is defined in Eq. (6.15).

For a weakly stationary normal process the mean function is constant and the covariance function, $\text{Cov}[x(t_i), x(t_j)]$, depends only on time difference $(t_j - t_i)$.

The normal process plays a significant role in stochastic analysis of random phenomena observed in natural sciences, since many random phenomena can be approximately represented by a normal process. As an example, we may consider wind-generated waves observed in the ocean. The wave profile recorded during a storm in the ocean shows that it varies randomly with time as demonstrated in Figure 9.8; however, the deviation from its mean line can be approximately represented by the normal probability distribution. It is interpreted that the wind-generated waves consist of many sinusoidal waves of different frequencies and phases superimposed in random fashion. The profile at time t, $x(t)$, may be considered as the sum of independent identically distributed random variables; hence, it follows by the central limit theorem given in Theorem 6.1 that $x(t)$ is approximately normally distributed. Prediction of various

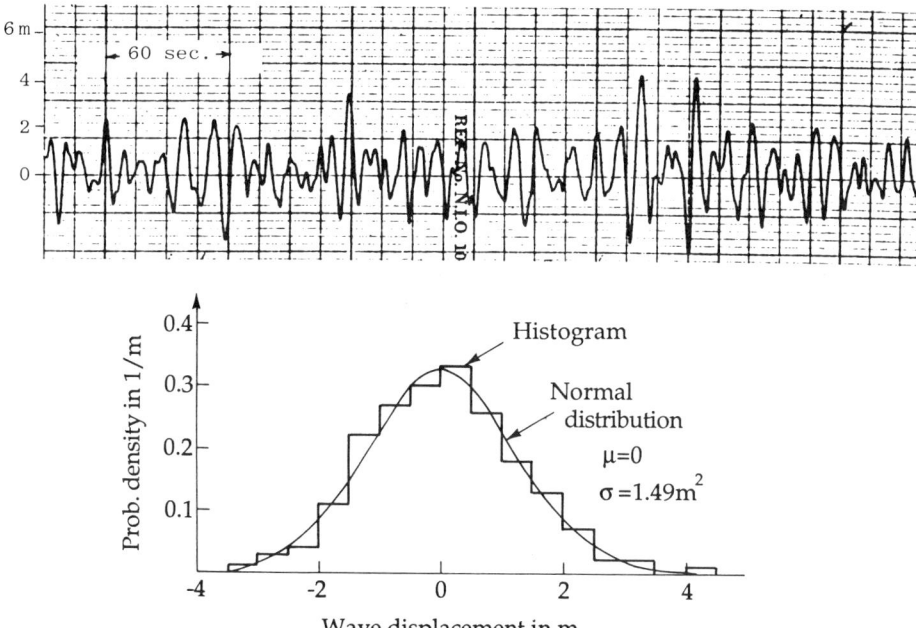

Figure 9.8 Wave profile in wind-generating seas and normal distribution.

statistical properties associated with the normal process will be discussed in Chapter 11.

9.3.2 *Wiener–Lévy Process*

This stochastic process is known as the *Brownian motion process* that describes the random movement exhibited by an extremely small particle immersed in a liquid or gas. Application of the Wiener–Lévy process, however, is not necessarily limited to the Brownian motion, but is also applied widely to other areas such as quantum mechanics and thermal noise in electric circuits. The process is defined as follows:

Definition 9.10. A stochastic process $x(t)$ is said to be a *Wiener–Lévy process* if

(i) $x(t)$ has stationary independent increment.

(ii) Every independent increment is normally distributed.

(iii) $E[x(t)] = 0$ for all time.

(iv) $x(0) = 0$.

From the conditions given in (i) and (iv), it can be proved that the variance of the Wiener–Lévy process increases linearly with time. If the mean value function, $E[x(t)]$, is not equal to zero but is μ (nonzero), then $x(t)$ is called a *Wiener–Lévy process with drift* μ. The Wiener–Lévy process will be discussed in detail in Chapter 13.

9.3.3 Poisson Process

Definition 9.11. A counting process $N(t)$ is said to be a *Poisson process* with mean rate (or intensity) ν if

(i) $N(t)$ has stationary independent increment.

(ii) $N(0) = 0$.

(iii) The number in any time interval of length τ is Poisson distributed with mean $\nu\tau$. That is,

$$\Pr\{N(t+\tau) - N(t) = k\} = e^{-\nu\tau}\frac{(\nu\tau)^k}{k!}, \qquad k = 0,1,2,\ldots \quad (9.11)$$

The process $N(t+\tau) - N(t)$, in item (iii) is called a *Poisson increment process*. Let us write $x(t) = N(t+\tau) - N(t)$. It was shown in Example 9.4 that for a stochastic process with independent increments, the autocovariance function becomes

$$\text{Cov}[x(t_1), x(t_2)] = \begin{cases} \text{Var}[N(t_1+\tau) - N(t_2)] & \text{for } 0 < t_2 - t_1 < \tau \\ 0 & \text{otherwise} \end{cases}$$
(9.12)

If $x(t)$ is Poisson distributed, then from knowledge of the variance of the Poisson distribution, we have

$$\text{Cov}[x(t_1), x(t_2)]$$
$$= \begin{cases} \nu(t_1 + \tau - t_2) = \nu\{\tau - (t_2 - t_1)\} & \text{for } 0 < t_2 - t_1 < \tau \\ 0 & \text{otherwise} \end{cases} \quad (9.13)$$

Thus, it can be proved that the Poisson increment process is covariance stationary.

The Poisson process is one of the most frequently employed counting processes. The process defined above is one of the most fundamental forms having a constant intensity v. There are several generalized forms of the Poisson process that are also extremely important in applied stochastic analysis techniques. Details of these various processes are discussed in Chapter 17.

9.3.4 Bernoulli Process

Definition 9.12. Consider a series of independent repeated trials with two outcomes: success and failure, rain and no rain, and so on. A counting process X_n is called a *Bernoulli process* if X_n represents the number of successes in n trials.

By letting p be the probability of success in each trial, the probability of k successes in n trials is given by the following binomial distribution:

$$Pr\{X_n = k\} = \binom{n}{k} p^k q^{n-k}, \quad \text{where } q = 1 - p \quad (9.14)$$

9.3.5 Shot Noise Process

Definition 9.13. A stochastic process $x(t)$ is said to be a *shot noise process* if it is induced by a sequence of impulses applied to a system at random times τ_k. In general, process $x(t)$ may be expressed in the form

$$x(t) = \sum_{k=1}^{N(t)} A_k w(t, \tau_k) \quad (9.15)$$

where $w(t, \tau_k)$ is the response of a system at time t resulting from an impulse A_k at time τ_k. A_k is a set of independent identically distributed random variables. $N(t)$ is a counting process with interarrival time τ_k, often assumed to be Poisson process.

It is assumed in Eq. (9.15) that the system's response is linear so that the effects of the individual impulses can be superimposed. One of the most commonly known examples of the shot noise process is the random fluctuation of an anode current in a vacuum-tube diode resulting from a series of impulses at every emission of an electron from the heated cathode. The shot noise process will be discussed in Section 17.3 in connection with the Poisson process.

CHAPTER 10

Spectral Analysis of Stochastic Processes

Prior to discussing the details of spectral analysis, it may be well to outline briefly the essence of the analysis in order to facilitate understanding of the principle.

We consider a continuous state stochastic process, $x(t)$, which is weakly stationary with mean value zero. From the time history of a single record, we define the autocorrelation function and the spectral density function, the latter being a function of frequency. It can be shown that these two functions are a Fourier transform pair that is called the Wiener–Khintchine theorem, and that the area under the spectral density function is equal to the variance of the process $x(t)$. This leads to the probability function through which various statistical properties of the process $x(t)$ can be evaluated as will be discussed in Chapter 11. The concept of spectral analysis also enables us to predict the statistical characteristics of the response of a system induced by a random excitation as will be discussed in Chapter 14. Thus, spectral analysis of stochastic processes carried out in the frequency domain is a prerequisite for predicting properties of stochastic processes.

In this chapter, the concept of transformation from a stochastic process in the time domain to the spectrum in the frequency domain is explained for both single and dual stochastic processes. Then, spectral analysis of integrated and differentiated processes and squared random processes that have extremely important application is discussed. Finally, extension of spectral analysis into a two-dimensional frequency space is outlined.

10.1 SPECTRAL ANALYSIS FOR A SINGLE RANDOM PROCESS

10.1.1 Autocorrelation Function

The autocorrelation function defined below plays a significant role in spectral analysis; therefore, we will give the definition first in a general form referring to the ensemble of a stochastic process.

Definition 10.1. The ensemble average of the product of a stochastic process at times t_1 and t_2 is defined as the *autocorrelation function* denoted by $R_{xx}(t_1, t_2)$. That is,

$$R_{xx}(t_1, t_2) = E[\mathbf{x}(t_1)\mathbf{x}(t_2)]$$
$$= \frac{1}{n}\sum_{k=1}^{n} {}^k x(t_1) {}^k x(t_2) \qquad (10.1)$$

We may recall that the ensemble average by subtracting the mean value from ${}^k x(t_1)$ and ${}^k x(t_2)$ is defined as the autocovariance function $C_{xx}(t_1, t_2)$ in Chapter 9. That is,

$$C_{xx}(t_1, t_2) = E[(\mathbf{x}(t_1) - E[\mathbf{x}(t_1)])(\mathbf{x}(t_2) - E[\mathbf{x}(t_2)])]$$
$$= \frac{1}{n}\sum_{k=1}^{n} \{{}^k x(t_1) - \bar{x}(t_1)\}\{{}^k x(t_2) - \bar{x}(t_2)\} \qquad (10.2)$$

From Eqs. (10.1) and (10.2), it is understood that the autocorrelation function is equal to the autocovariance function if the mean value of a stochastic process is zero.

The autocorrelation function is denoted by $R_{xx}(t_1, t_2)$ in order to distinguish it from the cross-correlation function between two random processes $x(t)$ and $y(t)$ that is denoted by $R_{xy}(t_1, t_2)$.

For the weakly stationary stochastic process defined in the previous chapter, the autocorrelation function $R_{xx}(t_1, t_2)$ becomes a function of the time difference only. That is, by writing $t_2 = t_1 + \tau$, we have

$$R_{xx}(\tau) = E[{}^k x(t) {}^k x(t+\tau)]$$
$$= \frac{1}{n}\sum_{k=1}^{n} {}^k x(t) {}^k x(t+\tau) \qquad (10.3)$$

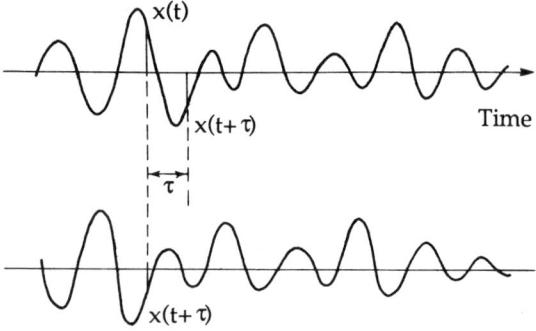

Figure 10.1 Definition of autocorrelation $E[x(t)x(t + \tau)]$.

Furthermore, by assuming a stochastic process $x(t)$ is ergodic, the ensemble average can be evaluated by the time average of a single record by shifting the record by time τ (see Figure 10.1). That is,

$$R_{xx}(\tau) = E[x(t)x(t + \tau)]$$

$$= \lim_{T \to \infty} \frac{1}{2T} \int_{-T}^{T} x(t)x(t + \tau) \, dt \qquad (10.4)$$

Accordingly, the autocovariance function for an ergodic random process may be written as

$$C_{xx}(t, t + \tau) = E[\{x(t) - E[x(t)]\}\{x(t + \tau) - E[x(t + \tau)]\}]$$

$$= \lim_{T \to \infty} \left[\frac{1}{2T} \int_{-T}^{T} x(t)x(t + \tau) \, dt - \left(\frac{1}{2T} \int_{-T}^{T} x(t) \, dt \right)^2 \right]$$

$$(10.5)$$

Several properties of the autocorrelation function $R_{xx}(\tau)$ are summarized below.

1. $R_{xx}(\tau)$ is an even function. That is,

$$R_{xx}(-\tau) = R_{xx}(\tau) \qquad (10.6)$$

SPECTRAL ANALYSIS FOR A SINGLE RANDOM PROCESS

Proof

$$R_{xx}(-\tau) = \lim_{T \to \infty} \frac{1}{2T} \int_{-T}^{T} x(t) x(t-\tau) \, dt$$

$$= \lim_{T \to \infty} \frac{1}{2T} \int_{-T}^{T} x(u+\tau) x(u) \, du = R(\tau)$$

2. $R_{xx}(\tau)$ is maximum at $\tau = 0$. That is,

$$R(0) \geq R(\tau) \tag{10.7}$$

Proof

$$\lim_{T \to \infty} \frac{1}{2T} \int_{-T}^{T} \{x(t) - x(t+\tau)\}^2 \, dt$$

$$= \lim_{T \to \infty} \frac{1}{2T} \int_{-T}^{T} \{x(t)\}^2 \, dt$$

$$- 2 \lim_{T \to \infty} \frac{1}{2T} \int_{-T}^{T} x(t) x(t+\tau) \, dt$$

$$+ \lim_{T \to \infty} \frac{1}{2T} \int_{-T}^{T} \{x(t+\tau)\}^2 \, dt$$

$$= R(0) - 2R(\tau) + R(0) \geq 0$$

Thus, $R(0) \geq R(\tau)$.

3. The magnitude of the autocorrelation function for $\tau = 0$, namely $R_{xx}(0)$, is equal to the variance of the random process with zero mean, and it represents the time average of the power or energy of the random process. By letting $\tau = 0$ in Eq. (10.4), we have

$$R_{xx}(0) = E[x^2(t)] = \lim_{T \to \infty} \frac{1}{2T} \int_{-T}^{T} \{x(t)\}^2 \, dt \tag{10.8}$$

An example of the autocorrelation function computed from the time history of a typical Gaussian random process is shown in Figure 10.2.

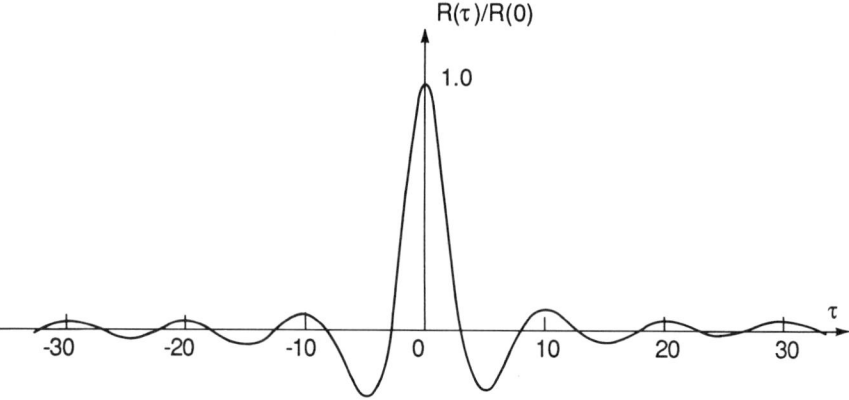

Figure 10.2 Example of autocorrelation function $R(\tau)$.

Example 10.1. An ergodic steady-state random process $x(t)$ is given by $x(t) = a\cos(\omega t + \varepsilon)$, and a and ω are constants and ε is a random variable. The autocorrelation function of $x(t)$ becomes

$$R_{xx}(\tau) = \lim_{T \to \infty} \frac{1}{2T} \int_{-T}^{T} a^2 \cos(\omega t + \varepsilon)\{\cos \omega(t + \tau) + \varepsilon\}\, dt$$

$$= \lim_{T \to \infty} \frac{1}{2T} \int_{-T}^{T} \frac{a^2}{2} \{\cos \omega\tau + \cos(2\omega t + \omega\tau + 2\varepsilon)\}\, dt$$

$$= \frac{a^2}{2} \cos \omega\tau \qquad \blacksquare$$

The result derived above can be easily extended to a random process composed of an accumulation of sinusoids with various amplitudes, frequencies, and phases.

Example 10.2. The random process $x(t)$ consists of many sinusoids expressed by

$$x(t) = \sum_{i=1}^{n} a_i \cos(\omega_i t + \varepsilon_i)$$

SPECTRAL ANALYSIS FOR A SINGLE RANDOM PROCESS   223

The autocorrelation is given by

$$R_{xx}(\tau) = \sum_{i=1}^{n} \frac{a_i^2}{2} \cos \omega_i \tau$$ ∎

Example 10.3. Let us consider the general expression of a periodic function $x(t)$ in the form of a complex Fourier series (see Appendix A: Fourier Transform). That is,

$$x(t) = \sum_{n=-\infty}^{\infty} c_n e^{in\omega_0 t}, \qquad n = 0, \pm 1, \pm 2, \ldots$$

where

$$c_n = \lim_{T \to \infty} \frac{1}{2T} \int_{-T}^{T} x(t) e^{-in\omega_0 t} \, dt$$

Then, we have

$$\begin{aligned} R_{xx}(\tau) &= \lim_{T \to \infty} \frac{1}{2T} \int_{-T}^{T} x(t) x(t+\tau) \, dt \\ &= \lim_{T \to \infty} \frac{1}{2T} \int_{-T}^{T} x(t) \sum_{n=-\infty}^{\infty} c_n e^{in\omega_0 (t+\tau)} \, dt \\ &= \sum_{n=-\infty}^{\infty} c_n e^{in\omega_0 \tau} \left[\lim_{T \to \infty} \frac{1}{2T} \int_{-T}^{T} x(t) e^{in\omega_0 t} \, dt \right] \\ &= \sum_{n=-\infty}^{\infty} c_n c_n^* e^{in\omega_0 \tau} = \sum_{n=-\infty}^{\infty} |c_n|^2 e^{in\omega_0 \tau} \end{aligned}$$ ∎

Example 10.4. Let a random process $x(t)$ be the sum of two statistically independent processes $x_1(t)$ and $x_2(t)$. The autocorrelation function of $x(t)$ becomes

$$\begin{aligned} R_{xx}(\tau) &= E[\{x_1(t) + x_2(t)\}\{x_1(t+\tau) + x_2(t+\tau)\}] \\ &= E[x_1(t)x_1(t+\tau) + x_1(t)x_2(t+\tau) \\ &\quad + x_2(t)x_1(t+\tau) + x_2(t)x_2(t+\tau)] \end{aligned}$$

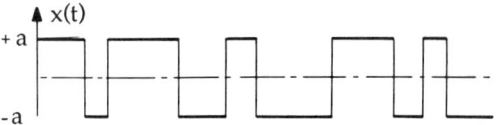

Figure 10.3   Random telegraph signal.

Since $x_1(t)$ and $x_2(t)$ are statistically independent,

$$R_{xx}(\tau) = E[x_1(t)x_1(t+\tau)] + E[x_1(t)]E[x_2(t+\tau)]$$
$$+ E[x_2(t)]E[x_1(t+\tau)] + E[x_2(t)x_2(t+\tau)]$$

Thus, if the mean value of either $x_1(t)$ or $x_2(t)$ is zero, $R_{xx}(\tau)$ becomes

$$R_{xx}(\tau) = R_{x_1 x_1}(\tau) + R_{x_2 x_2}(\tau) \qquad \blacksquare$$

Example 10.5.   Random telegraph signal is a random process $x(t)$ that takes the value either $+a$ or $-a$ and its occurrence follows the Poisson distribution with the average number per unit time μ (see Figure 10.3). The autocorrelation function can be evaluated following the relation $R_{xx}(\tau) = E[x(t)x(t+\tau)]$. However, it should be noted that the expected value of $x(t)x(t+\tau)$ depends on the number of sign changes in the time interval $(t, t+\tau)$. The product $x(t)x(t+\tau)$ is equal to $+a^2$ for an even number of changes in τ, while it is $-a^2$ for an odd number of changes. Therefore, $E[x(t)x(t+\tau)]$ becomes

$$E[x(t)x(t+\tau)] = a^2 \, Pr\{\text{even number of changes}\}$$
$$+ (-a^2) \, Pr\{\text{odd number of changes}\}$$
$$= a^2 [Pr\{0\} - Pr\{1\} + Pr\{2\} - Pr\{3\} + \cdots]$$
$$= a^2 e^{-2\mu\tau} \qquad (\text{see Exercise 5.7})$$

Thus, we have

$$R_{xx}(\tau) = a^2 e^{-2\mu\tau} \qquad \blacksquare$$

10.1.2   *Spectral Density Function*

It was shown in Eq. (10.8) that the magnitude of the autocorrelation function for $\tau = 0$, namely $R(0)$, represents the time average of the power

SPECTRAL ANALYSIS FOR A SINGLE RANDOM PROCESS

or energy of a random process $x(t)$. Let us consider this property in more detail.

Suppose the random process $x(t)$ is an electric current, then the integral involved in Eq. (10.8) can be considered as the power that is dissipated in a resistance of 1 Ω; hence, the right side of Eq. (10.8) represents the time average of the power during a sufficiently long time interval. Similarly, if $x(t)$ represents a wave profile in the ocean, the right side of Eq. (10.8) can be regarded as the average wave energy neglecting the factor ρg, where ρ = mass density of water and g = gravitational constant. Thus, by denoting the time average of the power or energy of $x(t)$ by $\overline{P}_x$, we have

$$\overline{P}_x = \lim_{T \to \infty} \frac{1}{2T} \int_{-T}^{T} \{x(t)\}^2 \, dt \tag{10.9}$$

Next, let us consider the average energy of $x(t)$ in the frequency domain by applying the Fourier transform. For this, it is necessary to examine the condition required for the existence of the Fourier transform of $x(t)$. In general, the Fourier transform of a function $x(t)$ exists if the integration of $x(t)$ with respect to time over its entire domain, namely $-\infty < t < \infty$, is finite (see Appendix A: Fourier Transform). In order to strictly satisfy this condition, it is necessary to consider the following truncated function $x_T(t)$ defined as (see Figure 10.4)

$$x_T(t) = \begin{cases} x(t) & \text{for } -T < t < T \\ 0 & \text{otherwise} \end{cases} \tag{10.10}$$

For the truncated function $x_T(t)$ thus defined, the average energy of $x_T(t)$ may be written as follows:

$$\overline{P}_x = \lim_{T \to \infty} \frac{1}{2T} \int_{-\infty}^{\infty} \{x_T(t)\}^2 \, dt \tag{10.11}$$

Next, let us express the average energy, $\overline{P}_x$, in terms of frequency by applying the following theorem:

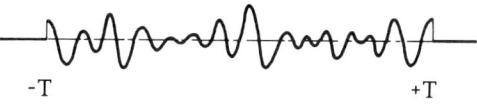

Figure 10.4 Truncated function x_T.

Theorem 10.1. Let the Fourier transform of a real function $x(t)$ be $X(\omega)$. Then we have

$$\int_{-\infty}^{\infty} \{x(t)\}^2 \, dt = \frac{1}{2\pi} \int_{-\infty}^{\infty} \{X(\omega)\}^2 \, d\omega \qquad (10.12)$$

Equation (10.12) is called the *Parseval theorem*.

Proof. Following the definition of the Fourier transform given in the Appendix A, we have

$$\int_{-\infty}^{\infty} \{x(t)\}^2 \, dt = \int_{-\infty}^{\infty} x(t) \left\{ \frac{1}{2\pi} \int_{-\infty}^{\infty} X(\omega) e^{i\omega t} \, d\omega \right\} dt$$

$$= \frac{1}{2\pi} \left\{ \int_{-\infty}^{\infty} x(t) e^{i\omega t} \, dt \right\} X(\omega) \, d\omega$$

$$= \frac{1}{2\pi} \int_{-\infty}^{\infty} X^*(\omega) X(\omega) \, d\omega$$

$$= \frac{1}{2\pi} \int_{-\infty}^{\infty} |X(\omega)|^2 \, d\omega$$

If the Fourier transform is carried out in terms of the frequency f in cycles per second (cps) instead of ω in radians per second (rps), then Parseval theorem may be written as

$$\int_{-\infty}^{\infty} \{x(t)\}^2 \, dt = \int_{-\infty}^{\infty} |X(f)|^2 \, df \qquad (10.13)$$

With the aid of the Parseval theorem, the average energy given in Eq. (10.10) can be written as follows:

$$\overline{P}_x = \begin{cases} \displaystyle\lim_{T \to \infty} \frac{1}{4\pi T} \int_{-\infty}^{\infty} |X_T(\omega)|^2 \, d\omega & \text{for frequency } \omega \quad (10.14a) \\ \displaystyle\lim_{T \to \infty} \frac{1}{2T} \int_{-\infty}^{\infty} |X_T(f)|^2 \, df & \text{for frequency } f \quad (10.14b) \end{cases}$$

Definition 10.2. The *spectral density function* of a random process $x(t)$ is given by

$$S_{xx}(\omega) = \lim_{T \to \infty} \frac{1}{2\pi T} |X_T(\omega)|^2 \qquad (10.15a)$$

SPECTRAL ANALYSIS FOR A SINGLE RANDOM PROCESS   227

In terms of frequency f, it is given by

$$S_{xx}(f) = \lim_{T \to \infty} \frac{1}{T} |X_T(f)|^2 \qquad (10.15b)$$

As is the case for the autocorrelation function, $S_{xx}(\omega)$ is called the autospectral density function in order to distinguish it from the cross-spectral density function $S_{xy}(\omega)$ defined for two random processes $x(t)$ and $y(t)$.

From Eqs. (10.14) and (10.15), and taking into consideration that the spectral density function is an even function, the average energy of $x(t)$ can be expressed as

$$\overline{P}_x = \begin{cases} \dfrac{1}{2} \int_{-\infty}^{\infty} S_{xx}(\omega)\,d\omega = \int_{0}^{\infty} S_{xx}(\omega)\,d\omega & (10.16a) \\[2mm] \dfrac{1}{2} \int_{-\infty}^{\infty} S_{xx}(f)\,df = \int_{0}^{\infty} S_{xx}(f)\,df & (10.16b) \end{cases}$$

Thus, the area under the spectral density function defined in Eq. (10.15) is equal to the average energy of a random process with respect to time.

Several remarks are in order here in regard to the spectral density function.

1. The spectral density function may also be defined as the Fourier transform of the auto-correlation function based on the Wiener-Khintchine theorem as presented in the next section.

2. The spectral density function in the frequency domain ω is often defined as follows:

$$S_{xx}(\omega) = \lim_{T \to \infty} \frac{1}{2T} |X_T(\omega)|^2 \qquad (10.17)$$

It should be noted that from this definition the area under the spectral density function does not represent the average energy; instead, it is π times the average energy.

3. $S_{xx}(\omega)$ is often called the *power spectrum*. This definition is proper for the random phenomena in communication engineering. However, $S_{xx}(\omega)$ can be evaluated for any random process such as waves in the ocean and motions and accelerations of ships and airplanes. For these cases, the term

"power" has no physical meaning; hence it may best be designated as a spectral density function

4. Justification for defining $S_{xx}(\omega)$ as a spectral "density function" may be clarified with the aid of the linear system concept discussed in Chapter 14. It will be shown in Section 14.1.2 that in the frequency domain, the output of a linear system to an input is given by

$$Y(\omega) = X(\omega) \cdot H(\omega) \qquad (10.18)$$

where

$X(\omega)$ = Fourier transform of input $x(t)$

$Y(\omega)$ = Fourier transform of output $y(t)$

$H(\omega)$ = frequency response function.

Let us consider an ideal bandpass filter as the frequency response function. That is,

$$H(\omega) = \begin{cases} 1 & \text{for } 0 < \omega_1 < \omega < \omega_2 \\ 0 & \text{otherwise} \end{cases} \qquad (10.19)$$

Then, the average output energy is given by

$$\lim_{T \to \infty} \frac{1}{2T} \int_{-\infty}^{\infty} \{y(t)\}^2 \, dt = \lim_{T \to \infty} \frac{1}{4\pi T} \int_{-\infty}^{\infty} |Y(\omega)|^2 \, d\omega$$

$$= \lim_{T \to \infty} \frac{1}{2\pi T} \int_{0}^{\infty} |Y(\omega)|^2 \, d\omega$$

$$= \lim_{T \to \infty} \frac{1}{2\pi T} \int_{0}^{\infty} |X(\omega)|^2 \cdot |H(\omega)|^2 \, d\omega$$

$$= \lim_{T \to \infty} \frac{1}{2\pi T} \int_{\omega_1}^{\omega_2} |X(\omega)|^2 \, d\omega$$

$$= \int_{\omega_1}^{\omega_2} S_{xx}(\omega) \, d\omega \qquad (10.20)$$

The above equation indicates that the average energy in the frequency domain $\omega_1 < \omega < \omega_2$ is given by the integration of $S_{xx}(\omega)$ over this interval of frequencies. Therefore, $S_{xx}(\omega)$ may be considered as representing energy density.

As will be shown later, the spectral density function plays a significant role in predicting statistical characteristics of a random process. Hence, it may be well to summarize the definitions and terminologies associated with the spectral density function in order to facilitate their usage in further analyses.

1. The kth *moment* of the spectral density function, denoted by m_k, is defined as

$$m_k = \begin{cases} \int_0^\infty \omega^k S(\omega)\, d\omega & \text{for frequency } \omega \quad (10.21\text{a}) \\ \int_0^\infty f^k S(f)\, df & \text{for frequency } f \quad (10.21\text{b}) \end{cases}$$

Care must be taken that m_k evaluated based on the frequency f is not equal to that evaluated based on the frequency ω except for the zeroth moment. The relationship between them is given by

$$m_k(\omega) = (2\pi)^k m_k(f), \quad k = 1, 2, 3, \ldots \quad (10.22)$$

2. The frequency where the spectral density function peaks is called the *modal frequency* and denoted by ω_m (or f_m).

3. The *mean frequency* of the spectral density function, denoted by $\bar{\omega}$ is defined as

$$\bar{\omega} = \frac{\int_0^\infty \omega S(\omega)\, d\omega}{\int_0^\infty S(\omega)\, d\omega} = \frac{m_1}{m_0} \quad (10.23)$$

A similar definition of the mean frequency $\bar{f}$ can also be given based on the frequency f. In this case, we have $\bar{f} = \bar{\omega}/2\pi$. It is to be noted that the mean frequency is not equal to the average value of the various frequencies involved in a random process; instead, it represents the mathematical mean of the spectral density function.

4. The *kth moment about the mean* of the spectral density function, denoted by μ_k, is defined as

$$\mu_k = \begin{cases} \int_0^\infty (\omega - \bar{\omega})^k S(\omega)\, d\omega & \text{for frequency } \omega \quad (10.24\text{a}) \\ \int_0^\infty (f - \bar{f})^k S(f)\, df & \text{for frequency } f \quad (10.24\text{b}) \end{cases}$$

and

$$\mu_k(\omega) = (2\pi)^k \mu_k(f), \qquad k = 2, 3, \ldots \quad (10.25)$$

The relationship between the moments m_k and μ_k evaluated in terms of frequency ω is as follows:

$$\begin{aligned} \mu_0 &= m_0 \\ \mu_1 &= 0 \\ \mu_2 &= m_2 - (m_1^2/m_0) \end{aligned} \quad (10.26)$$

5. The parameter ε defined as

$$\varepsilon = \sqrt{1 - \frac{m_2^2}{m_0 m_4}} \quad (10.27)$$

is called the *bandwidth parameter* of the spectrum (Cartwright and Longuet-Higgins, 1956). The ε value represents the spectral bandwidth for two extreme cases, namely $\varepsilon = 0$ for a random process with a narrow-band spectrum and $\varepsilon = 1$ for a random process having a wide-band spectrum.

6. The parameter ν defined as

$$\nu = \sqrt{\frac{\mu_2}{\mu_0} \frac{1}{\bar{\omega}}} \quad (10.28)$$

is called the *spectral width parameter* (Longuet-Higgins, 1975). This parameter plays an important role in the joint probability distribution of amplitudes and periods of a random process.

7. The parameter Q_p defined as

$$Q_p = \frac{\int_0^\infty \omega \{S(\omega)\}^2 d\omega}{\left\{\int_0^\infty S(\omega)\, d\omega\right\}^2} = \frac{\int_0^\infty f\{S(f)\}^2 df}{\left\{\int_0^\infty S(f)\, df\right\}^2} \quad (10.29)$$

is called the *spectral peakedness parameter* (Goda, 1970). It is a general trend that the value of the parameter Q_p increases with increasing sharpness of the spectrum.

10.1.3 Wiener–Khintchine Theorem

In this section, the relationship between the autocorrelation function of a random process that is defined in the time domain and its energy spectral density function defined in the frequency domain will be clarified.

For convenience, we assume that the Fourier transform of a random process $x(t)$ exists over the entire time domain. Hence the truncated function $x_T(t)$ as defined in Eq. (10.10) will not be considered in the following analysis. With this in mind, let us consider the Fourier transform of the autocorrelation function $R_{xx}(\tau)$ for the frequency ω. It is given by

$$\int_{-\infty}^{\infty} R_{xx}(\tau) e^{-i\omega\tau} d\tau = \lim_{T \to \infty} \frac{1}{2T} \int_{-\infty}^{\infty} \int_{-\infty}^{\infty} x(t) x(t+\tau) e^{-i\omega\tau} dt\, d\tau$$

$$= \lim_{T \to \infty} \frac{1}{2T} \int_{-\infty}^{\infty} \int_{-\infty}^{\infty} x(t) x(t+\tau) e^{-i\omega(t+\tau)} e^{i\omega t} dt\, d\tau$$

$$= \lim_{T \to \infty} \frac{1}{2T} X(\omega) X^*(\omega) = \lim_{T \to \infty} \frac{1}{2T} |X(\omega)|^2 \quad (10.30)$$

From the definition of the spectral density function given in Eq. (10.15a), we have

$$S_{xx}(\omega) = \frac{1}{\pi} \int_{-\infty}^{\infty} R_{xx}(\tau) e^{-i\omega\tau} d\tau \quad (10.31)$$

and hence

$$\int_{-\infty}^{\infty} R_{xx}(\tau) e^{-i\omega\tau} d\tau = \pi S_{xx}(\omega) \quad (10.32)$$

By taking the inverse Fourier transform, we have

$$R_{xx}(\tau) = \frac{1}{2} \int_{-\infty}^{\infty} S_{xx}(\omega) e^{i\omega\tau} d\omega \quad (10.33)$$

It can be seen from Eqs. (10.31) and (10.33) that the autocorrelation function, $R_{xx}(\tau)$, and the spectral density function, $S_{xx}(\omega)$, are a Fourier transform pair. This leads to the following theorem:

Theorem 10.2. For a weakly steady-state ergodic random process $x(t)$, its autocorrelation function, $R_{xx}(\tau)$, and the spectral density function, $S_{xx}(\omega)$, are related by the Fourier transform. That is,

$$S_{xx}(\omega) = \frac{1}{\pi}\int_{-\infty}^{\infty} R_{xx}(\tau) e^{-i\omega\tau}\, d\tau$$

$$R_{xx}(\tau) = \frac{1}{2}\int_{-\infty}^{\infty} S_{xx}(\omega) e^{i\omega\tau}\, d\omega$$

where

$$S_{xx}(\omega) = \lim_{T\to\infty} \frac{1}{2\pi T}|X(\omega)|^2$$

and

$$R_{xx}(\tau) = \lim_{T\to\infty} \frac{1}{2T}\int_{-T}^{T} x(t)x(t+\tau)\, dt$$

This is called the *Wiener–Khintchine Theorem*.

Since the autocorrelation function and the spectral density function are both real and even function, the Wiener–Khintchine theorem can be written as follows:

$$S_{xx}(\omega) = \frac{1}{\pi}\int_{-\infty}^{\infty} R_{xx}(\tau) e^{-i\omega\tau}\, d\tau$$

$$= \frac{2}{\pi}\int_{0}^{\infty} R_{xx}(\tau) \cos\omega t\, d\tau$$

$$R_{xx}(\tau) = \frac{1}{2}\int_{-\infty}^{\infty} S_{xx}(\omega) e^{i\omega\tau}\, d\omega$$

$$= \int_{0}^{\infty} S_{xx}(\omega) \cos\omega\tau\, d\omega \qquad (10.34)$$

It is noted that if the definition of the spectral density function given in Eq. (10.17) is used in Eq. (10.30), then the Wiener–Khintchine theorem can

SPECTRAL ANALYSIS FOR A SINGLE RANDOM PROCESS

be written as

$$S_{xx}(\omega) = \int_{-\infty}^{\infty} R_{xx}(\tau) e^{-i\omega\tau} d\tau$$

$$= 2\int_{0}^{\infty} R_{xx}(\tau) \cos \omega\tau \, d\tau$$

$$R_{xx}(\tau) = \frac{1}{2\pi} \int_{-\infty}^{\infty} S_{xx}(\omega) e^{-i\omega\tau} d\omega$$

$$= \frac{1}{\pi} \int_{0}^{\infty} S_{xx}(\omega) \cos \omega\tau \, d\omega \tag{10.35}$$

For the spectral density function thus evaluated, the area under the spectral density function does not represent the average energy of $x(t)$; instead, it represents π times the average energy as stated in connection with the definition given in Eq. (10.17).

Next, if the Fourier transform of the autocorrelation function is carried out in terms of the frequency f, we have

$$\int_{-\infty}^{\infty} R_{xx}(\tau) e^{-i2\pi f\tau} d\tau$$

$$= \lim_{T \to \infty} \frac{1}{2T} \int_{-\infty}^{\infty} \int_{-\infty}^{\infty} x(t) x(t+\tau) e^{-i2\pi f(t+\tau)} e^{i2\pi ft} \, dt \, d\tau$$

$$= \lim_{T \to \infty} \frac{1}{2T} |X(f)|^2 \tag{10.36}$$

From the definition of the spectral density function $S_{xx}(f)$ given in Eq. (10.14b), we have

$$S_{xx}(f) = 2\int_{-\infty}^{\infty} R_{xx}(\tau) e^{-i2\pi f\tau} d\tau \tag{10.37}$$

and the inverse Fourier transform yields

$$R_{xx}(\tau) = \frac{1}{2} \int_{-\infty}^{\infty} S_{xx}(f) e^{i2\pi f\tau} df \tag{10.38}$$

Thus, the Wiener–Khintchine theorem may be written in terms of frequency f as follows:

$$S_{xx}(f) = 2\int_{-\infty}^{\infty} R_{xx}(\tau) e^{-i2\pi f \tau}\, d\tau$$

$$= 4\int_{0}^{\infty} R_{xx}(\tau) \cos 2\pi f \tau\, d\tau$$

$$R_{xx}(\tau) = \frac{1}{2}\int_{-\infty}^{\infty} S_{xx}(f) e^{i2\pi f \tau}\, df$$

$$= \int_{0}^{\infty} S_{xx}(f) \cos 2\pi f \tau\, df \qquad (10.39)$$

We now summarize all functional relationships associated with spectral analysis of a random process $x(t)$ that are derived in the preceding three sections. Figure 10.5 shows an explanatory sketch indicating the relationships. That is, we first evaluate the autocorrelation function, $R_{xx}(\tau)$, from the time history of a random process $x(t)$. Next, by taking the Fourier transform of the autocorrelation function, the spectral density function, $S_{xx}(\omega)$, can be evaluated. Care must be given to the definition of the spectral density function considered in the transform. Then, from Eqs. (10.8), (10.9), and (10.34) or (10.39), we have the relationship that the area under the spectral density function is equal to the time average of the energy of the random process $x(t)$ as well as the value of the autocorrelation function for $\tau = 0$, and this in turn equals the variance of the random process if its mean value is zero. That is,

$$\left.\begin{array}{l}\int_{0}^{\infty} S_{xx}(\omega)\, d\omega \\ \int_{0}^{\infty} S_{xx}(f)\, df\end{array}\right\} = \bar{P}_x = R_{xx}(0) = \text{Var}[x(t)] \qquad \text{if } E[x(t)] = 0$$

$$(10.40)$$

Because of the relationship given in the above equation, the spectral density function $S_{xx}(\omega)$ is often called the *variance spectrum*.

It is noted again that the area under the spectral density function may or may not represent the average energy of a random process depending on the definition of the spectral density function. Table 10.1 summarizes the

SPECTRAL ANALYSIS FOR A SINGLE RANDOM PROCESS

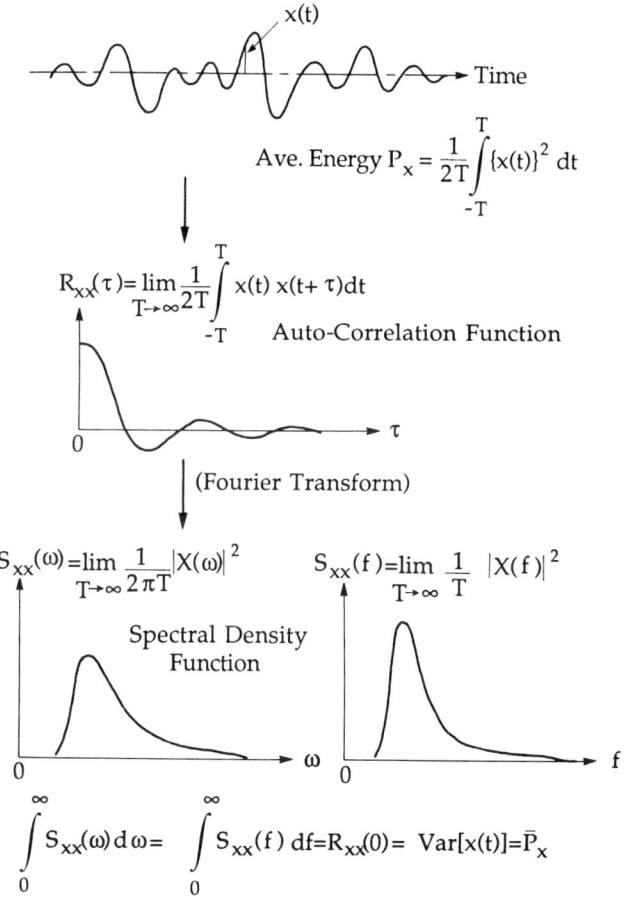

Figure 10.5 Principle and procedure of spectral analysis of random process $x(t)$ for $E[x(t)] = 0$.

Wiener–Khintchine theorem and functional relationship with the average energy for frequencies ω and f.

Example 10.6. The autocorrelation function for a spectral density function $S(\omega) = A$, where A is a constant and $-\infty < \omega < \infty$ is obtained as follows: The Fourier transform of the unit impulse function $\delta(t)$ is 1 (see Appendix C). Hence, we have $\mathscr{F}\{A\delta(t)\} = A$. However, by following the definition of the Wiener–Khintchine theorem given in Eq. (10.34), $\mathscr{F}\{\pi A\delta(\tau)\} = A$. Thus, $R_{xx}(\tau) = \pi A\delta(\tau)$. On the other hand, the autocor-

Table 10.1
Wiener–Khintchine Theorem and Relationship to Average Energy of Random Process $x(t)$ for $E[x(t)] = 0$

Frequency	Definition of auto-spectral density function	Wiener–Khintchine theorem $R_{xx}(\tau) \leftrightarrow \begin{cases} S_{xx}(\omega) \\ S_{xx}(f) \end{cases}$	Relation between spectral density function and average energy
ω in radians per sec.	$S_{xx}(\omega) = \lim_{T\to\infty} \dfrac{1}{2\pi T}\|X(\omega)\|^2$	$S_{xx}(\omega) = \dfrac{1}{\pi}\int_{-\infty}^{\infty} R_{xx}(\tau)e^{-i\omega\tau}\,d\tau = \dfrac{2}{\pi}\int_{0}^{\infty} R_{xx}(\tau)\cos\omega\tau\,d\tau$ $R_{xx}(\tau) = \dfrac{1}{2}\int_{-\infty}^{\infty} S_{xx}(\omega)e^{i\omega\tau}\,d\omega = \int_{0}^{\infty} S_{xx}(\omega)\cos\omega\tau\,d\omega$	$\int_{0}^{\infty} S_{xx}(\omega)\,d\omega = R_{xx}(0)$ $= \mathrm{Var}[x(t)] = \overline{P}_x$
ω in radians per sec.	$S_{xx}(\omega) = \lim_{T\to\infty} \dfrac{1}{2T}\|X(\omega)\|^2$	$S_{xx}(\omega) = \int_{-\infty}^{\infty} R_{xx}(\tau)e^{-i\omega\tau}\,d\tau = 2\int_{0}^{\infty} R_{xx}(\tau)\cos\omega\tau\,d\tau$ $R_{xx}(\tau) = \dfrac{1}{2\pi}\int_{-\infty}^{\infty} S_{xx}(\omega)e^{i\omega\tau}\,d\omega = \dfrac{1}{\pi}\int_{0}^{\infty} S_{xx}(\omega)\cos\omega\tau\,d\omega$	$\int_{0}^{\infty} S_{xx}(\omega)\,d\omega = \pi R_{xx}(0)$ $= \pi\,\mathrm{Var}[x(t)] = \pi\overline{P}_x$
f in cycles per sec.	$S_{xx}(f) = \lim_{T\to\infty} \dfrac{1}{T}\|X(f)\|^2$	$S_{xx}(f) = 2\int_{-\infty}^{\infty} R_{xx}(\tau)e^{-i2\pi f\tau}\,d\tau = 4\int_{0}^{\infty} R_{xx}(\tau)\cos 2\pi f\tau\,d\tau$ $R_{xx}(\tau) = \dfrac{1}{2}\int_{-\infty}^{\infty} S_{xx}(f)e^{i2\pi f\tau}\,df = \int_{0}^{\infty} S_{xx}(f)\cos 2\pi f\tau\,df$	$\int_{0}^{\infty} S_{xx}(f)\,df = R_{xx}(0)$ $= \mathrm{Var}[x(t)] = \overline{P}_x$

SPECTRAL ANALYSIS FOR A SINGLE RANDOM PROCESS

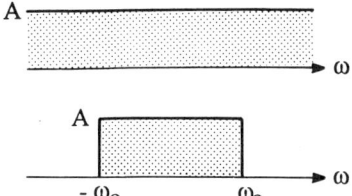

Figure 10.6 White noise spectrum and band-limited white noise spectrum.

relation function of $S(\omega) = A$, where $-\omega_0 < \omega < \omega_0$, is given by

$$R(\tau) = \frac{1}{2}\int_{-\omega_0}^{\omega_0} A e^{i\omega\tau}\, d\omega = \frac{A}{\tau}\sin\omega_0\tau \qquad \blacksquare$$

Two spectral density functions, $S(\omega) = A$ for $-\infty < \omega < \infty$ and $S(\omega) = A$ for $-\omega_0 < \omega < \omega_0$, shown in Figure 10.6 are called the *white noise spectrum* and the *band-limited white noise spectrum*, respectively. The former is not physically realizable since the time averaged energy (which is represented by the area under the spectral density function) becomes infinite; nevertheless, the concept is extremely useful for stochastic analysis of some problems. For example, the solution of a nonlinear system by applying the Fokker–Planck equation can be derived by assuming the excitation is a Gaussian random process with a white noise spectrum (see Chapter 15).

Example 10.7. A random process $x(t)$ is given by $x(t) = a + b\cos(\omega_0 t + \varepsilon)$, where a, b, and ω_0 are constants. The autocorrelation function of $x(t)$ is given by

$$R(\tau) = a^2 + \frac{b^2}{2}\cos\omega_0\tau \qquad \text{(see Example 10.1)}$$

Hence, the spectral density function becomes

$$S(\omega) = \frac{1}{\pi}\int_{-\infty}^{\infty}\left(a^2 + \frac{b^2}{2}\cos\omega_0\tau\right) e^{-i\omega\tau}\, d\tau$$

$$= \frac{a^2}{\pi}\int_{-\infty}^{\infty} e^{-i\omega\tau}\, d\tau + \frac{b^2}{4\pi}\int_{-\infty}^{\infty}\left(e^{i\omega_0\tau} + e^{-i\omega_0\tau}\right) e^{-i\omega\tau}\, d\tau$$

$$= 2a^2\delta(\omega) + \frac{b^2}{2}\{\delta(\omega - \omega_0) + \delta(\omega + \omega_0)\} \qquad \blacksquare$$

Example 10.8. $x(t)$ is a periodic function given by

$$x(t) = \sum_{n=-\infty}^{\infty} c_n e^{in\omega_0 t}$$

The autocorrelation function of $x(t)$ was derived in Example 10.3 as

$$R(\tau) = \sum_{n=-\infty}^{\infty} |c_n|^2 e^{in\omega_0 \tau}$$

Hence, the spectral density function can be written by

$$S(\omega) = \frac{1}{\pi} \int_{-\infty}^{\infty} \sum_{n=-\infty}^{\infty} |c_n|^2 e^{in\omega_0 \tau} e^{-i\omega\tau} d\tau$$

Since the Fourier transform of $e^{in\omega_0 \tau}$ is $2\pi\delta(\omega - n\omega_0)$, we have

$$S(\omega) = \sum_{n=-\infty}^{\infty} |c_n|^2 2\delta(\omega - n\omega_0)$$ ∎

Example 10.9. The autocorrelation function of a random process is given by

$$R(\tau) = \begin{cases} 1 - |\tau|/T & \text{for } |\tau| < T \\ 0 & \text{otherwise (see Figure 10.7)} \end{cases}$$

The spectral density function $S(\omega)$ can be evaluated as

$$S(\omega) = \frac{2}{\pi} \int_0^{\infty} R(\tau) \cos \omega\tau \, d\tau$$

$$= \frac{2}{\pi} \int_0^{\infty} \left(1 - \frac{\tau}{T}\right) \cos \omega\tau \, d\tau$$

$$= \frac{4}{\pi\omega^2 T} \left(\sin \frac{\omega T}{2}\right)^2$$ ∎

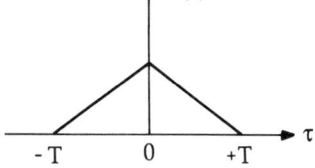

Figure 10.7 Autocorrelation function $R(\tau)$ in Example 10.9.

10.2 SPECTRAL ANALYSIS OF TWO RANDOM PROCESSES

The spectral analysis techniques developed for a single random process can be carried out on two random processes, $x(t)$ and $y(t)$. This is called *cross-spectral analysis* in contrast to autospectral analysis for a single random process. By carrying out cross-spectral analysis between two random processes $x(t)$ and $y(t)$, we can obtain various additional information on stochastic characteristics of the random processes. These include (1) the correlation and phase relationship between $x(t)$ and $y(t)$, (2) the linearity of a system if $x(t)$ and $y(t)$ represent the input and output of the system, respectively, and (3) stochastic characteristics of the random process compound of $x(t)$ and $y(t)$, and so on. Furthermore, cross-spectral analysis enables us to find the directional properties of a random process when its energy is propagating into various directions.

10.2.1 Cross-Correlation Function

Definition 10.3. The ensemble average of the product of two random processes $\mathbf{x}(t)$ and $\mathbf{y}(t)$ at times t_1 and t_2 is defined as the *cross-correlation function* denoted by $R_{xy}(t_1, t_2)$. That is,

$$R_{xy}(t_1, t_2) = E[\mathbf{x}(t_1)\mathbf{y}(t_2)]$$

$$= \frac{1}{n} \sum_{k=1}^{n} {}^k x(t_1) {}^k y(t_2) \qquad (10.41)$$

In particular, for a steady-state ergodic random process, we may write $t_2 = t_1 + \tau$, and the cross-correlation function can be written as

$$R_{xy}(\tau) = E[x(t)y(t+\tau)]$$

$$= \lim_{T \to \infty} \frac{1}{2T} \int_{-T}^{T} x(t)y(t+\tau)\, dt \qquad (10.42)$$

In evaluating the cross-correlation function $R_{xy}(\tau)$, as defined in Eq. (10.42), the record $y(t)$ should be shifted backward by time τ as shown in Figure 10.8. If $y(t)$ is shifted forward by time τ, then the cross-correlation function is denoted by $R_{yx}(\tau)$.

Several properties of the cross-correlation function between two random processes $x(t)$ and $y(t)$ are summarized below:

1. $R_{yx}(\tau) = R_{xy}(-\tau).$ \hfill (10.43)

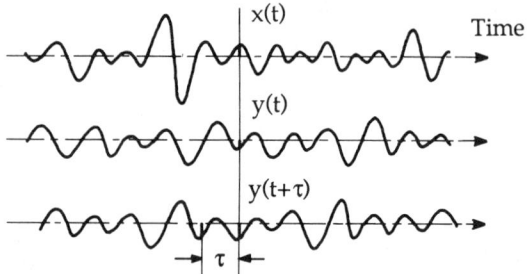

Figure 10.8 Definition of cross-correlation function $E[x(t)y(t+\tau)]$.

Proof. By definition, we have

$$R_{yx}(\tau) = \lim_{T \to \infty} \frac{1}{2T} \int_{-T}^{T} y(t)x(t+\tau)\, dt$$

By letting $t = u - \tau$,

$$R_{yx}(\tau) = \lim_{T \to \infty} \frac{1}{2T} \int_{-T}^{T} y(u-\tau)x(u)\, du = R_{xy}(-\tau)$$

2. $|R_{xy}(\tau)| \leq \sqrt{R_{xx}(0)R_{yy}(0)} \leq \tfrac{1}{2}\{R_{xx}(0) + R_{yy}(0)\}$ \hfill (10.44)

Proof. In general for real numbers a and b (where $b \neq 0$), we can write

$$\{ax(t) + by(t+\tau)\}^2 \geq 0$$

By taking the time average as defined by the autocorrelation and crosscorrelation functions, each term of the above equation can be written as

$$a^2 R_{xx}(0) + 2ab R_{xy}(\tau) + b^2 R_{yy}(0) \geq 0$$

By dividing by b^2,

$$R_{xx}(0)\left\{\left(\frac{a}{b}\right)^2 + 2\left(\frac{a}{b}\right)\frac{R_{xy}(\tau)}{R_{xx}(0)} + \frac{R_{yy}(0)}{R_{xx}(0)}\right\}$$

$$= R_{xx}(0)\left[\left\{\frac{a}{b} + \frac{R_{xy}(\tau)}{R_{xx}(0)}\right\}^2 + \frac{R_{yy}(0)}{R_{xx}(0)} - \left\{\frac{R_{xy}(\tau)}{R_{xx}(0)}\right\}^2\right] \geq 0$$

SPECTRAL ANALYSIS OF TWO RANDOM PROCESSES   241

In order to satisfy the above equation, the last two terms must be nonnegative. Hence, we have

$$\{R_{xy}(\tau)\}^2 \leq R_{xx}(0)R_{yy}(0)$$

Next, consider the following equation:

$$\{R_{xx}(0) - R_{yy}(0)\}^2 \geq 0$$

Then, by adding a nonnegative quantity $4R_{xx}(0)R_{yy}(0)$ to both sides of the equation,

$$\{R_{xx}(0) + R_{yy}(0)\}^2 \geq 4R_{xx}(0)R_{yy}(0)$$

From the above equation, the desired relationship can be derived.

3. $R_{xy}(\tau)$ is not always maximum at $\tau = 0$. Also, $R_{xy}(0)$ has no significant meaning in contrast to $R_{xx}(0)$, which is equal to the variance of the random process $x(t)$ if its mean value is zero.

Example 10.10. A random process $z(t)$ is given by $z(t) = ax(t) + y(t)$, where $x(t)$ and $y(t)$ are weakly steady-state random processes and a is a constant. The autocorrelation function of $z(t)$ is given by

$$R_{zz}(\tau) = E[z(t)z(t+\tau)]$$
$$= E[\{ax(t) + y(t)\} \cdot \{ax(t+\tau) + y(t+\tau)\}]$$
$$= a^2 R_{xx}(\tau) + a\{R_{xy}(\tau) + R_{yx}(\tau)\} + R_{yy}(\tau) \quad \blacksquare$$

Example 10.11. Let $x(t)$ and $y(t)$ be periodic functions having the same frequency ω_0. We may write

$$x(t) = \sum_{n=-\infty}^{\infty} c_n e^{in\omega_0 t}$$

and

$$y(t) = \sum_{n=-\infty}^{\infty} d_n e^{in\omega_0 t}$$

where

$$c_n = \frac{1}{2T}\int_{-T}^{T} x(t)e^{-in\omega_0 t}\,dt$$

$$d_n = \frac{1}{2T}\int_{-T}^{T} y(t)e^{-in\omega_0 t}\,dt$$

The cross-correlation function $R_{xy}(\tau)$ becomes

$$R_{xy}(\tau) = \frac{1}{2T}\int_{-T}^{T} x(t)y(t+\tau)\,dt$$

$$= \frac{1}{2T}\int_{-T}^{T} x(t)\sum_{n=-\infty}^{\infty} d_n e^{in\omega_0(t+\tau)}\,dt$$

$$= \sum_{n=-\infty}^{\infty}\left[d_n e^{in\omega_0\tau}\frac{1}{2T}\int_{-T}^{T} x(t)e^{in\omega_0 t}\,dt\right]$$

$$= \sum_{n=-\infty}^{\infty} (c_n^* d_n)e^{in\omega_0\tau}$$

It is understood from the above that $R_{xy}(\tau)$ and $(c_n^* d_n)$ form Fourier transforms of each other. ∎

10.2.2 Cross-Spectral Density Function

Analogous to the definition of the autospectral density function for a single random process $x(t)$, the cross-spectral density function of two random processes $x(t)$ and $y(t)$ is defined as follows:

Definition 10.4. The frequency function

$$S_{xy}(\omega) = \lim_{T\to\infty}\frac{1}{2\pi T}X^*(\omega)Y(\omega) \quad \text{for frequency } \omega \quad (10.45a)$$

$$S_{xy}(f) = \lim_{T\to\infty}\frac{1}{T}X^*(f)Y(f) \quad \text{for frequency } f \quad (10.45b)$$

is called the *cross-spectral density function* of random processes $x(t)$ and $y(t)$, where $X^*(\omega)$ and $X^*(f)$ are the conjugate functions of $X(\omega)$ and $X(f)$, respectively.

SPECTRAL ANALYSIS OF TWO RANDOM PROCESSES   243

The Wiener–Khintchine theorem given in Theorem 10.2 is also applied to the cross-correlation function and cross-spectral density function.

Theorem 10.3. For weakly steady-state ergodic random processes $x(t)$ and $y(t)$, the cross-correlation function $R_{xy}(\tau)$ and the cross-spectral density function $S_{xy}(\tau)$ are a Fourier transform pair. That is,

$$S_{xy}(\omega) = \frac{1}{\pi} \int_{-\infty}^{\infty} R_{xy}(\tau) e^{-i\omega\tau} d\tau$$

$$R_{xy}(\tau) = \frac{1}{2} \int_{-\infty}^{\infty} S_{xy}(\omega) e^{i\omega\tau} d\omega \tag{10.46}$$

It should be noted that the cross-spectral density function is a complex function in contrast to a real-valued function for the autospectral density function. Let us evaluate $S_{xy}(\omega)$ in detail following the Wiener–Khintchine theorem and clarify how the imaginary part of the function is introduced in the cross-spectral density function. That is,

$$\begin{aligned}
S_{xy}(\omega) &= \frac{1}{\pi} \int_{-\infty}^{\infty} R_{xy}(\tau) e^{-i\omega\tau} d\tau \\
&= \frac{1}{\pi} \left\{ \int_{-\infty}^{0} R_{xy}(\tau) e^{-i\omega\tau} d\tau + \int_{0}^{\infty} R_{xy}(\tau) e^{-i\omega\tau} d\tau \right\} \\
&= \frac{1}{\pi} \left\{ \int_{0}^{\infty} R_{yx}(\tau) e^{i\omega\tau} d\tau + \int_{0}^{\infty} R_{xy}(\tau) e^{-i\omega\tau} d\tau \right\} \\
&= \frac{1}{\pi} \int_{0}^{\infty} \{ R_{xy}(\tau) + R_{yx}(\tau) \} \cos \omega\tau \, d\tau \\
&\quad + i \frac{1}{\pi} \int_{0}^{\infty} \{ -R_{xy}(\tau) + R_{yx}(\tau) \} \sin \omega\tau \, d\tau \\
&= C_{xy}(\omega) + i Q_{xy}(\omega)
\end{aligned} \tag{10.47}$$

where

$$C_{xy}(\omega) = \frac{1}{\pi} \int_{0}^{\infty} \{ R_{xy}(\tau) + R_{yx}(\tau) \} \cos \omega\tau \, d\tau$$

$$Q_{xy}(\omega) = \frac{1}{\pi} \int_{0}^{\infty} \{ -R_{xy}(\tau) + R_{yx}(\tau) \} \sin \omega\tau \, d\tau \tag{10.48}$$

As can be seen in Eq. (10.47), the cross-spectral density function carries an imaginary part since the cross-correlation function $R_{xy}(\omega)$ is not equal to $R_{yx}(\omega)$. The real part $C_{xy}(\omega)$ given in Eq. (10.48) is referred to as the *cospectrum*, while the imaginary part $Q_{xy}(\omega)$ is referred to as the *quadrature spectrum*.

From Eq. (10.47), the amplitude spectrum of $S_{xy}(\omega)$ becomes

$$S_{xy}(\omega) = \sqrt{\{C_{xy}(\omega)\}^2 + \{Q_{xy}(\omega)\}^2} \tag{10.49}$$

and the phase spectrum is given by

$$\varepsilon(\omega) = \tan^{-1}\left\{\frac{Q_{xy}(\omega)}{C_{xy}(\omega)}\right\} \tag{10.50}$$

The cospectrum $C_{xy}(\omega)$ is an even function, while the quadrature spectrum $Q_{xy}(\omega)$ is an odd function. Various properties of the co- and quadrature spectra are summarized below.

$$C_{xy}(-\omega) = C_{xy}(\omega)$$
$$C_{yx}(\omega) = C_{xy}(\omega) \tag{10.51}$$
$$C_{yx}(-\omega) = C_{xy}(-\omega) = C_{xy}(\omega)$$

$$Q_{xy}(-\omega) = -Q_{xy}(\omega)$$
$$Q_{yx}(\omega) = -Q_{xy}(\omega) \tag{10.52}$$
$$Q_{yx}(-\omega) = -Q_{xy}(-\omega) = Q_{xy}(\omega)$$

By using the properties given in the above equations, we can derive the following relationship:

$$S_{xy}(-\omega) = C_{xy}(-\omega) + iQ_{xy}(-\omega)$$
$$= C_{xy}(\omega) - iQ_{xy}(\omega) = S_{xy}^*(\omega) \tag{10.53}$$

and

$$S_{yx}(\omega) = C_{yx}(\omega) + iQ_{yx}(\omega)$$
$$= C_{xy}(\omega) - iQ_{xy}(\omega) = S_{xy}^*(\omega) \tag{10.54}$$

where $S_{xy}^*(\omega)$ is the complex conjugate of $S_{xy}(\omega)$.

SPECTRAL ANALYSIS OF TWO RANDOM PROCESSES

In terms of the frequency f in cps, the cross-spectral density function can be written following the same procedure as used for the derivation of Eq. (10.47) as

$$S_{xy}(f) = 2\int_{-\infty}^{\infty} R_{xy}(\tau) e^{-i2\pi f\tau} d\tau = C_{xy}(f) + iQ_{xy}(f) \quad (10.55)$$

where

$$C_{xy}(f) = 2\int_{0}^{\infty} \{R_{xy}(\tau) + R_{yx}(\tau)\} \cos 2\pi f\tau \, d\tau$$

$$Q_{xy}(f) = 2\int_{0}^{\infty} \{-R_{xy}(\tau) + R_{yx}(\tau)\} \sin 2\pi f\tau \, d\tau$$

From information on the autospectral and cross-spectral density functions of two random processes, we can define an important parameter called the coherency function.

Definition 10.5. The following parameter indicating the correlation between two random processes is defined as the *coherency function*:

$$\gamma(\omega) = \frac{\{C_{xy}(\omega)\}^2 + \{Q_{xy}(\omega)\}^2}{S_{xx}(\omega) S_{yy}(\omega)} \quad (10.56)$$

If $x(t)$ and $y(t)$ are the input and output of a random vibration system, and the coherency function plays a significant role in clarifying the linearity of the system. This subject will be discussed in detail in Chapter 14 where the linear system is discussed.

10.2.3 Application—Directional Spectral Analysis

As an example of the application of cross-spectral analysis, let us consider the directional spectral analysis of a random phenomenon whose energy is not transmitted in a specific direction, but instead spreads into various directions. Wind-generated ocean waves subject to gravity force are a typical example of this sort of random phenomena. The energy transferred from the wind to the sea has spatial spread, and hence waves propagate in different directions although the predominant wave energy is in line with the wind direction.

Let us assume we measure the wave profile at a location in the sea and obtain the wave spectrum therefrom. This spectrum is often called the point spectrum and it provides information on the energy of all waves coming into that location from various directions, but does not provide information as to the direction in which the energy is propagating. Therefore, if the statistical prediction of wave characteristics such as wave amplitude and period is carried out from the point spectrum, it is presumed that all wave energy is propagating in one direction. This is certainly not true in reality. Although the energy-spreading mechanism of wind-generated waves is extremely complicated as it depends on the strength and time duration of the wind, the fetch, the water depth, and so on, the wave-spreading characteristics can be evaluated by obtaining the wave time histories at several locations and by carrying out autospectral as well as cross-spectral analysis (Panicker and Borgman, 1971; Panicker, 1974; among others).

The spreading characteristics of wave energy can also be found by measuring the wave profile and slopes at one location and by carrying out autospectral and cross-spectral analyses (Longuet-Higgins et al., 1961; Cartwright and Smith, 1964). The principle of this approach is outlined below.

Let us consider the source of the disturbance (wind in this example) to be along the axis shown in Figure 10.9, and derive the directional spectral density function, denoted by $S(\omega, \theta)$ as a function of the angle θ. For this, $S(\omega, \theta)$ is expanded into a Fourier series with respect to θ;

$$S(\omega, \theta) = \frac{a_0}{2} + \sum_{n=1}^{\infty} (a_n \cos n\theta + b_n \sin n\theta) \qquad (10.57)$$

where

$$a_n = \frac{1}{\pi} \int_{-\pi}^{\pi} S(\omega, \theta) \cos n\theta \, d\theta$$

$$b_n = \frac{1}{\pi} \int_{-\pi}^{\pi} S(\omega, \theta) \sin n\theta \, d\theta$$

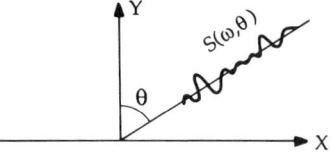

Figure 10.9 Directional random process with angle θ.

SPECTRAL ANALYSIS OF TWO RANDOM PROCESSES

The unknowns a_n and b_n will be expressed in terms of autospectral and cross-spectral density functions that can be evaluated from the displacement η, and slopes in the X and Y directions $\partial \eta/\partial x$ and $\partial \eta/\partial y$, respectively. For brevity, these three random processes are denoted as 1, 2, and 3, respectively.

In evaluating the autospectral and cross-spectral density functions, it is extremely convenient to express the progressive random waves in the vector form. That is, the displacement can be written as

$$\eta(t) = \text{Re} \int_{-\pi}^{\pi} \int_0^{\infty} e^{i(\mathbf{k}\cdot\mathbf{r} - \omega t + \varepsilon)} \, dA(\omega, \theta) \tag{10.58}$$

where

$$\mathbf{k} = k(\cos\theta \cdot \mathbf{i} + \sin\theta \cdot \mathbf{j})$$

$$\mathbf{r} = x \cdot \mathbf{i} + y \cdot \mathbf{j}$$

$$k = \omega^2/g$$

$$\varepsilon = \text{random phase}$$

Then, the cross-correlation function between η and $\partial \eta/\partial x$, denoted by $R_{12}(\tau)$, can be derived as

$$\begin{aligned}
R_{12}(\tau) &= \lim_{T \to \infty} \frac{1}{2T} \int_{-T}^{T} \eta(t) \frac{\partial}{\partial x} \eta(t + \tau) \, dt \\
&= \lim_{T \to \infty} \frac{1}{2T} \int_{-T}^{T} \left[\text{Re} \int_{-\pi}^{\pi} \int_0^{\infty} e^{i(\mathbf{k}\cdot\mathbf{r} - \omega t + \varepsilon)} \, dA(\omega, \theta) \right] \\
&\quad \times \left[\text{Re} \int_{-\pi}^{\pi} \int_0^{\infty} ik \cos\theta \, e^{i(\mathbf{k}\cdot\mathbf{r} - \omega(t+\tau) + \varepsilon)} \, dA(\omega, \theta) \right] dt \\
&= \int_{-\pi}^{\pi} ik \cos\theta \, R(\tau, \theta) \, d\theta
\end{aligned} \tag{10.59}$$

where $R(\tau, \theta)$ is the directional autocorrelation function.

Hence, the cross-spectral density function between η and $\partial \eta/\partial x$, denoted by $S_{12}(\omega)$, can be obtained from Eq. (10.59) as

$$S_{12}(\omega) = C_{12}(\omega) + iQ_{12}(\omega)$$

$$= \int_{-\pi}^{\pi} ik \cos\theta \, S(\omega, \theta) \, d\theta \tag{10.60}$$

where $S_{11}(\omega, \theta)$ = autospectral density function of η.

From Eq. (10.60), we have the following cospectrum and quadrature spectrum between η and $\partial \eta / \partial x$:

$$C_{12}(\omega) = 0$$

$$Q_{12}(\omega) = \int_{-\pi}^{\pi} k \cos \theta \, S(\omega, \theta) \, d\theta \qquad (10.61)$$

Computations of autospectral and cross-spectral density functions similar to that shown in the above are carried out for all three random processes, η, $\partial \eta / \partial x$, $\partial \eta / \partial y$, denoted as 1, 2, 3, respectively. The results are as follows:

$$C_{11}(\omega) = \int_{-\pi}^{\pi} S(\omega, \theta) \, d\theta \qquad Q_{11}(\omega) = 0$$

$$C_{12}(\omega) = 0 \qquad Q_{12}(\omega) = \int_{-\pi}^{\pi} k \cos \theta \, S(\omega, \theta) \, d\theta$$

$$C_{13}(\omega) = 0 \qquad Q_{13}(\omega) = \int_{-\pi}^{\pi} k \sin \theta \, S(\omega, \theta) \, d\theta$$

$$C_{22}(\omega) = \int_{-\pi}^{\pi} k^2 \cos^2 \theta \, S(\omega, \theta) \, d\theta \qquad Q_{22}(\omega) = 0$$

$$C_{23}(\omega) = \int_{-\pi}^{\pi} k^2 \sin \theta \cos \theta \, S(\omega, \theta) \, d\theta \qquad Q_{23}(\omega) = 0$$

$$C_{33}(\omega) = \int_{-\pi}^{\pi} k^2 \sin^2 \theta \, S(\omega, \theta) \, d\theta \qquad Q_{33}(\omega) = 0$$

$$(10.62)$$

With the aid of Eq. (10.62), the coefficients a_n and b_n in Eq. (10.57) can be evaluated as follows:

$$a_0 = (1/\pi) C_{11}(\omega)$$

$$a_1 = (1/\pi k) Q_{12}(\omega)$$

$$a_2 = (1/\pi k^2) \{ C_{22}(\omega) - C_{33}(\omega) \} \qquad (10.63)$$

$$b_1 = (1/\pi k) Q_{13}(\omega)$$

$$b_2 = (2/\pi k^2) C_{23}(\omega)$$

Thus, the spectral density function in an arbitrary direction can be evaluated from Eqs. (10.57) and (10.63). It is noted that because of the limitation in the number of measured components, the Fourier expansion is made for n up to $n = 2$ in this analysis. In order to let the partial sum of the Fourier expansion be equivalent to the sum of infinite terms, Longuet-Higgins et al. (1961) suggest that the Fourier expansion should be modified as

$$S(\omega, \theta) = \frac{a_0}{2} + \sum_{n=1}^{2} w_n(a_n \cos n\theta + b_n \sin n\theta) \qquad (10.64)$$

and let $w_1 = 2/3$ and $w_3 = 1/6$.

It is also noted that by additional measurements of the curvatures, $\partial^2 \eta / \partial x^2$, $\partial^2 \eta / \partial y^2$, and $\partial^2 \eta / \partial x \, \partial y$, it is possible to carry out the Fourier expansion up to $n = 4$ in $S(\omega, \theta)$.

10.3 INTEGRATED AND DIFFERENTIATED RANDOM PROCESSES

It is often of great importance to obtain the stochastic characteristics of the integrated or differentiated random processes $[\int x(t) \, dt$ or $\dot{x}(t)]$ from knowledge of a random process $x(t)$. These include evaluation of (1) $x(t)$ from $\dot{x}(t)$ such as the displacement of a particle in free Brownian motion from its velocity, (2) autocorrelation function of $\dot{x}(t)$ from the autocorrelation function of $x(t)$, and (3) spectral density function of the displacement $S_{xx}(\omega)$, from the spectrum of acceleration of the random process, $S_{\ddot{x}\ddot{x}}(\omega)$, and so on.

In the following, we may assume that the integration, $\int x(t) \, dt$, always exists. This assumption appears to be appropriate for almost all phenomena in natural sciences. On the other hand, care has to be taken in regard to the differentiability of a random process since derivatives do not always exist.

10.3.1 Mean, Variance, and Covariance

Let us write the integrated random process as

$$z(t) = \int_0^t x(t) \, dt \qquad (10.65)$$

and in the following $x(t)$ is assumed to be not necessarily a stationary random process with zero mean. The mean, variance, and covariance of

$z(t)$ can be evaluated as follows:

$$E[z(t)] = \int_0^t E[x(t)]\, dt \tag{10.66}$$

$$\operatorname{Var}[z(t)] = E\big[\{z(t)\}^2\big] - (E[z(t)])^2$$

$$= \int_0^t \int_0^t E[x(u)x(v)]\, du\, dv - \int_0^t \int_0^t E[x(u)]E[x(v)]\, du\, dv$$

$$= \int_0^t \int_0^t \{E[x(u)x(v)] - E[x(u)]E[x(v)]\}\, du\, dv$$

$$= \int_0^t \int_0^t C_{xx}(u,v)\, du\, dv \tag{10.67}$$

where $C_{xx}(u,v)$ is an autocovariance function defined in Eq. (10.5). Since $C_{xx}(u,v)$ is a symmetric function, we may write (see Figure 10.10)

$$\operatorname{Var}[z(t)] = 2\int_0^t \int_0^v C_{xx}(u,v)\, du\, dv \tag{10.68}$$

$$C_{zz}(t_1, t_2) = \operatorname{Cov}\left[\int_0^{t_1} x(u)\, du,\, \int_0^{t_2} x(v)\, dv\right]$$

$$= \int_0^{t_2} \int_0^{t_1} C_{xx}(u,v)\, du\, dv \tag{10.69}$$

As an example of the application of these formulae we consider the integrated Wiener–Lévy process, a nonstationary random process that will be discussed in detail in Chapter 13.

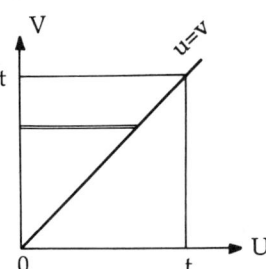

Figure 10.10 Integration domain for Eq. (10.67).

Example 10.12. Let $x(t)$ be a Wiener–Lévy process for which we have $E[x(t)] = 0$ and $C_{xx}(t_1, t_2) = E[x(t_1)x(t_2)] = \sigma^2 t_1$ for $t_1 < t_2$. Then, for the integrated process, we have

$$E[z(t)] = E\left[\int_0^t x(t)\, dt\right] = 0$$

$$\text{Var}[z(t)] = 2\int_0^t \int_0^v C_{xx}(u, v)\, du\, dv = 2\int_0^t \int_0^v \sigma^2 u\, du\, dv = \frac{\sigma^2 t^3}{3}$$

Care must be taken in the integration domain for evaluating the covariance function of $z(t)$ as shown in Figure 10.11. As shown in the figure, we have to consider $C_{xx}(u, v)$ for $u < v$ and for $u > v$ separately.

$$\text{Cov}[z(t_1), z(t_2)] = \int_0^{t_2}\int_0^{t_1} C_{xx}(u, v)\, du\, dv$$

$$= \int_0^{t_2}\int_0^{t_1} \sigma^2 u\, du\, dv + \int_0^{t_2}\int_0^{t_1} \sigma^2 v\, du\, dv$$

The integration of the first term becomes

$$\sigma^2 \int_0^{t_1}\int_0^v u\, du\, dv + \sigma^2 \int_{t_1}^{t_2}\int_0^{t_1} u\, du\, dv = \sigma^2\left\{\frac{t_1^3}{6} + \frac{t_1^2}{2}(t_2 - t_1)\right\}$$

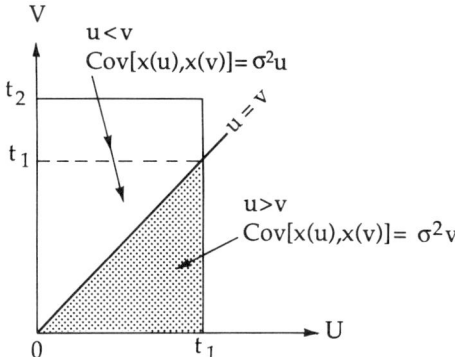

Figure 10.11 Integration domain for evaluating $\text{Cov}[z(t_1), z(t_2)]$ in Example 10.12.

The integration of the second term becomes

$$\sigma^2 \int_0^{t_1} \int_0^u v \, dv \, du = \frac{\sigma^2}{6} t_1^3$$

Thus, the covariance $[z(t_1), z(t_2)]$ becomes

$$C_{zz}(t_1, t_2) = \frac{\sigma^2}{6} t_1^2 (3t_2 - t_1)$$

By letting $t_1 = t_2$, we have $\text{Var}[z(t)] = \sigma^2 t^3 / 3$. ∎

The sample mean of a random process $x(t)$ is defined as

$$\bar{x}(t) = \frac{1}{T} \int_0^T x(t) \, dt \qquad (10.70)$$

For a stationary random process $x(t)$, the mean and variance of the sample mean can be evaluated from Eqs. (10.66) and (10.68) as follows:

$$E[\bar{x}(t)] = \frac{1}{T} \int_0^T E[x(t)] \, dt \qquad (10.71)$$

$$\text{Var}[\bar{x}(t)] = \text{Var}\left[\frac{1}{T} \int_0^T x(t) \, dt \right]$$

$$= \frac{2}{T^2} \int_0^T \int_0^v C_{xx}(u, v) \, du \, dv \qquad (10.72)$$

Next, the mean and variance of the differentiated random process is discussed. Assuming that a random process has finite second moment, the derivative of the random process is defined as

$$x'(t) = \lim_{\Delta t \to 0} \frac{1}{\Delta t} \{ x(t + \Delta t) - x(t) \} \qquad (10.73)$$

where the limit exists in the sense of *convergence in the mean square* given by

$$\lim_{\Delta t \to 0} E\left[\left(\frac{x(t + \Delta t) - x(t)}{\Delta t} - x'(t) \right)^2 \right] = 0 \qquad (10.74)$$

It can be proved that the above equation is satisfied if and only if the following two limits exist:

(i) $\lim\limits_{\Delta t \to 0} E\left[\dfrac{x(t+\Delta t) - x(t)}{\Delta t}\right]$

(ii) $\lim\limits_{\Delta t \to 0, \Delta t' \to 0} \text{Cov}\left[\dfrac{x(t+\Delta t) - x(t)}{\Delta t}, \dfrac{x(t+\Delta t') - x(t)}{\Delta t'}\right]$ (10.75)

Within these limits, a stationary stochastic process $x(t)$ is said to be *differentiable in mean square*. The mean, variance, and covariance of the derivatives of $x(t)$ are given as follows:

$$E[\dot{x}(t)] = \frac{d}{dt} E[x(t)] \qquad (10.76)$$

$$\text{Var}[\dot{x}(t)] = E[\{\dot{x}(t)\}^2] - (E[\dot{x}(t)])^2$$

$$= \frac{d^2}{dt^2} \text{Var}[x(t)] \qquad (10.77)$$

$$\text{Cov}[\dot{x}(t_1), \dot{x}(t_2)] = \text{Cov}\left[\frac{d}{dt_1} x(t_1), \frac{d}{dt_2} x(t_2)\right]$$

$$= \frac{\partial^2}{\partial t_1 \partial t_2} C_{xx}(t_1, t_2) \qquad (10.78)$$

Similarly, the covariance of $x(t_1)$ and $\dot{x}(t_2)$ can be written as

$$\text{Cov}[x(t_1), \dot{x}(t_2)] = \frac{d}{dt_2} C_{xx}(t_1, t_2) \qquad (10.79)$$

Example 10.13. The autocorrelation of a random process $x(t)$ is given by

$$R_{xx}(\tau) = \sigma^2 e^{-a\tau^2}$$

By letting $E[x(t)] = 0$ and $t_1 - t_2 = \tau$ in Eq. (10.78), the autocorrelation of its derivative $\dot{x}(t)$ becomes

$$R_{\dot{x}\dot{x}}(\tau) = -\frac{d^2}{d\tau^2} R_{xx}(\tau) = 2\sigma^2 a(1 - 2a\tau^2) e^{-a\tau^2} \qquad ∎$$

Example 10.14. A Wiener–Lévy process is not differentiable in mean square. Note that the covariance function of the Wiener–Lévy process is given as $C_{xx}(t_1, t_2) = \sigma^2 t_1$ for $t_1 < t_2$. Hence, the derivative

$$\frac{\partial}{\partial t_2} \frac{\partial}{\partial t_1} C_{xx}(t_1, t_2) = \frac{\partial}{\partial t_2} \sigma^2$$

does not exist. Although the process is not differentiable in mean square, we may write the derivative of the covariance function as

$$\frac{\partial}{\partial t_2}\left(\frac{\partial}{\partial t_1} C_{xx}(t_1, t_2)\right) = \frac{\partial}{\partial t_2} \sigma^2 \mathsf{U}(t_2 - t_1) = \sigma^2 \delta(t_2 - t_1)$$

where $\mathsf{U}(\ )$ and $\delta(\ )$ are the unit step function and unit impulse function, respectively. Thus, the derivative of the Wiener–Lévy process can be considered as a white noise process. ∎

10.3.2 Autocorrelation Function and Spectral Density Function of Derived Random Processes

Let $x(t)$ be a steady-state random process. Assuming that the time derivatives of the random process such as velocity $\dot{x}(t)$ and acceleration $\ddot{x}(t)$ exist, we will evaluate the autocorrelation function and the spectral density function of velocity and acceleration from the displacement autocorrelation function $R_{xx}(\tau)$, and the spectral density function $S_{xx}(\omega)$.

We first obtain the autocorrelation function. For this, it may be easier to consider the cross-correlation function of $x(t_1)$ and $\dot{x}(t_2)$. That is,

$$\begin{aligned} R_{x\dot{x}}(t_1, t_2) &= E[x(t_1)\dot{x}(t_2)] \\ &= E\left[x(t_1)\frac{x(t_2 + \Delta t) - x(t_2)}{\Delta t}\right] \\ &= \frac{1}{\Delta t}\{R_{xx}(t_1, t_2 + \Delta t) - R_{xx}(t_1, t_2)\} \end{aligned} \quad (10.80)$$

For $\Delta t \to 0$, and for a steady-state random process for which $t_1 - t_2 = \tau$, we have

$$R_{x\dot{x}}(\tau) = \frac{\partial}{\partial t_2} R_{xx}(t_1, t_2) = -\frac{d}{d\tau} R_{xx}(\tau) \quad (10.81)$$

INTEGRATED AND DIFFERENTIATED RANDOM PROCESSES

Similarly, we can derive

$$R_{\ddot{x}\ddot{x}}(\tau) = \frac{\partial^2}{\partial t_1 \partial t_2} R_{xx}(t_1, t_2) = -\frac{d^2}{d\tau^2} R_{xx}(\tau) \qquad (10.82)$$

and

$$R_{\ddot{x}\ddot{x}}(\tau) = \frac{d^4}{d\tau^4} R_{xx}(\tau) \qquad (10.83)$$

Thus, for a random process with zero mean, we have

$$\text{Var}[\dot{x}(t)] = -\frac{d^2}{d\tau^2} R_{xx}(0)$$

$$\text{Var}[\ddot{x}(t)] = \frac{d^4}{d\tau^4} R_{xx}(0) \qquad (10.84)$$

In order to obtain the spectral density functions of $\dot{x}(t)$ and $\ddot{x}(t)$ from the displacement spectral density function $S_{xx}(\omega)$, we may apply the property regarding the Fourier transform of the time derivative of a function given in Appendix A. That is, if we write the Fourier transform of a random process $x(t)$ as

$$\mathscr{F}\{x(t)\} = X(\omega)$$

then, we have

$$\mathscr{F}\{\dot{x}(t)\} = \dot{X}(\omega) = i\omega X(\omega)$$

and

$$\mathscr{F}\{\ddot{x}(t)\} = \ddot{X}(\omega) = -\omega^2 X(\omega)$$

Hence, by definition, the spectral density functions of velocity, $S_{\dot{x}\dot{x}}(\omega)$, and acceleration, $S_{\ddot{x}\ddot{x}}(\omega)$, can be written as follows:

$$S_{\dot{x}\dot{x}}(\omega) = \lim_{T \to \infty} \frac{1}{2\pi T} |\dot{X}(\omega)|^2$$

$$= \omega^2 \lim_{T \to \infty} \frac{1}{2\pi T} |X(\omega)|^2 = \omega^2 S_{xx}(\omega)$$

$$S_{\ddot{x}\ddot{x}}(\omega) = \lim_{T \to \infty} \frac{1}{2\pi T} |\ddot{X}(\omega)|^2$$

$$= \omega^4 \lim_{T \to \infty} \frac{1}{2\pi T} |X(\omega)|^2 = \omega^4 S_{xx}(\omega) \qquad (10.85)$$

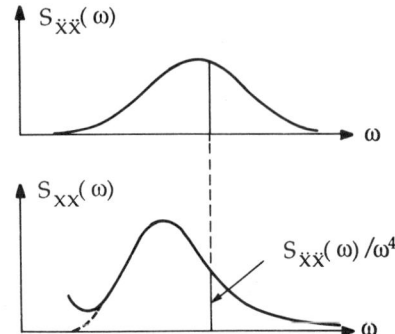

Figure 10.12 Evaluation of displacement spectrum $S_{\ddot{x}\ddot{x}}(\omega)$ from acceleration spectrum $S_{xx}(\omega)$.

It is clear from Eq. (10.85) that the velocity and acceleration spectral density functions can be evaluated from the displacement spectral density function $S_{xx}(\omega)$ without knowledge of the time history of velocity and acceleration. Inversely, the displacement spectral density function can be evaluated from the velocity or acceleration spectral density function. As a practical example, the displacement spectrum of ocean waves is usually obtained from the wave acceleration measured by a floating buoy.

In converting the acceleration spectrum, $S_{\ddot{x}\ddot{x}}(\omega)$, to the displacement spectrum, $S_{xx}(\omega)$, by applying Eq. (10.85), there are always very large values in the low frequency range of $S_{xx}(\omega)$ as illustrated in Figure 10.12. This is because $S_{\ddot{x}\ddot{x}}(\omega)/\omega^4$ becomes very large for small ω. It is customary to bring the spectral density function to zero ignoring the unusually large values of $S_{xx}(\omega)$ for small ω.

If the spectral density functions are expressed in terms of the frequency f in cps, the relationships given in Eq. (10.85) can be written as follows:

$$S_{\dot{x}\dot{x}}(f) = (2\pi f)^2 S_{xx}(f)$$
$$S_{\ddot{x}\ddot{x}}(f) = (2\pi f)^4 S_{xx}(f) \tag{10.86}$$

Example 10.15. The magnitude of the spectral density function of the displacement of a random vibration system is a in m²-sec for the frequency range $\omega_1 < \omega < \omega_2$. The spectral density function of the acceleration of this system can be obtained by applying Eq. (10.85) in g units as $a\omega^4/(9.8)^2$ as shown in Figure 10.13. ∎

Example 10.16. The spectral density function of the wave acceleration is obtained in g units as a function of frequency ω in rps. In order to obtain the spectral density function of wave displacement in metric units as a function of frequency f in cps, it is necessary to convert the frequency scale to $f = \omega/2\pi$ and the acceleration scale to metric units. Then, the scale

SQUARED RANDOM PROCESSES

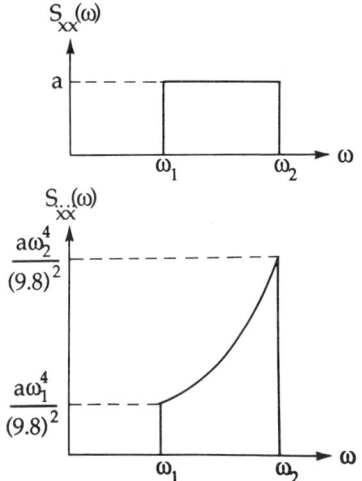

Figure 10.13 Evaluation of acceleration spectrum $S_{\ddot{x}\ddot{x}}(\omega)$ from displacement spectrum $S_{xx}(\omega)$ in Example 10.15.

of the displacement spectral density function becomes

$$S_{\ddot{x}\ddot{x}}(\omega) \cdot (9.8)^2(2\pi)/(2\pi f)^4 \quad \text{(see Figure 10.14)}. \qquad \blacksquare$$

10.4 SQUARED RANDOM PROCESSES

Let $x(t)$ be a stationary normal process with zero mean and let us consider the square of $x(t)$. This random process is often called the *square-law*

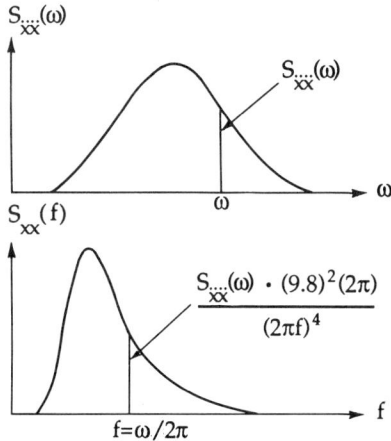

Figure 10.14 Evaluation of displacement spectrum $S_{xx}(\omega)$ from acceleration spectrum $S_{\ddot{x}\ddot{x}}(\omega)$ in Example 10.16.

detector in electrical engineering, but we shall refer to it here as the *squared random process*. The squared random process is frequently encountered in diverse areas of engineering and physics. For example, in the evaluation of the hydrodynamic impact forces or aerodynamic wind forces on a structure, these forces are usually proportional to the square of the velocity, where the velocity can be considered as a Gaussian random process with zero mean.

We first derive the mean and variance of the squared random process $y(t) = ax^2(t)$, where a is a positive constant, assuming that $x(t)$ is a weakly stationary normal process with zero mean and an autocorrelation function $R_{xx}(\tau)$. The mean of $y(t)$ can be obtained by

$$E[y(t)] = E[ax^2(t)] = aR_{xx}(0) \qquad (10.87)$$

In order to obtain the variance of $y(t)$, let us derive the autocorrelation function of the process $y(t)$, denoted by $R_{yy}(\tau)$. It is given by

$$R_{yy}(\tau) = a^2 E[y(t)y(t+\tau)] = a^2 E[x^2(t)x^2(x+\tau)] \qquad (10.88)$$

By applying the result derived in Example 3.16, we may write

$$E[x^2(t)x^2(t+\tau)] = E[x^2(t)]E[x^2(t+\tau)] + 2(E[x(t)x(t+\tau)])^2 \qquad (10.89)$$

Hence, the autocorrelation function $R_{yy}(\tau)$ becomes

$$R_{yy}(\tau) = a^2\{R_{xx}(0)\}^2 + 2a^2\{R_{xx}(\tau)\}^2 \qquad (10.90)$$

By letting $\tau = 0$, we have

$$E[y^2(t)] = 3a^2\{R_{xx}(0)\}^2 \qquad (10.91)$$

Since the mean value $E[y(t)] = aR_{xx}(0)$ as shown in Eq. (10.87), we can obtain the variance of $y(t)$ as

$$\text{Var}[y(t)] = E[y^2(t)] - (E[y(t)])^2 = 2a^2\{R_{xx}(0)\}^2 \qquad (10.92)$$

The probability density function of $y(t)$ can be obtained by applying the techniques for the transformation of random variables given in Eq. (7.6). Since $x(t)$ is normally distributed with zero mean and variance $R_{xx}(0)$, the probability density function of $y(t)$ becomes

$$f(y) = 2\left[\frac{1}{\sqrt{2\pi}\sqrt{R_{xx}(0)}} \exp\left\{-\frac{x^2}{2R_{xx}(0)}\right\}\right]_{x=\sqrt{y/a}} \left|\frac{1}{2\sqrt{ay}}\right|$$

$$= \frac{1}{\sqrt{2\pi aR_{xx}(0)}\sqrt{y}} \exp\left\{-\frac{y}{2aR_{xx}(0)}\right\}, \quad 0 \leq y < \infty \quad (10.93)$$

Note that the probability density function of $y(t)$ is the gamma distribution shown in Eq. (6.18) with $m = 1/2$, $\lambda = 1/\{2aR_{xx}(0)\}$.

Note that by letting $y/\{aR_{xx}(0)\} = z$, Eq. (10.93) becomes the χ^2 distribution with one degree of freedom.

In the field of engineering and physics, the squared random process is often given in the form of $y(t) = ax(t)|x(t)|$. For example, the hydrodynamic (or aerodynamic) drag force acting on a fixed body in an oscillating flow is proportional to the square of the fluid velocity, but its direction is opposite that of the fluid velocity; hence, the drag force is in the form of $ax(t)|x(t)|$. In this case, the sample space of $y(t)$ is $(-\infty, \infty)$. The probability density function becomes one-half that given in Eq. (10.93) and y should be replaced by $|y|$.

The spectral density function of $y(t)$ can be obtained by applying the Wiener–Khintchen theorem to the autocorrelation function given in Eq. (10.90). The first term of Eq. (10.90) is a constant $a^2\{R_{xx}(0)\}^2$, and its Fourier transform is given by $2\pi a^2\{R_{xx}(0)\}^2\delta(\omega)$ (see Appendix A). However, following the definition of the Wiener–Khintchen theorem given in Eq. (10.32), we may write the first term of the spectral density function of $S_{yy}(\omega)$ as $2a^2\{R_{xx}(0)\}^2\delta(\omega)$ so that the area under the spectral density function is equal to the second moment of $y(t)$. The Fourier transform of the second term of Eq. (10.90) is given by the convolution integral $(1/2\pi)(2a^2)S_{xx}(\omega)*S_{xx}(\omega)$. Again, by following the definition given in Eq. (10.32), we may write $(a^2/\pi^2)S_{xx}(\omega)*S_{xx}(\omega)$. Thus we have

$$S_{yy}(\omega) = a^2\left[2\{R_{xx}(0)\}^2\delta(\omega) + \frac{1}{\pi^2}S_{xx}(\omega)*S_{xx}(\omega)\right] \quad (10.94)$$

10.5 HIGHER ORDER SPECTRAL ANALYSIS

The spectral analysis introduced so far deals with autospectral density function, which represents the contribution of each frequency to the time average of the total energy of a random process. The energy contribution of each frequency is assumed to be independent; hence, in the time domain, the random process can be expressed by superposition of independent frequency components. This may not be the case, however, for some physical phenomena that has nonlinear characteristics. For nonlinear random processes, the frequency components of the energy are not necessarily independent; that is, interaction of energy may take place between frequencies. In order to clarify the stochastic properties of these random phenomena, it is necessary to obtain the higher order moments of the processes. This can be achieved by carrying out higher order spectral analysis.

Let us elaborate on the above statement. Many random phenomena are commonly assumed to be a Gaussian random process whose stochastic properties can be expressed in terms of the mean value and variance (second moment for the zero mean), which is equal to the area under the spectral density function. If a phenomenon cannot be assumed to be a Gaussian random process, the stochastic properties can no longer be expressed by the first two moments. The third moment (or higher moments) may be required to adequately describe the stochastic properties. Here, the third moment can be evaluated from the two-dimensional spectral density function, called the bispectrum, and fourth moment can be obtained from the three-dimensional spectral density function, called the trispectrum, and so on.

Bispectral analysis was first carried out by Hasselman et al. (1962) to clarify the nonlinear interaction of ocean waves. Since then the analysis has been applied to problems in many diverse fields such as studies on fluid turbulence (Yeh and Van Atta, 1973; Lii et al., 1976), growth mechanism of wind-generated waves (Liu and Green, 1978), internal waves in the ocean (McComas and Briscoe, 1980), plasma-wave studies (Kim and Powers, 1978), ship rolling motion (Yamanouchi and Ohtsu, 1972), noise in mechanical gear trains (Sato et al., 1980), economic time series (Godfrey, 1965), and others.

The following discussion is limited to second-order spectral analysis (bispectrum) of an ergodic random process; however, the principles and techniques for bispectral analysis are equally valid for higher order spectral analysis.

Analogous to the autocorrelation function defined in Eq. (10.4), the two-dimensional autocorrelation function of a random process, denoted by $M_{xx}(\tau_1, \tau_2)$ is defined by

$$M_{xx}(\tau_1, \tau_2) = E[x(t)x(t+\tau_1)x(t+\tau_2)]$$

$$= \lim_{T \to \infty} \frac{1}{2T} \int_{-T}^{T} x(t)x(t+\tau_1)x(t+\tau_2)\, dt \quad (10.95)$$

HIGHER ORDER SPECTRAL ANALYSIS

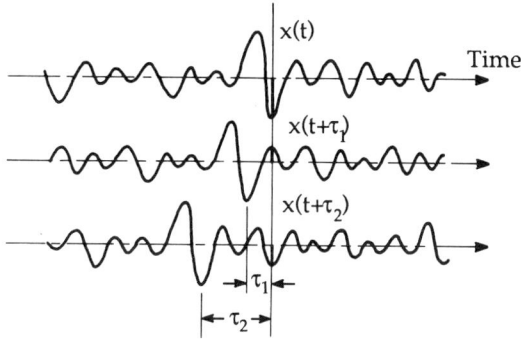

Figure 10.15 Two-dimensional autocorrelation function $M_{xx}(\tau_1, \tau_2) = E[x(t)x(t+\tau_1)x(t+\tau_2)]$.

Two-dimensional auto-correlation function, therefore, can be evaluated by shifting the time history of $x(t)$ by time τ_1 and τ_2 as shown in Figure 10.15, and then by integrating the product of $x(t)x(t+\tau_1)x(t+\tau_2)$.

From the definition given in Eq. (10.95), we can derive the following symmetry relations of $M_{xx}(\tau_1, \tau_2)$:

$$M_{xx}(\tau_1, \tau_2) = M_{xx}(\tau_2, \tau_1)$$
$$= M_{xx}(-\tau_2, \tau_1 - \tau_2) = M_{xx}(\tau_1 - \tau_2, -\tau_2)$$
$$= M_{xx}(-\tau_1, \tau_2 - \tau_1) = M_{xx}(\tau_2 - \tau_1, -\tau_1) \quad (10.96)$$

Proof of the above relationship is simple. For example, in order to prove $M_{xx}(\tau_1, \tau_2) = M_{xx}(\tau_1 - \tau_2, -\tau_2)$, replace τ_1 and τ_2 in Eq. (10.95) by $(\tau_1 - \tau_2)$ and $(-\tau_2)$, respectively, and then let $t - \tau_2 = u$.

Analogous to the definition of the unispectrum $S_{xx}(\omega)$, the bispectrum of a random process $x(t)$ is defined as follows:

Definition 10.6. The quantity

$$B_{xx}(\omega_1, \omega_2) = \lim_{T \to \infty} \frac{1}{2\pi T} X(\omega_1) X(\omega_2) X^*(\omega_1 + \omega_2)$$

$$= \lim_{T \to \infty} \frac{1}{2\pi T} X(\omega_1) X(\omega_2) X(\omega_3) \quad (10.97)$$

is defined as the *bispectrum* of a random process $x(t)$, where $X(\omega)$ is the

Fourier transform of $x(t)$, $X^*(\omega)$ is the conjugate of $X(\omega)$, and $\omega_1 + \omega_2 + \omega_3 = 0$.

The definition given in Eq. (10.97) is of a somewhat different form from that given in Eq. (10.15) for the unispectrum $S_{xx}(\omega)$; however, they are essentially the same. This is because the spectral density function $S_{xx}(\omega)$ may be written in the following form:

$$S_{xx}(\omega) = \lim_{T \to \infty} \frac{1}{2\pi T} |X(\omega)|^2 = \lim_{T \to \infty} \frac{1}{2\pi T} X(\omega) X^*(\omega)$$

$$= \lim_{T \to \infty} \frac{1}{2\pi T} X(\omega_1) X(\omega_2) \qquad (10.98)$$

where $\omega_1 + \omega_2 = 0$.

The Wiener–Khintchine theorem can also be applied to the two-dimensional analysis of a random process. That is, the bispectrum, $B_{xx}(\omega_1, \omega_2)$ is a Fourier transform (a two-dimensional transform in this case) of the autocorrelation function, which can be written as

$$B_{xx}(\omega_1, \omega_2) = \frac{1}{\pi^2} \int_{-\infty}^{\infty} \int_{-\infty}^{\infty} M_{xx}(\tau_1, \tau_2) e^{-i(\omega_1 \tau_1 + \omega_2 \tau_2)} d\tau_1 d\tau_2 \qquad (10.99)$$

where

$$M_{xx}(\tau_1, \tau_2) = E[x(t)x(t+\tau_1)x(t+\tau_2)]$$

$$= \frac{1}{2^2} \int_{-\infty}^{\infty} \int_{-\infty}^{\infty} B_{xx}(\omega_1, \omega_2) e^{i(\omega_1 \tau_1 + \omega_2 \tau_2)} d\omega_1 d\omega_2 \qquad (10.100)$$

By letting $\tau_1 = \tau_2 = 0$ in Eq. (10.100), we have

$$M_{xx}(0,0) = E[x^3(t)] = \lim_{T \to \infty} \frac{1}{2T} \int_{-\infty}^{\infty} \{x(t)\}^3 dt$$

$$= \int_0^{\infty} \int_0^{\infty} B(\omega_1, \omega_2) d\omega_1 d\omega_2 \qquad (10.101)$$

Thus, it is clear that the integral volume of the bispectrum $B_{xx}(\omega_1, \omega_2)$ with respect to ω_1 and ω_2 represents the third moment of the random process $x(t)$.

It was shown in Eq. (10.96) that there are six two-dimensional autocorrelation functions for a given τ_1 and τ_2. Since the Wiener–Khintchine

HIGHER ORDER SPECTRAL ANALYSIS   263

theorem holds for each of these autocorrelation functions, there are six bispectra for a given ω_1 and ω_2 having the same value. These are,

$$B_{xx}(\omega_1, \omega_2) = B_{xx}(\omega_2, \omega_1) = B(\omega_1, -\omega_1 - \omega_2)$$
$$= B_{xx}(-\omega_1 - \omega_2, \omega_1) = B_{xx}(\omega_2, -\omega_1 - \omega_2)$$
$$= B_{xx}(-\omega_1 - \omega_2, \omega_2) \qquad (10.102)$$

Furthermore, the bispectrum is a complex function as defined in Eq. (10.97), and hence a conjugate exists for each bispectrum,

$$B_{xx}(\omega_1, \omega_2) = B^*(-\omega_1, -\omega_2) \qquad (10.103)$$

Therefore, if we consider the absolute value of the bispectrum, then an additional six conjugates of the bispectra given in Eq. (10.103) must be included. This implies that the absolute values of all of 12 bispectra are equal.

Figure 10.16 shows a pictorial sketch indicating the symmetric characteristics of the bispectrum. As can be seen in the figure, the six bispectra given

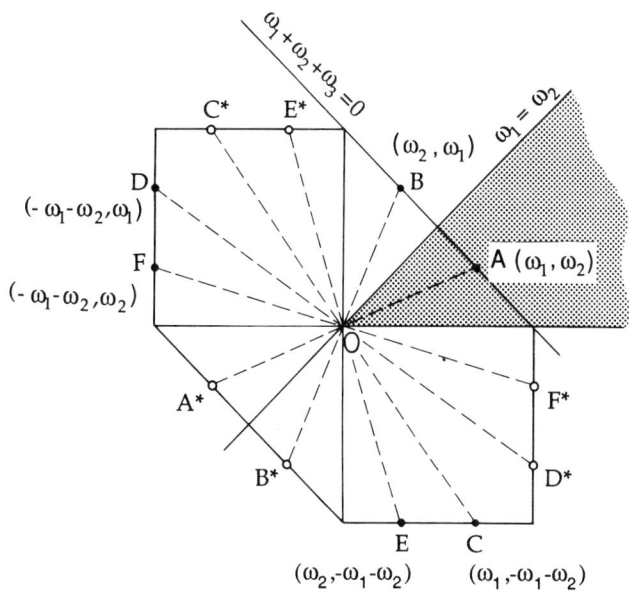

Figure 10.16 Pictorial sketch indicating symmetric characteristics of bispectrum.

in Eq. (10.102), denoted by A, B, C, and so on, are pairwise symmetric with respect to the line $\omega_1 = \omega_2$ shown in the figure. The conjugates, denoted by A*, B*, and so on in the figure, are symmetric with respect to the origin, the point zero in the figure. Because of these symmetries, the bispectrum A in the figure can be considered as a representative of all other 11 bispectra. Therefore, it is sufficient to evaluate the bispectrum only in the fundamental region, which is the octant, namely, the domain defined by $0 \leqslant \omega_2 < \omega_1$ and $0 \leqslant \omega_1 < \infty$. However, it should be remembered in evaluating the bispectrum that the bispectrum is defined based on three frequencies, ω_1, ω_2, and $(\omega_1 + \omega_2)$, as given in Eq. (10.97), and thereby we have the following relationship:

$$0 \leqslant \omega_2 < \omega_1 \leqslant |\omega_1 + \omega_2| \tag{10.104}$$

Figure 10.17 shows an example of the bispectrum of ocean wave record obtained by Hasselmann et al. (1962). The unispectrum of waves is shown

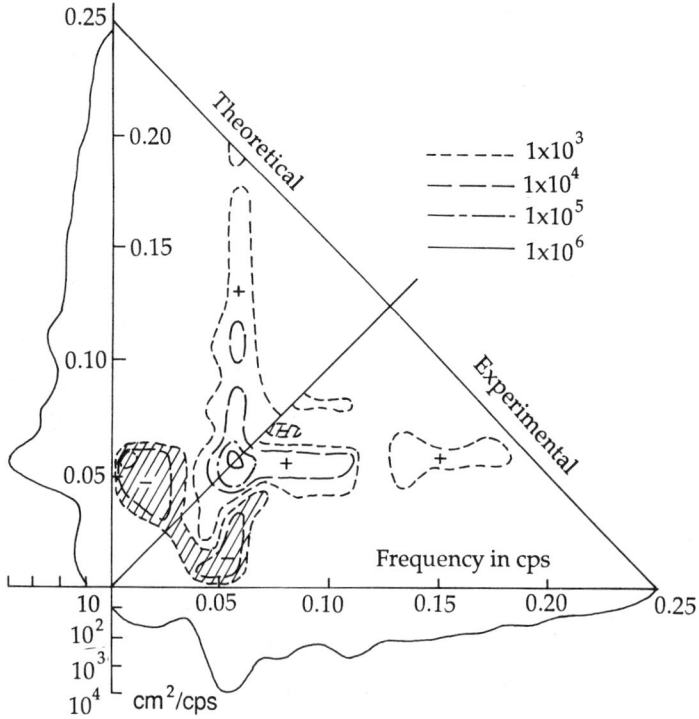

Figure 10.17 Example of bispectrum analysis of an ocean wave record (from Hasselmann et al., 1962).

along the two axes in the figure. Contours are drawn in the figure in cm^3-sec^2 representing the absolute value of the bispectrum per unit frequency band squared. Included also in the upper half of the figure is the bispectrum computed from a theoretical study by Hasselmann et al. (1962) on the nonlinear energy transfer in waves in finite water depth. In the case of perfect agreement between theory and experiment, these two bispectra would be symmetric with respect to the 45° line. It is also noted that if the random process is a perfect Gaussian random process, then the absolute value of the spectrum is zero everywhere except on the 45° line.

It can be seen in the figure that the bispectrum shows very large values in the neighborhood of f_1 and $f_2 = 0.055$, which is the peak frequency of the unispectrum. This implies that the energy interaction between frequencies in the proximity of the peak frequency is very large. It can also be seen that the interaction of energy between frequencies $f_1 = 0.110$ and $f_2 = 0.055$, is rather high.

Bispectral analysis also plays a significant role in finding the stochastic properties of a nonlinear system. In this case, cross-bispectral analysis is made between input and output of the system. This subject will be discussed in detail in Chapter 15.

EXERCISES

10.1 A random process $x(t)$ is given by

$$x(t) = \sin(\omega t + \varepsilon)$$

where ω is a positive constant and ε is a random variable uniformly distributed over the interval 0 and π. Is this a weakly steady-state random process?

10.2 Are the following autocorrelation functions permissible? Give the reason(s).

(a) $R(\tau) = a(\cos \tau + \sin \tau)$, where $a > 0$

(b) $R(\tau) = \{1 + |\tau|\} e^{-|\tau|}$

(c) $R(\tau) = |\tau| e^{-|\tau|}$.

10.3 Obtain the spectral density function of the random process $x(t)$ given by

$$x(t) = \sum_{i=1}^{n} a_i \cos(\omega_i t + \varepsilon_i)$$

where a_i are constants.

10.4 Evaluate the spectral density function for the following autocorrelation function:

$$R(\tau) = \tfrac{1}{4}\{1 + e^{-2a|\tau|}\}$$

10.5 Let a random process $z(t)$ be the sum of two steady-state random processes given by

$$z(t) = ax(t) + by(t)$$

where a and b are constants and $x(t)$ and $y(t)$ are correlated. Obtain the spectral density function of $z(t)$ in terms of the spectral density functions associated with $x(t)$ and $y(t)$.

10.6 The autocorrelation function of a random process $x(t)$ is given by

$$R(\tau) = ae^{-b^2\tau^2}$$

where a and b are constants. Evaluate the spectral density function.

10.7 Let $\dot{x}(t)$ be the time derivative of the random process $x(t)$ given in the previous problem. Evaluate the cross-spectral density function of $x(t)$ and $\dot{x}(t)$.

10.8 The cross-correlation function $R_{xy}(\tau)$ is given by

$$R_{xy}(\tau) = \begin{cases} A & \text{for } 0 < \tau < a \\ 0 & \text{for } \tau = 0, \ a > \tau, \text{ and } \tau < -a \\ -A & \text{for } -a < \tau < 0 \end{cases}$$

Evaluate the cross-spectral density function $S_{xy}(f)$.

CHAPTER 11

Amplitudes and Periods of Gaussian Random Processes

Throughout this chapter we consider a Gaussian random process and obtain statistical characteristics of amplitudes and periods of the process under different conditions. First, methods to statistically predict the magnitude of amplitudes (or crests) from the given spectral density function of a random process are discussed. Then, the joint probability distribution of amplitudes and periods is presented followed by the distribution of periods. The last section discusses the estimation of extreme amplitude (or crest) expected to occur in a specified number of observations or time. In all cases, the probability density functions are developed based on the concept of narrow-band as well as non-narrow-band random processes.

11.1 DISTRIBUTION OF AMPLITUDES FOR NARROW-BAND PROCESSES

11.1.1 *Probability Density Function of Amplitudes*

Let us assume a random process $x(t)$ to be a Gaussian random process with a narrow-band spectrum. From the narrow-band random process assumption defined in Chapter 9, we can write $x(t)$ as

$$x(t) = A(x)\cos\{\omega_0 t + \varepsilon(t)\}$$
$$= A(t)\{\cos\varepsilon(t)\cos\omega_0 t - \sin\varepsilon(t)\sin\omega_0 t\} \quad (11.1)$$

268 AMPLITUDES AND PERIODS OF GAUSSIAN RANDOM PROCESSES

Furthermore, by assuming $x(t)$ to be a normal random process with zero mean and variance σ^2, we can express $x(t)$ in the following form:

$$x(t) = \sum_{n=1}^{\infty} (a_n \cos n\omega t + b_n \sin n\omega t) \quad (11.2)$$

where

$$a_n = \frac{2}{T}\int_0^T x(t)\cos n\omega t\, dt$$

$$b_n = \frac{2}{T}\int_0^T x(t)\sin n\omega t\, dt \quad (11.3)$$

and the coefficients a_n and b_n are normally distributed with zero mean and variance σ^2. By writing $n\omega t$ in Eq. (11.2) as $(n\omega - \omega_0)t + \omega_0 t$, we can express Eq. (11.2) as follows:

$$x(t) = x_c(t)\cos \omega_0 t - x_s(t)\sin \omega_0 t \quad (11.4)$$

where

$$x_c(t) = \sum_{n=1}^{\infty} \{a_n \cos(n\omega - \omega_0)t + b_n \sin(n\omega - \omega_0)t\}$$

$$x_s(t) = \sum_{n=1}^{\infty} \{a_n \sin(n\omega - \omega_0)t - b_n \cos(n\omega - \omega_0)t\} \quad (11.5)$$

From a comparison between Eqs. (11.1) and (11.4), we have

$$x_c(t) = A(t)\cos \varepsilon(t)$$

$$x_s(t) = A(t)\sin \varepsilon(t) \quad (11.6)$$

Let us define the random variables X_c and X_s referring to the value of $x_c(t)$ and $x_s(t)$ at time t. It is noted from Eq. (11.5) that X_c and X_s are normally distributed. In order to obtain the joint probability distribution of X_c and X_s, it is necessary to find the mean and variance of X_c and X_s as well as the correlation between them. The mean of both X_c and X_s is zero, since the random variables a_n and b_n all have zero mean. Thus, we have,

$$E[x_c] = E[x_s] = 0 \quad (11.7)$$

DISTRIBUTION OF AMPLITUDES FOR NARROW-BAND PROCESSES

Next, let us consider the covariance of the two random variables X_c and X_s. Since the mean values of X_c and X_s are zero, we can write the covariance from Eq. (11.5) as

$$\text{Cov}[X_c, X_s] = E[x_c x_s]$$

$$= E\left[\sum_{n=1}^{\infty}\{a_n \cos(n\omega - \omega_0)t + b_n \sin(n\omega - \omega_0)t\}\right.$$

$$\left.\times \sum_{n=1}^{\infty}\{a_m \sin(m\omega - \omega_0)t - b_m \cos(m\omega - \omega_0)t\}\right]$$

$$= \sum_{n=1}^{\infty}\sum_{m=1}^{\infty}\left[E[a_n a_m]\cos(n\omega - \omega_0)t \sin(m\omega - \omega_0)t\right.$$

$$+ E[b_n a_m]\sin(n\omega - \omega_0)t \sin(m\omega - \omega_0)t$$

$$- E[a_n b_m]\cos(n\omega - \omega_0)t \cos(m\omega - \omega_0)t$$

$$\left. - E[b_n b_m]\sin(n\omega - \omega_0)t \cos(m\omega - \omega_0)t\right]$$

(11.8)

The first term of the above equation can be evaluated from Eq. (11.3) as follows:

$$E[a_n a_m] = \frac{4}{T^2}\int_0^T\int_0^T E[x(t)x(s)]\cos n\omega t \cos m\omega s \, dt \, ds$$

$$= \frac{4}{T^2}\int_0^T\int_0^T R(t-s)\cos n\omega t \cos m\omega s \, dt \, ds \quad (11.9)$$

where $R(t-s)$ is the autocorrelation function as defined in Section 10.1. By writing $t - s = \tau$, we have

$$E[a_n a_m] = \frac{4}{T^2}\int_0^T\int_{-s}^{T-s} R(\tau)\cos n\omega(s+\tau)\cos m\omega s \, d\tau \, ds \quad (11.10)$$

Furthermore, by letting $u = s/T$ and $\omega T = 2\pi$,

$$E[a_n a_m] = \frac{4}{T}\int_{-Tu}^{T(1-u)}\int_0^1 R(\tau)\cos n\omega(Tu+\tau)\cos m\omega Tu\, du\, d\tau$$

$$= \frac{4}{T}\int_{-Tu}^{T(1-u)}\int_0^1 R(\tau)\{(\cos 2\pi nu \cos 2\pi mu)\cos n\omega\tau$$

$$- (\sin 2\pi nu \cos 2\pi mu)\sin n\omega\tau\}\, du\, d\tau \quad (11.11)$$

From the properties of the orthogonal function, we have

$$\int_0^1 \cos 2\pi nu \cos 2\pi mu\, du = \begin{cases} 1/2 & \text{if } m = n \\ 0 & \text{otherwise} \end{cases}$$

and

$$\int_0^1 \sin 2\pi nu \cos 2\pi mu\, du = 0 \quad \text{for all } m, n \quad (11.12)$$

Hence, Eq. (11.11) can be reduced to

$$E[a_n a_m] = \frac{2}{T}\int_{-Tu}^{T(1-u)} R(\tau)\cos n\omega\tau\, d\tau \quad \text{if } m = n \quad (11.13)$$

where $n \to \infty$ as $T \to \infty$ such that $n\omega = \omega_n$ remains constant. Thus, we can write

$$\lim_{T \to \infty} TE[a_n a_m] = 2 \lim_{T \to \infty} \int_{-Tu}^{T(1-u)} R(\tau)\cos \omega_n\tau\, d\tau \quad (11.14)$$

On the other hand, from the Wiener–Khintchine theorem given in Theorem 10.2, the integration in Eq. (11.14) for $T \to \infty$ is equal to $\pi S(\omega_n)$, where $S(\omega_n)$ is the spectral density function. Thus, Eq. (11.14) becomes

$$\lim_{T \to \infty} TE[a_n a_m] = \begin{cases} 2\pi S(\omega_n) & \text{for } n = m \\ 0 & \text{otherwise} \end{cases} \quad (11.15)$$

By applying the same procedure used in the derivation of the above equation to the other three terms of Eq. (11.8), we can prove that

$$\lim_{T \to \infty} E[a_n b_m] = 0$$

$$\lim_{T \to \infty} TE[b_n a_n] = 0$$

$$\lim_{T \to \infty} TE[b_n b_m] = \begin{cases} 2\pi S(\omega_n) & \text{for } n = m \\ 0 & \text{otherwise} \end{cases} \quad (11.16)$$

DISTRIBUTION OF AMPLITUDES FOR NARROW-BAND PROCESSES 271

Thus, from Eqs. (11.15) and (11.16), the covariance of two random variables X_c and X_s becomes,

$$\text{Cov}[x_c, x_s] = E[x_c x_s] = 0 \qquad (11.17)$$

This implies that the two random variables X_c and X_s are uncorrelated. Furthermore, by the same procedure as shown in the derivation of Eq. (11.17), the second moments of the random variables X_c and X_s can be evaluated as follows:

$$E[x_c^2] = E[x_s^2] = \sigma^2 = \lim_{T \to \infty} \sum_{n=1}^{\infty} \frac{2\pi}{T} S(\omega_n) = \int_0^{\infty} S(\omega) \, d\omega \qquad (11.18)$$

Thus, from the results obtained in Eqs. (11.17) and (11.18), it can be stated that the two random variables X_c and X_s are statistically independent, each as a normal probability distribution with zero mean and variance σ^2. Hence, the joint probability density function of X_c and X_s can be written as

$$f(x_c, x_s) = \frac{1}{2\pi\sigma^2} e^{-(1/2\sigma^2)(x_c^2 + x_s^2)}, \qquad -\infty < x_c < \infty, -\infty < x_s < \infty$$

$$(11.19)$$

Next, the joint probability density given in Eq. (11.19) will be transformed to the joint probability density function of the amplitude A and the phase ε using the relationship given in Eq. (11.6). (For the transform technique, see Chapter 7.) The transformation yields,

$$f(A, \varepsilon) = \frac{1}{2\pi\sigma^2} e^{-A^2/2\sigma^2}, \qquad 0 \leq A < \infty, 0 \leq \varepsilon \leq 2\pi \qquad (11.20)$$

Then, the marginal probability density function of the amplitude A can be written as

$$f(A) = \int_0^{2\pi} f(A, \varepsilon) \, d\varepsilon = \frac{A}{\sigma^2} e^{-A^2/2\sigma^2}, \qquad 0 \leq A < \infty \qquad (11.21)$$

The above probability density function is the Rayleigh probability distribution with parameter $2\sigma^2$ (see Section 6.3). It can be concluded therefore, that under the assumption of a narrow-band Gaussian process with zero mean and a variance σ^2, the amplitude has Rayleigh distribution with parameter $2\sigma^2$. Note that the variance σ^2 is equal to the area under the spectrum of the random process as proved in Eq. (10.40). Therefore, we can evaluate statistical characteristics of the amplitude of a narrow-band Gaussian random process from knowledge of the spectrum of the process.

The marginal probability density function of the phase ε can be derived from Eq. (11.20) as

$$f(\varepsilon) = \int_0^\infty f(A, \varepsilon)\, dA = \frac{1}{2\pi}, \qquad 0 \leqslant \varepsilon \leqslant 2\pi \qquad (11.22)$$

The above probability density function is the uniform distribution presented in Section 6.4. This implies that the phase of a narrow-band Gaussian random process can take on any value between 0 and 2π.

There is another way to derive the probability density function of the amplitude of a random process that is much more simple than the method presented above. In this approach, however, the probability distribution of amplitude is assumed to be equal to that of the envelope of the profile of random process shown in Figure 11.1. The derivation of the probability density function is as follows:

It was stated in Section 9.2 in connection with a narrow-band random process that the amplitude $A(t)$ and phase $\varepsilon(t)$ both vary slowly with time. Hence, if their time derivatives, $\dot{A}(t)$ and $\dot{\varepsilon}(t)$, are assumed to be negligibly small, the velocity process $\dot{x}(t)$ can be written as

$$\frac{\dot{x}(t)}{\omega_0} = -A(t)\sin\{\omega_0 t + \varepsilon(t)\} \qquad (11.23)$$

From the properties of a Gaussian random process, it can be proved that $\dot{x}(t)$ is statistically independent of $x(t)$ and that it is also a normal process with zero mean and variance $\sigma^2 \omega_0^2$ [see Eq. (10.85)]. Hence, for a given time t, $\dot{x}/\omega_0$ has a normal distribution with zero mean and variance σ^2. Then,

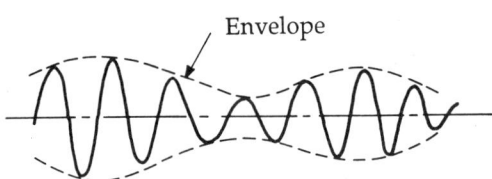

Figure 11.1 Narrow-band random process and its envelope.

the joint probability density function of x and $\dot{x}/\omega_0$ can be written as

$$f(x, \dot{x}/\omega_0) = \frac{1}{2\pi\sigma^2} e^{-(1/2\sigma^2)\{x^2 + (\dot{x}/\omega_0)^2\}},$$

$$-\infty < x < \infty, \; -\infty < \dot{x}/\omega < \infty \quad (11.24)$$

Note that the above joint probability density function is of the same form as that given in Eq. (11.19). Writing x and $\dot{x}/\omega_0$ in polar coordinates, as was done in Eq. (11.6), the marginal probability density function of amplitude A can be obtained by the same procedure as employed in the derivation of Eq. (11.21).

11.1.2 Distribution of Crest-to-Trough Excursions

In this section, the probability density function of the crest-to-trough excursions shown in Figure 11.2 will be derived. The distribution of crest-to-trough excursions may be obtained by assuming that the excursions are equal to twice the amplitude. In other words, by letting the random variable A in Eq. (11.21) representing the amplitude by $H/2$, where H represents the crest-to-trough excursions, the Jacobians for the transformation of random variables from A to H becomes $1/2$. Hence, the probability density function of H can simply be obtained as

$$f(H) = \frac{H}{4\sigma^2} e^{-H^2/8\sigma^2}, \quad 0 \leqslant H < \infty \quad (11.25)$$

The above probability density function is also the Rayleigh probability distribution given in Section 6.3 with the parameters $R = 8\sigma^2$. Although Eq. (11.25) is most commonly used for predicting the statistical characteristics of crest-to-trough excursions of a narrow-band Gaussian random process, it is a general trend that the distribution somewhat overestimates

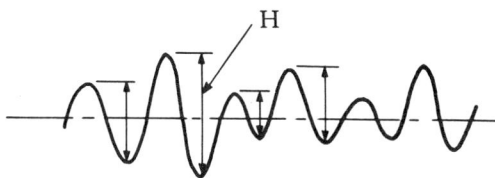

Figure 11.2 Crest-to-trough excursion of narrow-band random process.

the probability of occurrence of large excursion. It should be noted that the Rayleigh distribution is essentially applicable to the envelope of a random process that can be drawn as a pair of symmetrical curves by connecting crests and troughs, respectively, as shown in Figure 11.1.

If the spectrum is strictly narrow-banded, the envelopes are slowly varying so that the maxima and minima of the upper and lower envelopes coincide exactly with crests and troughs. However, this is not usually the case in practice. In evaluating the magnitude of crest-to-trough excursions, it may be more appropriate to consider the sum of the two values of the upper and lower envelopes separated by one-half of the average period rather than twice the magnitude of the value of the upper (or lower) envelope. A method to obtain the probability function for crest-to-trough excursions based on this concept is given by Tayfun (1981). In order to apply his method, it is necessary first to derive the joint probability density function of the two values of the envelope separated by the time interval τ. For this, we express the random process $x(t)$ in two components $x_c(t)$ and $x_s(t)$ as given in Eq. (11.4), and consider the joint probability density function of $x_{c1}(t)$, $x_{s1}(t)$, $x_{c2}(t + \tau)$, and $x_{s2}(t + \tau)$.

$x_{c1}(t)$ and $x_{s1}(t)$ can be expressed exactly by the same formula given in Eq. (11.5), while $x_{c2}(t + \tau)$ and $x_{s2}(t + \tau)$ can be given by

$$x_{c2}(t + \tau) = \sum_{m=1}^{\infty} \{a_m \cos(m\omega - \omega_0)(t + \tau) + b_m \sin(m\omega - \omega_0)(t + \tau)\}$$

$$x_{s2}(t + \tau) = \sum_{m=1}^{\infty} \{a_m \sin(m\omega - \omega_0)(t + \tau) - b_m \cos(m\omega - \omega_0)(t + \tau)\}$$

(11.26)

It then follows that analogous to Eq. (11.6), we have the following relationship between (x_c, x_s) and (A, ε), where A and ε are amplitude and phase, respectively:

$$x_{c1}(t) = A_1(t)\cos \varepsilon_1(t)$$

$$x_{s1}(t) = A_1(t)\sin \varepsilon_1(t)$$

$$x_{c2}(t + \tau) = A_2(t + \tau)\cos \varepsilon_2(t + \tau)$$

$$x_{s2}(t + \tau) = A_2(t + \tau)\sin \varepsilon_2(t + \tau) \qquad (11.27)$$

By writing $x_{c1}(t)$, $x_{s1}(t)$, $x_{c2}(t + \tau)$, and $x_{s2}(t + \tau)$, as x_{c1}, x_{s1}, x_{c2}, and x_{s2}, respectively, those random variables are jointly normally distributed

DISTRIBUTION OF AMPLITUDES FOR NARROW-BAND PROCESSES

with the following properties:

(i) $E[x_{c1}] = E[x_{s1}] = E[x_{c2}] = E[x_{s2}] = 0$ (11.28)

(ii) $E[x_{c1}x_{s1}] = E[x_{c2}x_{s2}] = 0$ (11.29)

(iii) $E[x_{c1}^2] = E[x_{s1}^2] = E[x_{c2}^2] = E[x_{s2}^2] = \sigma^2$ (11.30)

(iv) $E[x_{c1}x_{c2}] = E[x_{s1}x_{s2}]$

$$= \int_0^\infty S(\omega)\cos(\omega - \omega_0)\tau \, d\omega = \rho \quad (11.31)$$

(v) $E[x_{c1}x_{s2}] = -E[x_{s1}x_{c2}]$

$$= \int_0^\infty S(\omega)\sin(\omega - \omega_0)\tau \, d\omega = \lambda \quad (11.32)$$

where $S(\omega)$ is the spectrum of the random process.

As an example, the derivation of the formula for $\text{Cov}[x_{c1}, x_{c2}]$ is given in the following:

Because of the zero means, the covariance of x_{c1} and x_{c2} may be written as $E[x_{c1}, x_{c2}]$. From Eqs. (11.5) and (11.26), we have,

$$E[x_{c1}x_{c2}] = \sum_{n=1}^\infty \sum_{m=1}^\infty [E[a_na_m]\cos(n\omega - \omega_0)t\cos(m\omega - \omega_0)(t+\tau)$$
$$+ E[b_na_m]\sin(n\omega - \omega_0)t\cos(m\omega - \omega_0)(t+\tau)$$
$$+ E[a_nb_m]\cos(n\omega - \omega_0)t\sin(m\omega - \omega_0)(t+\tau)$$
$$+ E[b_nb_m]\sin(n\omega - \omega_0)t\sin(m\omega - \omega_0)(t+\tau)]$$

(11.33)

From the results obtained in Eqs. (11.15) and (11.16), the above equation becomes

$$E[x_{c1}x_{c2}] = \lim_{T\to\infty} \sum_{n=1}^\infty \frac{2\pi}{T} S(\omega_n)\cos(n\omega - \omega_0)\tau$$

$$= \int_0^\infty S(\omega)\cos(\omega - \omega_0)\tau \, d\omega = \rho$$

This is shown in Eq. (11.31). Thus, from Eqs. (11.29) through (11.32), the covariance matrix of x_{c1}, x_{s1}, x_{c2}, and x_{s2} can be written as

$$\Sigma = \begin{pmatrix} \sigma^2 & 0 & \rho & \lambda \\ 0 & \sigma^2 & -\lambda & \rho \\ \rho & -\lambda & \sigma^2 & 0 \\ \lambda & \rho & 0 & \sigma^2 \end{pmatrix} \tag{11.34}$$

Consequently, the joint normal probability density function of the random variables $(x_{c1}, x_{s1}, x_{c2}, x_{s2})$ with zero means can be expressed by using the formulation given in Section 6.1 as follows:

$$f(\mathbf{X}) = \frac{1}{(2\pi)^2 \sqrt{|\Sigma|}} e^{-\frac{1}{2}\mathbf{X}'\Sigma^{-1}\mathbf{X}} \tag{11.35}$$

where

$$\mathbf{X}' = (x_{c1}, x_{s1}, x_{c2}, x_{s2})$$

$$\sqrt{|\Sigma|} = \sigma^4 - \rho^2 - \lambda^2$$

and

$$\Sigma^{-1} = \frac{1}{\sqrt{|\Sigma|}} \begin{pmatrix} \sigma^2 & 0 & -\rho & -\lambda \\ 0 & \sigma^2 & \lambda & -\rho \\ -\rho & \lambda & \sigma^2 & 0 \\ -\lambda & -\rho & 0 & \sigma^2 \end{pmatrix}$$

Then, Eq. (11.35) yields

$$f(x_{c1}, x_{s1}, x_{c2}, x_{s2})$$
$$= \frac{1}{(2\pi)^2 \sqrt{|\Sigma|}} \exp\left\{ -\frac{1}{2\sqrt{|\Sigma|}} \left[\sigma^2 (x_{c1}^2 + x_{s1}^2 + x_{c2}^2 + x_{s2}^2) \right.\right.$$
$$\left.\left. - 2\rho (x_{c1}x_{c2} + x_{s1}x_{s2}) - 2\lambda (x_{c1}x_{s2} - x_{s1}x_{c2}) \right] \right\}$$

(11.36)

We now transform the random variables $(x_{c1}, x_{s1}, x_{c2}, x_{s2})$ to $(A_1, \varepsilon_1, A_2, \varepsilon_2)$ by using the functional relationship given in Eq. (11.27). By applying the transformation method presented in Chapter 7, the joint probability density function of A_1, ε_1, A_2, and ε_2 becomes

$$f(A_1, \varepsilon_1, A_2, \varepsilon_2)$$

$$= \frac{1}{4\pi^2\sqrt{|\Sigma|}} \exp\left\{-\frac{1}{2\sqrt{|\Sigma|}}\left[\sigma^2(A_1^2 + A_2^2) - 2A_1 A_2\{\rho\cos(\varepsilon_2 - \varepsilon_1)\right.\right.$$

$$\left.\left. + \lambda \sin(\varepsilon_2 - \varepsilon_1)\}\right]\right\}$$

$$0 \leqslant A_1 < \infty, 0 \leqslant A_2 < \infty, 0 \leqslant \varepsilon_1 \leqslant 2\pi, 0 \leqslant \varepsilon_2 \leqslant 2\pi \quad (11.37)$$

The joint probability distribution of amplitudes A_1 and A_2 can be obtained by integrating Eq. (11.37) with respect to ε_1 and ε_2. For this, we may write

$$\rho\cos(\varepsilon_2 - \varepsilon_1) + \lambda\sin(\varepsilon_2 - \varepsilon_1) = \sqrt{\rho^2 + \lambda^2}\cos\left(\varepsilon_2 - \varepsilon_1 - \tan^{-1}\frac{\lambda}{\rho}\right)$$

$$(11.38)$$

Then, we have

$$f(A_1, A_2)$$

$$= \frac{A_1 A_2}{4\pi^2\sqrt{|\Sigma|}} \exp\left[-\frac{1}{2\sqrt{|\Sigma|}}\sigma^2(A_1^2 + A_2^2)\right]$$

$$\times \int_0^{2\pi}\int_0^{2\pi} \exp\left\{\frac{A_1 A_2}{\sqrt{|\Sigma|}}\sqrt{\rho^2 + \lambda^2}\cos\left(\varepsilon_1 - \varepsilon_2 - \tan^{-1}\frac{\lambda}{\rho}\right)\right\} d\varepsilon_1 d\varepsilon_2$$

$$(11.39)$$

In carrying out the double integration in the above equation, let us first consider transformation of the two random variables ε_1 and ε_2 to a single random variable $(\varepsilon_1 - \varepsilon_2)$. By applying the transformation techniques shown in Section 7.2, it can easily be shown that the double integral in Eq. (11.39) is equivalent to the constant 2π times a single integral with respect to $(\varepsilon_1 - \varepsilon_2)$. Furthermore, $\cos(\varepsilon_2 - \varepsilon_1 - \tan^{-1}\lambda/\rho)$ is periodic. Hence, Eq.

(11.39) can be written as

$$f(A_1, A_2) = \frac{A_1 A_2}{\sqrt{|\Sigma|}} \exp\left[-\frac{1}{2\sqrt{|\Sigma|}} \sigma^2 (A_1^2 + A_2^2)\right]$$

$$\times \int_0^{2\pi} \frac{1}{2\pi} \exp\left(\frac{A_1 A_2}{\sqrt{|\Sigma|}} \sqrt{\rho^2 + \lambda^2} \cos\phi\right) d\phi$$

$$= \frac{A_1 A_2}{\sqrt{|\Sigma|}} \exp\left[-\frac{1}{2\sqrt{|\Sigma|}} \sigma^2 (A_1^2 + A_2^2)\right] I_0\left(\frac{A_1 A_2}{\sqrt{|\Sigma|}} \sqrt{\rho^2 + \lambda^2}\right) \quad (11.40)$$

where

$$\phi = \varepsilon_2 - \varepsilon_1 - \tan^{-1} \lambda/\rho,$$

$I_0(\ ) = $ modified Bessel function of zero order.

This is the joint probability density function of two amplitudes A_1 and A_2 that are separated by the time interval τ (see Rice, 1945).

We now consider the sum of amplitudes A_1 and A_2 that are separated by one-half the period, denoted by $T/2$. Here the period, T, is also a random variable that has the probability density function $f(T)$. By writing the crest-to-trough excursion separated by the time interval $T/2$ as $H = A_1 + A_2$, we can derive the probability density function of H from Eq. (11.40) with $\tau = T/2$ in Eqs. (11.31) and (11.32). Since this probability density function cannot be expressed in closed form, it may be written as

$$f(H|\tau = T/2) = \int_0^H f_{A_1 A_2}(H - A_2, A_2; \tau = T/2) \, dA_2 \quad (11.41)$$

where, $f_{A_1 A_2}(\ )$ is the joint probability density function of A_1 and A_2 given by Eq. (11.40).

By allowing the period T to take all possible values from 0 to ∞, the probability density function of crest-to-trough excursions can be written as

$$f(H) = \int_0^\infty \int_0^H f(T) f_{A_1 A_2}(H - A_2, A_2; \tau = T/2) \, dA_2 \, dT \quad (11.42)$$

where $f(T)$ is the probability density function of the period.

Computation of the probability density function given in Eq. (11.42), however, is extremely complicated. To simplify the problem, Tayfun (1981) assumes that the spectrum is concentrated around the mean frequency $\bar{\omega}$,

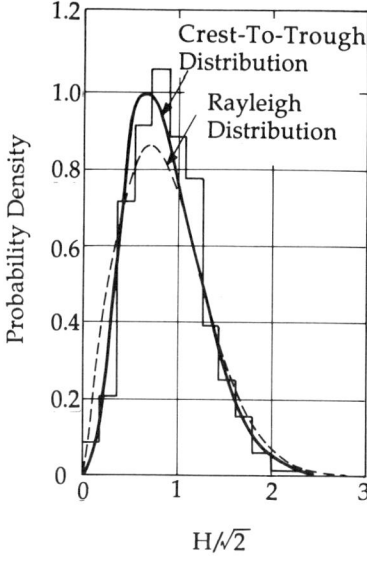

Figure 11.3 Comparison between histogram of crest-to-trough excursion, Rayleigh distribution, and the probability density function given in Eq. (11.42) (from Tayfun, 1981).

and thereby the probability density function of the period is sharply centered around the frequency $2\pi/\bar{\omega}$. Then, Eq. (11.40) can be substantially simplified in that $\lambda(\tau) = 0$, and hence $f(H)$ becomes

$$f(H) = 2\int_0^H f(H - A_2, A_2; \pi/\bar{\omega})\, dA_2 \qquad (11.43)$$

As an example of the application of this method, Figure 11.3 shows a comparison between the Rayleigh probability distribution and the probability density function given in Eq. (11.42) computed from the wave spectrum obtained from data measured in the open ocean. Included also in the figure is the histogram of wave height constructed from the measurements. It can be seen in the figure that the crest-to-trough distribution computed by Eq. (11.42) gives a higher probability density around the mean value than the Rayleigh distribution and that it agrees well with the observed data.

11.1.3 *Envelope Process*

It was shown in the previous section that the envelope of the profile of a random process plays a significant role in evaluating statistical properties of the process when the process has a narrow-band spectrum. In fact, many practical problems regarding the probability of occurrence of a random

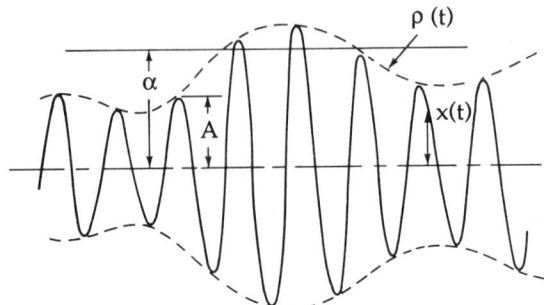

Figure 11.4 Envelope process $\rho(t)$.

process exceeding a specified level can be solved through consideration of the level crossing of the envelope. For example, evaluation of statistical properties associated with the group phenomenon of a random process in which a sequence of large amplitudes having nearly equal periods exceeds a certain level (see Figure 11.4) is extremely important in many engineering fields. The statistical properties of the group phenomenon are often evaluated through the envelope process, $\rho(t)$ in Figure 11.4, instead of dealing directly with the amplitude A. Therefore, accurate determination of the envelope of a stochastic process is required for the analysis.

The following discussion pertains to the mathematical relationship between a random process $x(t)$ and its envelope process $\rho(t)$ with the condition that $x(t)$ is narrow-banded.

We may write a narrow-band random process $x(t)$ as given in Eq. (11.4). That is,

$$x(t) = \rho(t)\cos\{\omega_0 t + \varepsilon(t)\}$$
$$= x_c(t)\cos \omega_0 t - x_s(t)\sin \omega_0 t$$

where

$$x_c(t) = \rho(t)\cos \varepsilon(t)$$
$$x_s(t) = \rho(t)\sin \varepsilon(t)$$

Let us express $x(t)$ in the form of a complex quantity as follows:

$$x(t) = \text{Re}\{x_c(t) + ix_s(t)\}e^{i\omega_0 t}$$
$$= \text{Re}\{\rho(t)e^{i\omega_0 t}\} = \text{Re}\{z(t)\} \qquad (11.44)$$

where

$$z(t) = \rho(t)e^{i\omega_0 t}$$

Next, let us consider the Hilbert transform of $x(t)$, denoted by $\tilde{x}(t)$. It is given by (see Appendix B)

$$\tilde{x}(t) = x_c(t)\sin \omega_0 t + x_s(t)\cos \omega_0 t \qquad (11.45)$$

Then, from Eqs. (11.44) and (11.45), we can derive

$$x(t) + i\tilde{x}(t) = \{x_c(t) + ix_s(t)\}e^{i\omega_0 t} = z(t) \qquad (11.46)$$

Thus, $z(t)$ can be evaluated from the record $x(t)$ and its Hilbert transform with the aid of Eq. (11.46), and therefrom, the envelope process $\rho(t)$ can be obtained as

$$\rho(t) = z(t)e^{-i\omega_0 t} \qquad (11.47)$$

11.1.4 Significant Value and Extreme Value

The significant value is a statistical measure that is often used in engineering problems to represent the severity of a random process. It is defined as follows:

Definition 11.1. The average of the highest one-third of the amplitudes (or the crest-to-trough excursions) of a random process is called the *significant amplitude* (or *significant crest-to-trough excursion*).

For example, the significant crest-to-trough excursion of a random process is evaluated by obtaining the one-third largest recorded excursions and by averaging the magnitude of those excursions. Therefore, in principle, the significant value is determined from the observed data; however, it can be evaluated from the spectral density function of a Gaussian random process assuming that the random process has a narrow-band spectrum and that the crest-to-trough excursions follow the Rayleigh probability distribution.

Let us write the probability density function of the Rayleigh distribution as

$$f(x) = \frac{2x}{R}e^{-x^2/R}, \quad 0 \leq x \leq \infty \qquad (11.48)$$

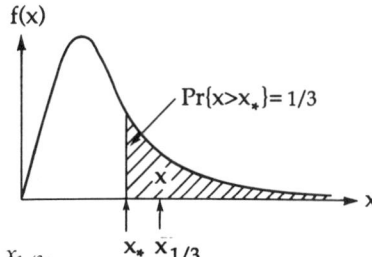

Figure 11.5 Definition of significant value $x_{1/3}$.

where x represents the crest-to-trough excursion (double amplitude) of a Gaussian random process with the variance σ^2. R is equal to $8\sigma^2$ as shown in Eq. (11.25). From the definition of the significant value, denoted by $x_{1/3}$, it is essentially equal to the center of gravity of the area containing the highest one-third values of the probability density function (see Figure 11.5). By denoting the lower limit of the highest one-third of the probability density function as x_*, we have

$$\Pr\{X \geq x_*\} = \int_{x_*}^{\infty} \frac{2x}{R} e^{-x^2/R}\, dx = \frac{1}{3} \qquad (11.49)$$

giving

$$x_* = \sqrt{R(\ln 3)} = 1.048\sqrt{R} \qquad (11.50)$$

The significant value, $x_{1/3}$, is then obtained by taking the moment about the origin so that

$$\frac{1}{3} x_{1/3} = \int_{x_*}^{\infty} x f(x)\, dx = x_* e^{-x_*^2/R} + \sqrt{\pi R} \left\{ 1 - \Phi\left(\sqrt{\frac{2}{R}}\, x_*\right) \right\} \qquad (11.51)$$

Substituting x_* of Eq. (11.50) in Eq. (11.51), we have

$$x_{1/3} = \left[\sqrt{\ln 3} + 3\sqrt{\pi}\left\{1 - \Phi\left(\sqrt{2(\ln 3)}\right)\right\}\right]\sqrt{R} \sim 1.42\sqrt{R} \qquad (11.52)$$

Since the parameter R of the Rayleigh distribution for peak-to-trough excursions is 8 times the variance, the significant value becomes

$$x_{1/3} = 1.42\sqrt{8m_0} = 4.01\sqrt{m_0} \qquad (11.53)$$

where m_0 = area under the spectral density function.

The formula for evaluating the significant value given in Eq. (11.52) can be generalized to evaluate the average of the highest $1/n$ observations,

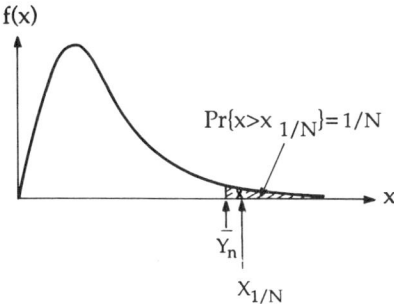

Figure 11.6 Definition of $\bar{y}_n$ and $x_{1/n}$.

denoted by $X_{1/n}$, based on the Rayleigh distribution. That is,

$$x_{1/n} = \left[\sqrt{\ln n} + n\sqrt{\pi}\left\{1 - \Phi(\sqrt{2\ln n})\right\}\right]\sqrt{R} \qquad (11.54)$$

The second term in Eq. (11.54) becomes negligibly small in comparison with the first term for large n, and hence Eq. (11.54) approximately becomes

$$x_{1/n} = \sqrt{\ln n}\sqrt{R} = 2\sqrt{2\ln n}\sqrt{m_0} \qquad (11.55)$$

It is noted that the formula given in Eq. (11.55) agrees with the probable extreme value expected in n observations (where n is large) derived in Chapter 8. We may recognize a contradiction that although the formula for evaluating the probable extreme value in n observations, denoted by $\bar{Y}_n$, in Chapter 8, agrees with that for the average of the highest $1/n$ observations, $X_{1/n}$, derived in Eq. (11.55), the definitions of these two parameters differ. That is, $\bar{Y}_n$ is defined as the value for which the probability of being exceeded is $1/n$ as shown in Figure 11.6. On the other hand, $X_{1/n}$ is defined as the average of the highest $1/n$ observations. This contradiction, however, is admissible for the present case since we are concerned with a large sample size. It is not difficult to see that $\bar{Y}_n$ and $X_{1/n}$ are approximately equal for large n.

11.2 DISTRIBUTION OF MAXIMA FOR NON-NARROW-BAND PROCESSES

In this section, the assumption of a narrow-band spectrum is no longer adhered to in predicting amplitudes of a random process. As shown in Figure 11.1, the narrow-band random process has a single peak during a

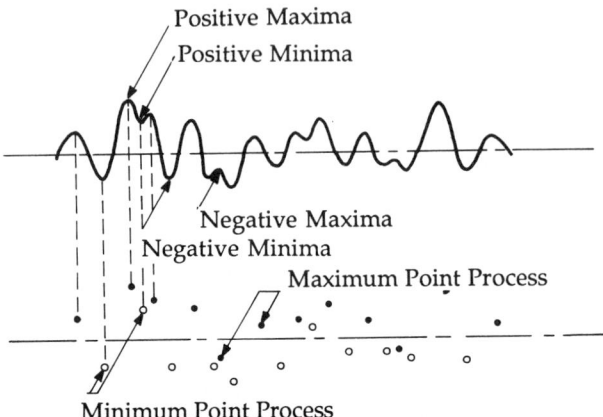

Figure 11.7 Explanatory sketch of maximum and minimum point processes.

half cycle, and the peak determines the amplitude. For a random process for which the spectrum is not narrow banded, there can be two or more local peaks occurring during a half cycle determined from successive zero crossing as illustrated in Figure 11.7.

As can be seen in the figure, the local crests and troughs located on the positive side (above the mean value) of a record are defined as the *positive maxima* and the *positive minima*, respectively, while those located on the negative side (below the mean value) are defined as the *negative maxima* and the *negative minima*, respectively. These local crests and troughs form a stationary random point process, called the *maximum point process* and the *minimum point process*. Table 11.1 summarizes the conditions for these point processes.

If a random process is assumed to be Gaussian with zero mean, then the maximum point process and the minimum point process follow the same probability law. That is, the positive maxima and the negative minima follow the same probability law, as do also the negative maxima and the positive minima.

11.2.1 *Expected Number of Maxima*

Prior to deriving the probability distribution of the maxima, let us first consider the expected (average) number of positive maxima exceeding a certain level per unit time. As will be shown in the next section, the probability density function of the positive maxima of a Gaussian random

DISTRIBUTION OF MAXIMA FOR NON-NARROW-BAND PROCESSES 285

Table 11.1
Conditions for the Maximum and Minimum Point Processes

	$x(t) \geq 0$ and $\dot{x}(t)=0$	$x(t) \leq 0$ and $\dot{x}(t)=0$	
$\ddot{x}(t)<0$ (Crests)	Positive maxima	Negative maxima	Maximum point process
$\ddot{x}(t)>0$ (Troughs)	Positive minima	Negative minima	Minimum point process

process $x(t)$ can be derived based on information of the expected number of positive maxima per unit time.

Let Z be a random variable representing the local positive maxima (see Figure 11.8), and let N_ζ be the number of times that the maxima Z exceeds a specified level ζ per unit time. We will obtain the expected value of N_ζ, denoted by $\overline{N}_\zeta$, assuming that the random process $x(t)$ is a Gaussian random process but its spectrum is no longer narrow banded. It is further assumed that time derivatives of the random process $x(t)$ exist.

By writing the joint probability density function of displacement, velocity, and acceleration at time t as $f(x, \dot{x}, \ddot{x})$, we have

$$f(x, \dot{x}, \ddot{x})\, dx\, d\dot{x}\, d\ddot{x}$$
$$= \Pr\{x < X \leq x + dx, \dot{x} < \dot{X} \leq \dot{x} + d\dot{x}, \ddot{x} < \ddot{X} \leq \ddot{x} + d\ddot{x}\} \quad (11.56)$$

If the random variable X represents the maxima, then the velocity $\dot{X} = 0$ and the acceleration $\ddot{X} < 0$. Hence, the quantity $f(x, 0, \ddot{x})\, dx\, d\ddot{x}$ is equal to the probability that X is between x and $x + dx$ with negative acceleration

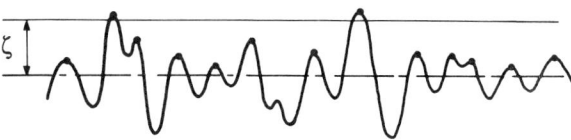

Figure 11.8 Positive maxima and a specified level ζ.

between $\ddot{x}$ and $\ddot{x} + d\ddot{x}$. This implies that for a unit total time, $f(x, 0, \ddot{x}) \, dx \, d\ddot{x}$ represents that fraction of time each peak lies within $(x, x + dx)$ with negative acceleration $(\ddot{x}, \ddot{x} + d\ddot{x})$.

Since the acceleration can assume any value between $(-\infty, 0)$ and since we are considering the positive maxima to be exceeding a certain level ζ, the expected (average) number of maxima per unit time can be written as

$$\overline{N}_\zeta = \int_\zeta^\infty \int_{-\infty}^0 |\ddot{x}| f(x, 0, \ddot{x}) \, d\ddot{x} \, dx \qquad (11.57)$$

Equation (11.57) was derived by Rice (1944, 1945). Middleton (1960), on the other hand, derived Eq. (11.57) by applying the unit impulse function. Since Middleton's approach is unique and may be applied for many other problems in stochastic processes, his method is also outlined below.

For the portion of the random process exceeding a certain level ζ, consider the Heaviside unit step function that can be written as $\mathsf{u}\{x(t) - \zeta\}$, where $\mathsf{u}\{\ \}$ denoted by the unit step function. That is,

$$\mathsf{u}\{x(t) - \zeta\} = \begin{cases} 1 & \text{for } x(t) - \zeta > 0 \\ 1/2 & \text{for } x(t) - \zeta = 0 \\ 0 & \text{for } x(t) - \zeta < 0 \end{cases} \qquad (11.58)$$

As illustrated in Figure 11.9, $\mathsf{u}\{x(t) - \zeta\}$ is a step function that becomes unity only in the domain where $x(t) > \zeta$. Next, consider the product $\ddot{x}(t) \cdot \delta\{x(t)\}$. Following the definition of the unit impulse function, the products are unit impulses with alternate signs. The product is negative when the process $x(t)$ peaks. Thus, the product $\ddot{x}(t) \cdot \delta\{x(t)\} \cdot \mathsf{u}\{x(t) - \zeta\}$ represents unit impulses with negative signs indicating the peaks above the level ζ.

Since the time integration of this quantity gives the number of positive maxima (peaks) above the level ζ, the quantity $\ddot{x}(t) \cdot \delta\{x(t)\} \cdot \mathsf{u}\{x(t) - \zeta\}$ can be considered to represent the number of maxima denoted by N_ζ, exceeding a specified level ζ per unit time. Therefore, the expected number

Figure 11.9 Unit step function $\mathsf{u}\{x(t) - \zeta\}$.

DISTRIBUTION OF MAXIMA FOR NON-NARROW-BAND PROCESSES

$\overline{N}_\zeta$ becomes

$$\overline{N}_\zeta = E\big[\ddot{x}(t)\cdot\delta\{\dot{x}(t)\}\mathsf{u}\{x(t)-\zeta\}\big]$$

$$= \int_0^\infty \int_0^\infty \int_{-\infty}^0 |\ddot{x}(t)|\delta\{\dot{x}(t)\}\mathsf{u}\{x(t)-\zeta\}f(x,\dot{x},\ddot{x})\,d\ddot{x}\,d\dot{x}\,dx \quad (11.59)$$

By using the following properties of the unit step function and the unit impulse function,

$$\int_0^\infty \mathsf{u}\{x(t)-\zeta\}f(x,\dot{x},\ddot{x})\,dx = \int_\zeta^\infty f(x,\dot{x},\ddot{x})\,dx \quad (11.60)$$

and

$$\int_0^\infty \delta\{\dot{x}(t)\}f(x,\dot{x},\ddot{x})\,d\dot{x} = f(x,0,\ddot{x}) \quad (11.61)$$

we have

$$\overline{N}_\zeta = \int_\zeta^\infty \int_{-\infty}^0 |\ddot{x}|f(x,0,\ddot{x})\,d\ddot{x}\,dx$$

This is the formula given in Eq. (11.57) but derived through a different approach.

Now, the joint probability density function $f(x,\dot{x},\ddot{x})$ with zero mean can be written as follows (see Section 6.1.3):

$$f(\mathbf{X}) = \frac{1}{(2\pi)^{3/2}\sqrt{|\Sigma|}}e^{-\frac{1}{2}\mathbf{X}'\Sigma^{-1}\mathbf{X}} \quad (11.62)$$

where

$$\mathbf{x}' = (x,\dot{x},\ddot{x})$$

Σ = covariance matrix of $x, \dot{x}, \ddot{x}$.

The elements of the covariance matrix Σ can be obtained by the same procedure as Eq. (11.8). That is, a narrow-band Gaussian random process with zero mean $x(t)$ can be presented as follows [see Eq. (11.4)]:

$$x(t) = x_c(t)\cos\omega_0 t - x_s(t)\sin\omega_0 t$$

Then, we may write the velocity process $\dot{x}(t)$ and acceleration process $\ddot{x}(t)$ as

$$\dot{x}(t) = -\omega_0\{x_c(t)\sin\omega_0 t + x_s(t)\cos\omega_0 t\}$$
$$\ddot{x}(t) = \omega_0^2\{-x_c(t)\cos\omega_0 t + x_s(t)\sin\omega_0 t\} \quad (11.63)$$

Since the mean values of $x(t)$ and $\dot{x}(t)$ are zero, we can write the covariance of $x(t)$ and $\dot{x}(t)$ as

$$\text{Cov}[x(t), \dot{x}(t)] = E[x(t)\dot{x}(t)]$$
$$= \omega_0\big(-E[x_c^2]\cos\omega_0 t \sin\omega_0 t + E[x_c x_s]\sin^2\omega_0 t$$
$$- E[x_c x_s]\cos^2\omega_0 t + E[x_s^2]\sin\omega_0 t \cos\omega_0 t\big) \quad (11.64)$$

As shown in Eqs. (11.8) through (11.18), we have

$$E[x_c x_s] = 0, \; E[x_c^2] = E[x_s^2] = \text{Var}[x(t)] = \sigma^2$$

Hence, it can be proved from Eq. (11.64) that

$$\text{Cov}[x(t), \dot{x}(t)] = 0 \quad (11.65)$$

It can also be proved that

$$\text{Cov}[\dot{x}(t), \ddot{x}(t)] = 0$$
$$\text{Cov}[x(t), \ddot{x}(t)] = -\omega^2 \sigma_0^2 \quad (11.66)$$

If we express the elements of the covariance matrix in terms of the moments of the spectral density function, then the covariance of a narrow-band Gaussian random processes $x(t)$, $\dot{x}(t)$, and $\ddot{x}(t)$ can be written as

$$\Sigma = \begin{pmatrix} m_0 & 0 & -m_2 \\ 0 & m_2 & 0 \\ -m_2 & 0 & m_4 \end{pmatrix} \quad (11.67)$$

where $m_j = \int \omega^j S(\omega)\,d\omega$.

DISTRIBUTION OF MAXIMA FOR NON-NARROW-BAND PROCESSES   289

We have $\mathbf{x}' = (x, 0, \ddot{x})$ for the present problem, and thereby $\mathbf{x}'\Sigma^{-1}\mathbf{x}$ can be evaluated from Eq. (11.67) as

$$\mathbf{X}'\Sigma^{-1}\mathbf{X} = \frac{m_4 x^2 + 2m_2 x\ddot{x} + m_0 \ddot{x}^2}{\Delta} \qquad (11.68)$$

where $\Delta = m_0 m_4 - m_2^2$.

Hence, the joint probability density function $f(x, 0, \ddot{x})$ becomes

$$f(x, 0, \ddot{x}) = \frac{1}{(2\pi)^{3/2}\sqrt{m_2 \Delta}} e^{-(m_4 x^2 + 2m_2 x\ddot{x} + m_0 \ddot{x}^2)/2\Delta} \qquad (11.69)$$

Then, the integration with respect to $\ddot{x}$ in Eq. (11.57) can be obtained as

$$\int_{-\infty}^{0} |\ddot{x}| f(x, 0, \ddot{x}) \, d\ddot{x}$$

$$= \frac{1}{2\pi\sqrt{m_0}} \sqrt{\frac{m_4}{m_2}} \left[\frac{\varepsilon}{\sqrt{2\pi}} \exp\left\{ -\frac{1}{2\varepsilon^2}\left(\frac{x}{\sqrt{m_0}}\right)^2 \right\} \right.$$

$$+ \sqrt{1-\varepsilon^2}\left(\frac{x}{\sqrt{m_0}}\right) \exp\left\{ -\frac{1}{2}\left(\frac{x}{\sqrt{m_0}}\right)^2 \right\}$$

$$\left. \times \left\{ 1 - \Phi\left(-\frac{\sqrt{1-\varepsilon^2}}{\varepsilon}\frac{x}{\sqrt{m_0}}\right) \right\} \right] \qquad (11.70)$$

where

$$\varepsilon = \sqrt{1 - \frac{m_2^2}{m_0 m_4}} \qquad \text{(see Equation 10.27)}$$

$$\Phi(u) = \frac{1}{\sqrt{2\pi}} \int_{-\infty}^{u} e^{-u^2/2} \, du$$

Further integration of Eq. (11.70) with respect to x is somewhat intricate. However, if we limit our discussion to a random process for which ε is less than 0.9, and consider the level ζ to be large, then we can assume

$$\Phi\left(-\frac{\sqrt{1-\varepsilon^2}}{\varepsilon}\frac{x}{\sqrt{m_0}}\right) \sim 0 \qquad (11.71)$$

With the aid of Eq. (11.71), the expected number of maxima per unit time given in Eq. (11.57) can be evaluated as follows:

$$\overline{N}_\zeta = \int_\zeta^\infty \int_{-\infty}^0 |\ddot{x}| f(x,0,\ddot{x})\, d\ddot{x}\, dx$$

$$= \frac{1}{2\pi}\sqrt{\frac{m_4}{m_2}}\left[\sqrt{1-\varepsilon^2}\, e^{-\zeta^2/2m_0} + \varepsilon^2\left\{1 - \Phi\left(\zeta/\sqrt{m_0}\right)\right\}\right]$$

$$= \frac{1}{2\pi}\sqrt{\frac{m_2}{m_0}}\, e^{-\zeta^2/2m_0} + \frac{1}{2\pi}\sqrt{\frac{m_4}{m_2}}\, \varepsilon^2\left\{1 - \Phi\left(\zeta/\sqrt{m_0}\right)\right\} \quad (11.72)$$

It is noted that for a narrow-band random process, namely for $\varepsilon = 0$, the first term gives the average number of maxima per unit time.

11.2.2 Probability Distribution of Maxima

The probability density function of the maxima of a random process can now be derived by applying the results obtained in the previous section. We first consider the derivation of the probability density function of the local positive maxima, denoted by Z.

The probability that Z will exceed a certain level ζ is equivalent to the average value of the ratio of the number of positive maxima above ζ per unit time, denoted by N_ζ, to the total number of positive maxima per unit time, denoted by N_{x+}. That is,

$$\Pr\{Z > \zeta\} = 1 - F(\zeta) = E\left[\frac{N_\zeta}{N_{x+}}\right] \quad (11.73)$$

where, $E[\ ]$ stands for the expected value, and $F(\zeta)$ is the cumulative distribution function of Z.

If we assume that (N_ζ/N_{x+}) and N_{x+} are statistically independent, then Eq. (11.73) becomes as follows (see Exercise 3.3):

$$\Pr\{Z > \zeta\} = \frac{E[N_\zeta]}{E[N_{x+}]} = \frac{\overline{N}_\zeta}{\overline{N}_{x+}} \quad (11.74)$$

The equation for the average number $\overline{N}_\zeta$ is given Eq. (11.57). Using a similar expression, we can write the equation for the average number $\overline{N}_{x+}$ as

$$\overline{N}_{x+} = \int_0^\infty \int_{-\infty}^0 |\ddot{x}| f(x,0,\ddot{x})\, d\ddot{x}\, dx \quad (11.75)$$

Then, the probability density function of the positive maxima, $f(\zeta)$, can be written from Eq. (11.74) with the aid of Eqs. (11.57) and (11.75) as

$$f(\zeta) = \frac{d}{d\zeta}\left\{1 - \frac{\overline{N}_\zeta}{\overline{N}_{x+}}\right\} = \frac{d}{d\zeta}\left\{1 - \frac{\int_\zeta^\infty \int_{-\infty}^0 |\ddot{x}| f(x, 0, \ddot{x})\, d\ddot{x}\, dx}{\int_0^\infty \int_{-\infty}^0 |\ddot{x}| f(x, 0, \ddot{x})\, d\ddot{x}\, dx}\right\}$$

$$= \frac{d}{d\zeta}\left\{1 - \frac{\int_{-\infty}^0 |\ddot{x}| F_x(\infty, 0, \ddot{x})\, d\ddot{x} - \int_{-\infty}^0 |\ddot{x}| F_x(\zeta, 0, \ddot{x})\, d\ddot{x}}{\int_0^\infty \int_{-\infty}^0 |\ddot{x}| f(x, 0, \ddot{x})\, d\ddot{x}\, dx}\right\}$$

$$= \frac{\int_{-\infty}^0 |\ddot{x}| f(\zeta, 0, \ddot{x})\, d\ddot{x}}{\int_0^\infty \int_{-\infty}^0 |\ddot{x}| f(x, 0, \ddot{x})\, d\ddot{x}\, dx} \tag{11.76}$$

where $F_x(x, 0, \ddot{x})$ is the marginal cumulative distribution function of $f(x, 0, \ddot{x})$.

The numerator in Eq. (11.76) is obtained in Eq. (11.70). In order to evaluate the denominator, we first carry out the integration of Eq. (11.69) with respect to x and then perform in the integration with respect to $\ddot{x}$. That is, by taking into account that $\ddot{x} < 0$ for the present case, Eq. (11.75) yields

$$\overline{N}_{x+} = \int_0^\infty \int_{-\infty}^0 |\ddot{x}| f(x, 0, \ddot{x})\, d\ddot{x}\, dx$$

$$= \frac{1}{(2\pi)\sqrt{m_2 m_4}} \int_{-\infty}^0 |\ddot{x}| e^{-\ddot{x}^2/2m_4}\left\{1 - \Phi\left(\frac{m_2}{\sqrt{m_4 \Delta}} \ddot{x}\right)\right\} d\ddot{x}$$

$$= \frac{1}{(2\pi)\sqrt{m_2 m_4}}\left\{\int_0^\infty \ddot{x} e^{-\ddot{x}^2/2m_4}\, d\ddot{x} + \int_{-\infty}^0 \ddot{x} e^{-\ddot{x}^2/2m_4} \Phi\left(\frac{m_2}{\sqrt{m_4 \Delta}} \ddot{x}\right) d\ddot{x}\right\}$$

$$= \frac{1}{4\pi\sqrt{m_2 m_4}}\left(m_4 + m_2\sqrt{\frac{m_4}{m_0}}\right) = \frac{1}{4\pi}\left(\sqrt{\frac{m_2}{m_0}} + \sqrt{\frac{m_4}{m_2}}\right)$$

$$= \frac{1}{4\pi}\left(\frac{1 + \sqrt{1 - \varepsilon^2}}{\sqrt{1 - \varepsilon^2}}\right)\sqrt{\frac{m_2}{m_0}} \tag{11.77}$$

Thus, from Eqs. (11.70) and (11.77), the probability density function for the positive maxima given in Eq. (11.76) is found to be

$$f(\zeta) = \frac{2/\sqrt{m_0}}{1 + \sqrt{1 - \varepsilon^2}} \left[\frac{\varepsilon}{\sqrt{2\pi}} \exp\left\{ -\frac{1}{2\varepsilon^2}\left(\frac{\zeta}{\sqrt{m_0}}\right)^2 \right\} \right.$$

$$\left. + \sqrt{1 - \varepsilon^2} \left(\frac{\zeta}{\sqrt{m_0}}\right) \exp\left\{ -\frac{1}{2}\left(\frac{\zeta}{\sqrt{m_0}}\right)^2 \right\} \Phi\left(\frac{\sqrt{1 - \varepsilon^2}}{\varepsilon} \frac{\zeta}{\sqrt{m_0}}\right) \right]$$

$$0 \leqslant \zeta < \infty \quad (11.78)$$

The above probability density function may be expressed in dimensionless form by defining a random variable, $\xi = \zeta/\sqrt{m_0}$. That is,

$$f(\xi) = \frac{2}{1 + \sqrt{1 - \varepsilon^2}} \left[\frac{\varepsilon}{\sqrt{2\pi}} e^{-\xi^2/2\varepsilon^2} + \sqrt{1 - \varepsilon^2}\, \xi e^{-\xi^2/2} \Phi\left(\frac{\sqrt{1 - \varepsilon^2}}{\varepsilon} \xi\right) \right]$$

$$(11.79)$$

Here we may consider the probability density function for two special cases, namely, $\varepsilon = 0$ and $\varepsilon = 1$. For a random process $x(t)$ with $\varepsilon = 0$ representing a narrow-band random process, the probability density function of the positive maxima is obtained from Eq. (11.79) as

$$f(\xi) = \xi e^{-\xi^2/2}, \quad 0 \leqslant \xi < \infty \quad (11.80)$$

which is the Rayleigh probability density function derived in Eq. (11.21) expressed in dimensional form by letting $\xi = A/\sigma$. On the other hand, for a random process with $\varepsilon = 1$ representing a wide-band random process, the probability density function of the positive maxima becomes

$$f(\xi) = \sqrt{\frac{2}{\pi}} e^{-\xi^2/2}, \quad 0 \leqslant \xi < \infty \quad (11.81)$$

which is a truncated normal distribution (truncated at $\xi = 0$). The probability density functions of the dimensionless positive maxima for various values of ε (where $0 \leqslant \varepsilon \leqslant 1$) are shown in Figure 11.10.

The probability density function derived in Eqs. (11.78) and (11.79) are for positive maxima. If negative maxima are included in the maximum

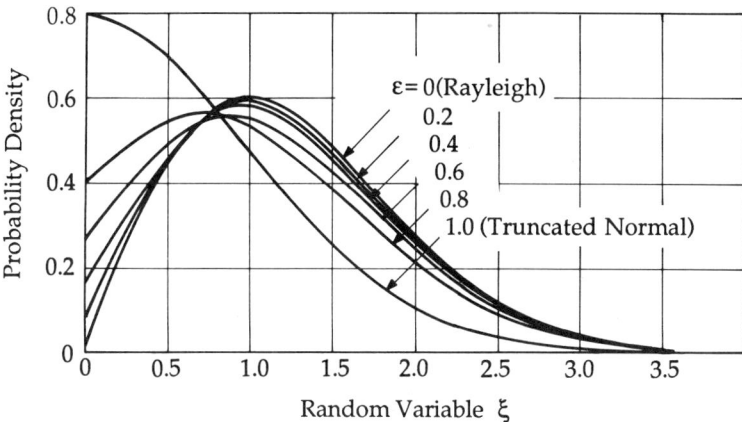

Figure 11.10 Probability density function of positive maxima as a function of bandwidth parameter ε.

point process as shown in Figure 11.11, then the probability density function of the maximum point process extends into the negative range. The procedure for deriving the probability density function of the maxima where both positive and negative maxima are included is the same as that of the positive maxima alone.

The probability density function of positive and negative maxima is given in the following form, which is similar to that shown in Eq. (11.76). That is,

$$f(\zeta) = \frac{\int_{-\infty}^{0} |\ddot{x}| f(\zeta, 0, \ddot{x}) \, d\ddot{x}}{\int_{-\infty}^{\infty} \int_{-\infty}^{0} |\ddot{x}| f(x, 0, \ddot{x}) \, d\ddot{x} \, dx}, \qquad -\infty < \zeta < \infty \quad (11.82)$$

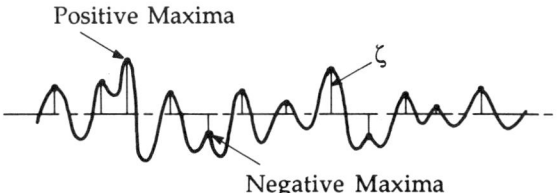

Figure 11.11 Positive maxima and negative minima.

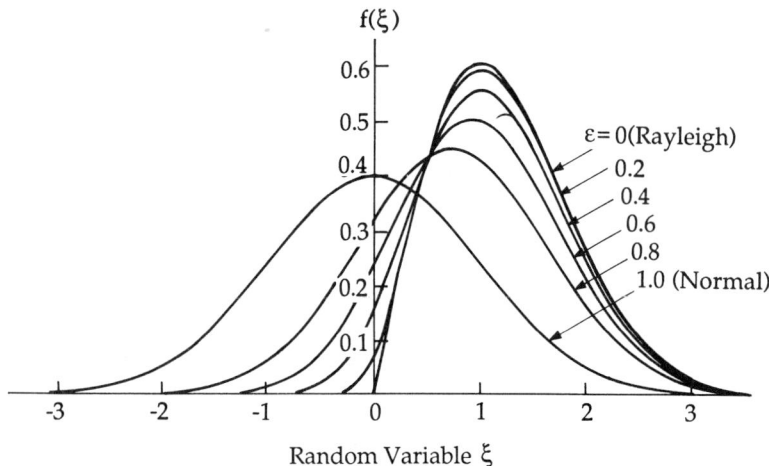

Figure 11.12 Probability density function of maximum point process including positive and negative maxima as a function of bandwidth parameter ε (from Cartwright and Longuet-Higgins, 1956).

The difference between Eqs. (11.76) and (11.82) is that the integration range of the displacement x is extended to $-\infty$. As a result, the probability density function in dimensionless form becomes (Cartwright and Longuet-Higgins, 1956)

$$f(\xi) = \frac{\varepsilon}{\sqrt{2\pi}} e^{-\frac{1}{2}(\xi^2/\varepsilon^2)} + \sqrt{1 - \varepsilon^2}\, \xi e^{-\xi^2/2} \Phi\left(\frac{\sqrt{1 - \varepsilon^2}}{\varepsilon} \xi\right), \quad -\infty < \xi < \infty$$

(11.83)

Figure 11.12 shows the probability density function given in Eq. (11.83). In this case, the probability density function becomes the Rayleigh distribution for $\varepsilon = 0$ and becomes the normal distribution for $\varepsilon = 1$.

11.3 JOINT PROBABILITY DISTRIBUTION OF AMPLITUDES AND PERIODS

In the previous two sections, methods to predict statistical properties of amplitudes (or maxima) from the spectral density function were considered. The prediction of amplitudes (or maxima) of a random process is certainly important for many problems in physical and engineering sciences. However, recent comprehensive advances in the application of the stochastic process approach in engineering design have revealed that consideration of amplitudes alone is not sufficient; instead, probabilistic information on amplitudes together with associated periods is extremely important.

As an example, let us consider the heaving (translation in the vertical plane) motion of a floating offshore structure in a seaway. The heaving motion is induced by waves, and hence a wave with large amplitudes results in large heaving motion, in general. However, the response characteristics of the structure's motion in a seaway are frequency dependent. If the wave frequency (and thereby wave length) is either sufficiently large or small in comparison with the natural heaving frequency of the structure, the heaving motion may not be serious even though the wave height is large. On the other hand, violent heaving motion may take place under the resonance condition in which the wave frequency is very close to the natural heaving frequency of the structure. Therefore, for the design of an offshore structure, it is very important to estimate the extreme wave height whose period is close to the natural motion period of the structure. Thus, the probabilistic knowledge of wave height and associated period will provide information vital to the safe operation and design of the offshore structure.

The joint probability distribution of amplitude and period has been developed for a narrow-band as well as for a non-narrow-band random process for predicting amplitudes and associated periods of ocean waves. However, application of the joint probability distribution functions presented in this section is not limited to ocean waves; the distribution functions are equally applicable to problems in various fields of physics and the engineering sciences.

11.3.1 *Joint Distribution for Narrow-Band Processes*

The joint probability distribution of amplitudes and periods specifically applicable to a random process with a narrow-band spectrum was developed by Longuet-Higgins (1957, 1975, 1983) as presented below.

Let us write the Gaussian random process as

$$x(t) = \sum_{n=1}^{\infty} a_n \cos(\omega_n t + \varepsilon_n)$$

$$= x_c(t)\cos \bar{\omega} t - x_s(t) \sin \bar{\omega} t \qquad (11.84)$$

where

$$x_c(t) = \sum_{n=1}^{\infty} a_n \cos\{(\omega_n - \bar{\omega})t + \varepsilon_n\}$$

$$x_s(t) = \sum_{n=1}^{\infty} a_n \sin\{(\omega_n - \bar{\omega})t + \varepsilon_n\}$$

Here it is assumed that the narrow-band spectrum is concentrated in the vicinity of the mean frequency, $\bar{\omega}$, defined in Eq. (10.23). It can be proved that the random variables x_c, $\dot{x}_c$, x_s, and $\dot{x}_s$ are statistically independent, all normally distributed with zero mean. The variance of x_c and x_s is μ_0, and the variance of $\dot{x}_c$ and $\dot{x}_s$ is μ_2, where μ_0 and μ_2 are the zeroth and second moment of the spectrum about its mean frequency, respectively. The proof is left as an exercise in this chapter.

It follows, therefore, that the joint probability density function of x_c, $\dot{x}_c$, x_s, $\dot{x}_s$ is a quadrivariate normal distribution with the mean

$$E[x_c] = E[\dot{x}_c] = E[x_s] = E[\dot{x}_s] = 0 \qquad (11.85)$$

and the covariance matrix

$$\Sigma = \begin{pmatrix} \mu_0 & 0 & 0 & 0 \\ 0 & \mu_2 & 0 & 0 \\ 0 & 0 & \mu_0 & 0 \\ 0 & 0 & 0 & \mu_2 \end{pmatrix} \qquad (11.86)$$

The reader may recognize the significant advantage of employing the mean frequency $\bar{\omega}$ for presenting a random process in the form shown in Eq. (11.84) in that the use of $\bar{\omega}$ yields the covariance matrix of $(x_c, \dot{x}_c, x_s, \dot{x}_s)$ as a simple diagonal matrix. The joint probability density function then can be written as

$$f(x_c, \dot{x}_c, x_s, \dot{x}_s) = \frac{1}{(2\pi)^2 \mu_0 \mu_2} e^{-(x_c^2 + x_s^2)/2\mu_0} e^{-(\dot{x}_c^2 + \dot{x}_s^2)/2\mu_2} \qquad (11.87)$$

By letting

$$x_c = A \cos \phi$$

$$x_s = A \sin \phi \qquad (11.88)$$

JOINT PROBABILITY DISTRIBUTION OF AMPLITUDES AND PERIODS

we transform the random variables from $(x_c, \dot{x}_c, x_s, \dot{x}_s)$ to $(A, \dot{A}, \phi, \dot{\phi})$, where A is the amplitude and ϕ is the phase. Then, the random process can now be expressed in the form

$$x(t) = \text{Re}\{A(t)e^{i\phi(t)}e^{i\bar{\omega}t}\} \tag{11.89}$$

where

$$A(t)e^{i\phi(t)} = \sum_{n=1}^{\infty} a_n e^{i\{(\omega_n - \bar{\omega})t + \varepsilon_n\}}$$

By applying the formula for changing random variables (see Chapter 7), the joint probability density function of A, ϕ, $\dot{A}$, and $\dot{\phi}$ becomes

$$f(A, \phi, \dot{A}, \dot{\phi}) = \frac{A^2}{(2\pi)^2 \mu_0 \mu_2} e^{-A^2/2\mu_0} e^{-(\dot{A}^2 + A^2\dot{\phi}^2)/2\mu_2},$$

$$0 \leqslant A < \infty, -\infty < \dot{A} < \infty, 0 \leqslant \phi \leqslant 2\pi, -\infty < \dot{\phi} < \infty$$

(11.90)

Then, the marginal joint probability density function of A and $\dot{\phi}$ can be obtained as

$$f(A, \dot{\phi}) = \frac{A^2}{\sqrt{2\pi}\, \mu_0 \sqrt{\mu_2}} e^{-A^2/2\mu_0} e^{-A^2\dot{\phi}^2/2\mu_2}, \quad 0 \leqslant A < \infty, -\infty < \dot{\phi} < \infty \tag{11.91}$$

Next, the phase velocity $\dot{\phi}$ may be expressed in terms of the period. Assuming that $\dot{\phi} \ll \bar{\omega}$, the period of $x(t)$ may be written as

$$T = 2\pi/(\bar{\omega} + \dot{\phi}) \tag{11.92}$$

Then, we have

$$\dot{\phi} = (2\pi/T) - \bar{\omega} \tag{11.93}$$

By using the relationship given in Eq. (11.93), the joint probability density function $f(A, \dot{\phi})$ can be converted to the joint probability density function of amplitude and period $f(A, T)$. In order to express the joint probability density function in dimensionless form we may define

$$\xi = A/\sqrt{m_0} = \text{dimensionless amplitude}$$

$$\eta = T/\overline{T} = \text{dimensionless period} \quad (11.94)$$

where

$$m_0 = \text{area under the spectral density function}$$

$$\overline{T} = \text{mean period} = 2\pi/\overline{\omega}$$

Then, the joint probability density function of amplitudes and periods (in dimensionless form) can be derived from Eqs. (11.91), (11.93), and (11.94) as

$$f(\xi, \eta) = \frac{1}{\sqrt{2\pi}\,\nu} \frac{\xi^2}{\eta^2} \exp\left[-\frac{\xi^2}{2}\left\{1 + \left(1 - \frac{1}{\eta}\right)^2 \frac{1}{\nu^2}\right\}\right] \quad (11.95)$$

where $\nu = \sqrt{\mu_2/(\mu_0 \overline{\omega})}$ is defined as the spectral width parameter in Eq. (10.28).

Since the period takes only positive values, the distribution must be truncated at $\eta = 0$. Furthermore, by normalizing the distribution so that $f(\xi, \eta)$ satisfies the condition that the integration over the entire sample space becomes unity, the joint probability density function becomes

$$f(\xi, \eta) = \frac{1}{\sqrt{2\pi}\,\nu}\left(1 + \frac{\nu^2}{4}\right)\frac{\varepsilon^2}{\eta^2} \exp\left[-\frac{\xi^2}{2}\left\{1 + \left(1 - \frac{1}{\eta}\right)^2 \frac{1}{\nu^2}\right\}\right], \quad (11.96)$$

$$0 \leqslant \xi < \infty, 0 \leqslant \eta < \infty$$

According to Longuet-Higgins, this joint probability density function appears to be valid for values of the spectral width parameter ν up to 0.6.

By letting $h = 2\xi$, the joint probability density function of the double amplitudes (crest-to-trough excursions) and periods is given in dimensionless form as follows:

$$f(h, \eta) = \frac{1}{8\sqrt{2\pi}\,\nu}\left(1 + \frac{\nu^2}{4}\right)\frac{h^2}{\eta^2} \exp\left[-\frac{h^2}{8}\left\{1 + \left(1 - \frac{1}{\eta}\right)^2 \frac{1}{\nu^2}\right\}\right], \quad (11.97)$$

$$0 \leqslant h < \infty, 0 \leqslant \eta < \infty$$

JOINT PROBABILITY DISTRIBUTION OF AMPLITUDES AND PERIODS

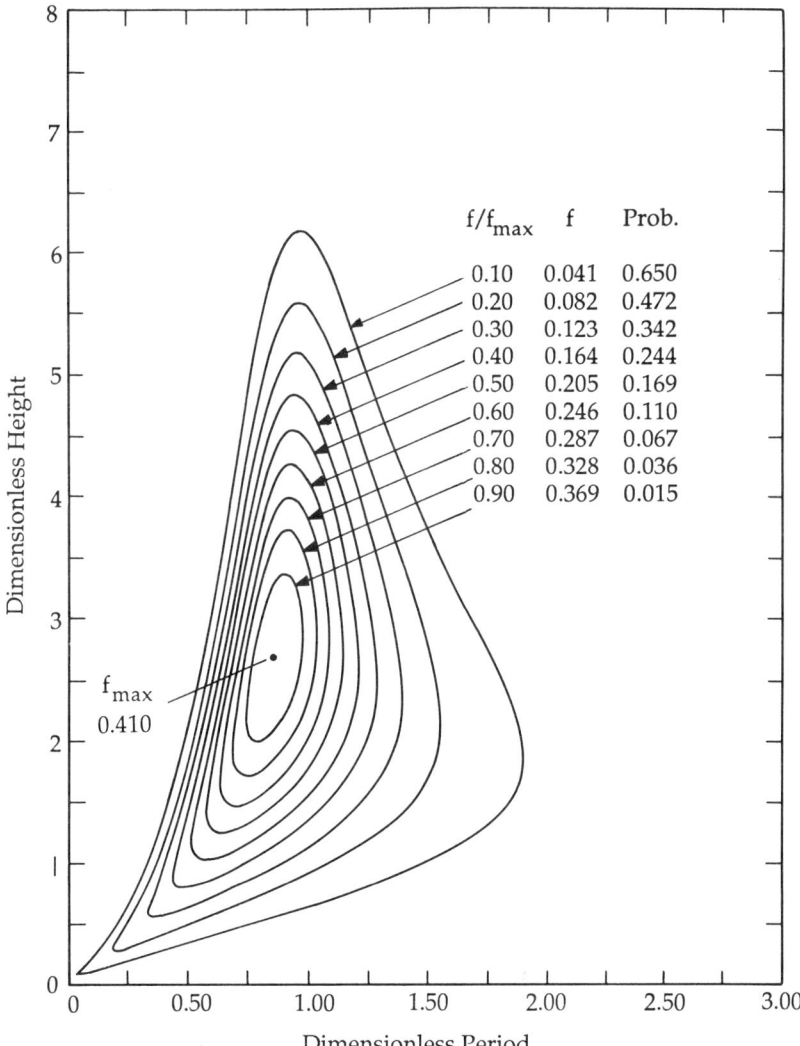

Figure 11.13 Contour lines of joint probability density function of height and period (dimensionless) given in Eq. (11.97) for $\nu = 0.40$ (Longuet-Higgens, 1983).

The joint probability density function $f(h, \eta)$ for the parameter $\nu = 0.40$ is shown in Figure 11.13. Included also in this figure is the probability of occurrence of all combinations of double amplitudes and periods within the contour line. The latter is computed by integrating the volume of the probability density function enclosed by the contour line.

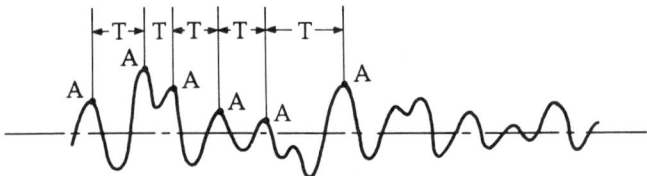

Figure 11.14 Positive maxima and associated time intervals for non-narrow-band random process.

11.3.2 *Joint Distribution for Non-Narrow-Band Processes*

It is extremely difficult to develop the joint probability distribution of amplitudes and periods for a random process with a non-narrow-band spectrum. This is because there are multiple crests (positive maxima) during a half-cycle of the random process, and hence, in the strict sense, the amplitude cannot be defined. The currently available joint probability distribution for a non-narrow-band random process is the joint distribution of positive maxima and associated time intervals as illustrated in Figure 11.14 (Cavanié et al., 1976). As shown in the figure, the positive maxima A do not necessarily represent the amplitudes nor do the time intervals T necessarily represent the period.

In order to derive the joint probability density function of the positive maxima (crests) and associated time intervals, let us first consider the probability that the positive maxima, denoted by Z, will exceed a specified level ζ with acceleration $\ddot{x}$. It can be obtained by the same approach as used in Eq. (11.74) as

$$\Pr\{Z > \zeta \text{ with acceleration } \ddot{x}\} = \frac{\overline{N}_{\zeta,\ddot{x}}}{\overline{N}_{x+}} \qquad (11.98)$$

where

$\overline{N}_{\zeta,\ddot{x}}$ = average number of times that positive maxima exceed the level ζ with acceleration $\ddot{x}$ per unit time

$\overline{N}_{x+}$ = average number of positive maxima per unit time

By letting $f(x, \dot{x}, \ddot{x})$ be the joint probability density function of displacement, velocity, and acceleration, $\overline{N}_{\zeta,\ddot{x}}$ and $\overline{N}_{x+}$ can be evaluated by

$$\overline{N}_{\zeta,\ddot{x}} = \int_{\zeta}^{\infty} |\ddot{x}| f(x, 0, \ddot{x})\, d\ddot{x}$$

$$\overline{N}_{x+} = \int_{0}^{\infty} \int_{-\infty}^{0} |\ddot{x}| f(x, 0, \ddot{x})\, d\ddot{x}\, dx \qquad (11.99)$$

From Eqs. (11.98) and (11.99), the joint probability density function of the positive maxima and acceleration can be written as follows:

$$f(\zeta, \ddot{x}) = \frac{d}{d\zeta}\left\{1 - \frac{\overline{N}_{\zeta, \ddot{x}}}{N_{x+}}\right\}$$

$$= \frac{|\ddot{x}| f(\zeta, 0, \ddot{x})}{\int_0^\infty \int_{-\infty}^0 |\ddot{x}| f(x, 0, \ddot{x}) \, d\ddot{x} \, dx} \qquad (11.100)$$

From the joint probability density function $f(x, 0, \ddot{x})$ given in Eq. (11.69), $f(\zeta, \ddot{x})$ can be evaluated as

$$f(\zeta, \ddot{x}) = \frac{2|\ddot{x}|}{\sqrt{2\pi}\sqrt{\Delta m_4}\left(1 + \sqrt{1 - \varepsilon^2}\right)} e^{-(m_0 \ddot{x} + 2m_2 \zeta \ddot{x} + m_4 \zeta^2)/2\Delta},$$

$$0 \leq \zeta < \infty, \, -\infty < \ddot{x} \leq 0 \qquad (11.101)$$

where

$m_j = j$th moment of the spectrum

$\Delta = m_0 m_4 - m_2^2$

$\varepsilon = \sqrt{1 - m_2^2/m_0 m_4}$ [see the definition given in Eq. (10.27)].

Next, let us transform the random variables $(\zeta, \ddot{x})$ to (ζ, T) by using the functional relationship given by

$$\ddot{x} = -\omega^2 \zeta = -(2\pi/T)^2 \zeta \qquad (11.102)$$

Then, the joint probability density function of the maxima, ζ, and associated time interval, T, can be obtained in dimensionless form as follows (Arhan et al. 1976; Cavanié et al., 1976):

$$f(\xi, \tau) = \sqrt{\frac{2}{\pi}} \frac{\alpha^3 \xi^2}{\varepsilon(1 - \varepsilon^2)\tau^5} \exp\left[-\frac{\xi^2}{2\varepsilon^2 \tau^4}\left\{(\tau^2 - \alpha^2)^2 + \alpha^4 \beta^2\right\}\right]$$

$$(11.103)$$

where

$$\xi = \zeta/\sqrt{m_0}$$

$$\tau = T/T_m$$

T_m = average time between two successive maxima

$$= \frac{2\pi}{\alpha}\sqrt{\frac{m_2}{m_4}} = 4\pi\left(\frac{\sqrt{1-\varepsilon^2}}{1+\sqrt{1-\varepsilon^2}}\right)\sqrt{\frac{m_0}{m_2}}$$

$$\alpha = \frac{1}{2}\left(1+\sqrt{1-\varepsilon^2}\right)$$

$$\beta = \varepsilon/\sqrt{1-\varepsilon^2}$$

The formula derived above needs to be modified in order to satisfy the condition required for a probability density function. For this let us evaluate the mean time interval by

$$\bar{\tau} = \int_0^\infty \int_0^\infty \tau f(\xi,\tau)\,d\xi\,d\tau \qquad (11.104)$$

Since the mean time interval is not unity as shown in Table 11.2, a correction is incorporated in the joint density function. For this, consider a new random variable given by

$$\tau_* = \tau/\bar{\tau} \qquad (11.105)$$

Table 11.2

Mean Time Interval $\bar{\tau}$ as a Function of the Bandwidth Parameter ε (from Arhan et al., 1976)

ε	$\bar{\tau}(\varepsilon)$	ε	$\bar{\tau}(\varepsilon)$
0.05	0.9988	0.55	0.9500
0.10	0.9963	0.60	0.9445
0.15	0.9928	0.65	0.9396
0.20	0.9886	0.70	0.9355
0.25	0.9838	0.75	0.9331
0.30	0.9787	0.80	0.9335
0.35	0.9732	0.85	0.9393
0.40	0.9675	0.90	0.9573
0.45	0.9617	0.95	1.0133
0.50	0.9558		

Then, the distribution of ξ and τ_* becomes

$$f(\xi, \tau_*) = \sqrt{\frac{2}{\pi}} \frac{\alpha^3 \xi^2}{\varepsilon(1-\varepsilon^2)\tau_*^5 \bar{\tau}^4}$$

$$\times \exp\left\{-\frac{\xi^2}{2\varepsilon^2 \tau_*^4 \bar{\tau}^4}\left[\left(\tau_*^2 \bar{\tau}^2 - \alpha^2\right)^2 + \alpha^4 \beta^2\right]\right\} \quad (11.106)$$

Furthermore, by letting $h = 2\xi$, the joint probability density function of h and τ_* is given by

$$f(h, \tau_*) = \frac{1}{4\sqrt{2\pi}} \frac{\alpha^3 h^2}{\varepsilon(1-\varepsilon^2)\tau_*^5 \bar{\tau}^4}$$

$$\times \exp\left\{-\frac{h^2}{8\varepsilon^2 \tau_*^4 \bar{\tau}^4}\left[\left(\tau_*^2 \bar{\tau}^2 - \alpha^2\right)^2 + \alpha^4 \beta^2\right]\right\},$$

$$0 \leq h < \infty, 0 \leq \tau_* < \infty \quad (11.107)$$

It is noted again that h is not the double amplitude in a strict sense; instead, it is the double excursion of the maxima. Example of the joint probability density functions, $f(h, \tau_*)$, for $\varepsilon = 0.6$ is shown in Figure 11.15.

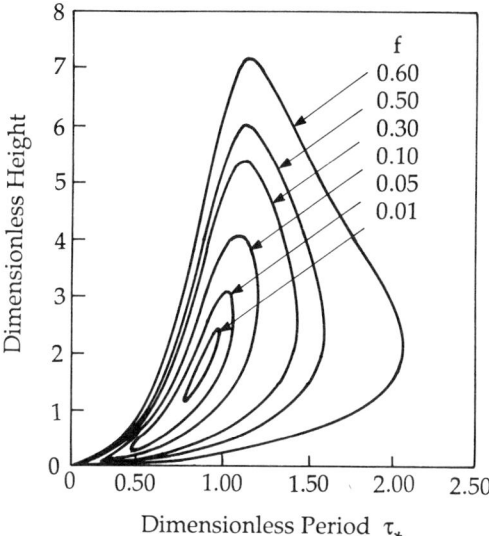

Figure 11.15 Contour lines of joint probability density function of positive maxima and associated time intervals (dimensionless) given in Eq. (11.107) for $\varepsilon = 0.6$ (Arhan et al., 1976; Cavanié et al., 1976).

11.4 DISTRIBUTION OF PERIODS

11.4.1 Expected Period

The problem of counting the number of times the random process exceeds a certain level within a given time duration is called the *level* (or *threshold*) *crossing problem* in stochastic processes. In particular, if the level is chosen to be zero, it is called the *zero-crossing problem*, and it is directly related to the period of a random process.

Definition 11.2. The *zero-crossing period* (or simply called *period*) of a random process is defined as the time interval between two upward (or downward) crossing of the zero line (or mean line).

Figure 11.16 illustrates the definition of the zero-crossing period based on the upward crossing of the zero line. As can be seen in the figure, the period thus defined depends solely on the crossings of the mean line (zero line) irrespective of the number of crests (or troughs) occurring during the successive zero crossings. Since the zero-crossing period varies from one cycle to the next, its average value is defined as the average zero-crossing period or simply called the average period.

The average (zero-crossing) period should not be confused with the mean period of random process, which is defined as the period associated with the mean frequency $\bar{\omega}$, given in Eq. (10.23), namely $\bar{T} = 2\pi/\bar{\omega}$. The mean period thus defined has no relationship with the average value of the period obtained by $(1/N)\sum_{i=1}^{N} T_i$. The mean period does not have much practical significance; however, the mean frequency $\bar{\omega}$ plays a significant role in stochastic analysis of a random process since a drastic simplification in the mathematical derivation of the distribution can be achieved by taking the moment of the spectrum about the mean frequency as shown in Section 11.3.1.

In the following, the prediction of the expected (average) zero-crossing period from knowledge of the spectrum will be presented. We first consider the level crossing for a specified level and evaluate the expected (average) number of crossings. Analogous to the concept for evaluating the expected

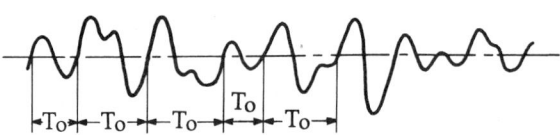

Figure 11.16 Definition of zero-crossing period.

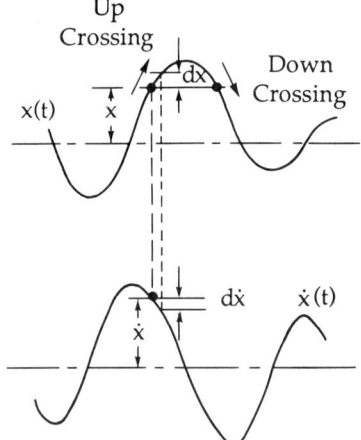

Figure 11.17 Explanatory sketch of level crossing problem.

number of maxima per unit times discussed in Section 11.2.1, the expected number of crossings the level x with velocity $\dot{x}$ per unit time (including both positive and negative crossings) is given by

$$\overline{N}_x = \int_{-\infty}^{\infty} |\dot{x}| f(x, \dot{x}) \, d\dot{x} \qquad (11.108)$$

The expected number of crossings per unit time with a positive slope (up-crossing in Figure 11.17) is then given by

$$\overline{N}_{x+} = \frac{1}{2} \int_{-\infty}^{\infty} |\dot{x}| f(x, \dot{x}) \, d\dot{x} \qquad (11.109)$$

In particular, for a specified level α, we have,

$$\overline{N}_{\alpha+} = \frac{1}{2} \int_{-\infty}^{\infty} |\dot{x}| f(\alpha, \dot{x}) \, d\dot{x} \qquad (11.110)$$

Let $\alpha = 0$. Then, the average number of zero crossings with positive slope becomes

$$\overline{N}_{0+} = \frac{1}{2} \int_{-\infty}^{\infty} |\dot{x}| f(0, \dot{x}) \, d\dot{x} \qquad (11.111)$$

To evaluate Eqs. (11.110) and (11.111), the joint probability density function $f(x, \dot{x})$ is necessary. By assuming that the random process $x(t)$ is Gaussian with zero mean, it is known that $x(t)$ and $\dot{x}(t)$ are statistically independent and that the variances are m_0 and m_2, respectively, where m_0

is the zeroth moment and m_2 is the second moment of the spectral density function. Hence, (11.110) becomes

$$\overline{N}_\alpha = \frac{1}{2\pi\sqrt{m_0 m_2}} e^{-\alpha^2/2m_0} \int_0^\infty \dot{x} e^{-\dot{x}^2/2m_2} \, d\dot{x} = \frac{1}{2\pi} \sqrt{\frac{m_2}{m_0}} e^{-\alpha^2/2m_0} \quad (11.112)$$

By letting $\alpha = 0$, the expected number of zero crossings per unit time becomes

$$\overline{N}_{0+} = \frac{1}{2\pi} \sqrt{m_2/m_0} \quad (11.113)$$

Thus, the average zero-crossing period and average frequency, denoted by $\overline{T}_0$ and $\overline{\omega}_0$, respectively, can be evaluated from the spectral density function as

$$\overline{T}_0 = 2\pi\sqrt{m_0/m_2}$$

$$\overline{\omega}_0 = \sqrt{m_2/m_0} \quad (11.114)$$

It is noted that the moments of the spectral density function m_0 and m_2 in Eq. (11.114) are evaluated for a spectrum expressed in terms of the frequency ω. If the spectral density function is given in terms of the frequency f, then the average zero-crossing period becomes

$$\overline{T}_0 = \sqrt{m_{0f}/m_{2f}} \quad (11.115)$$

where m_{0f}, m_{2f} are the zeroth and second moments, respectively, of the spectral density function $S(f)$.

11.4.2 Probability Density Function of Periods

In the previous section, the discussion pertained to the expected (average) period. For some engineering problems, however, it is often important to know the distribution of the periods of a random process. For example, for the design of offshore structures floating in a seaway, there is a great need for information on the frequency of occurrence of wavelengths (and thereby the wave periods) that may cause a resonant motion of the structure.

The probability density function of the periods of a random phenomena was first derived by Rice (1944, 1945) in connection with the level crossing problem. That is, in order to derive the probability density function of the periods, we first consider the general problem of evaluating the expected number of up-crossings of the level $(\alpha, \alpha + d\alpha)$ with velocity $(\dot{x}_1, \dot{x}_1 + d\dot{x}_1)$ and also of down-crossings of the same level with velocity $(\dot{x}_2, \dot{x}_2 + d\dot{x}_2)$. This is simply an extension of the concept used in the derivation of Eq. (11.110) to the two-dimensional case. The expected number per unit time is given by

$$\overline{N}_{\alpha+,\alpha-} = \int_0^\infty \dot{x}_1 \int_{-\infty}^0 \dot{x}_2 f(\alpha, \dot{x}_1, \alpha, \dot{x}_2) \, d\dot{x}_2 \, d\dot{x}_1 \qquad (11.116)$$

On the other hand, the expected number of crossings at the level α with positive slope per unit time is given in Eq. (11.112). Hence, the ratio of Eq. (11.116) to Eq. (11.112) gives approximately the probability of crossings at the level α with a positive slope followed by a crossing at the same level with a negative slope. This ratio may be interpreted as the amount of time the random process spends above the level α. By letting the time difference between the up- and down-crossings be τ, the probability density function of time is given by

$$f(\tau) = \frac{\int_0^\infty \dot{x}_1 \int_{-\infty}^0 \dot{x}_2 f(\alpha, \dot{x}_1, \alpha, \dot{x}_2) \, d\dot{x}_2 \, d\dot{x}_1}{\frac{1}{2\pi}\sqrt{m_2/m_0} \exp\{-\alpha^2/2m_0\}} \qquad (11.117)$$

Let $\alpha = 0$, then Eq. (11.117) yields the following probability density function of a half period:

$$f(\tau) = \frac{\int_0^\infty \dot{x}_1 \int_{-\infty}^0 \dot{x}_2 f(0, \dot{x}_1, 0, \dot{x}_2) \, d\dot{x}_2 \, d\dot{x}_1}{\frac{1}{2\pi}\sqrt{m_2/m_0}} \qquad (11.118)$$

Equation (11.118) was further evaluated by Rice (1945) for a Gaussian process with zero mean. The following is an excerpt of Rice's work. For the present problem, the covariance matrix of the random variables $(x_1, \dot{x}_1, x_2, \dot{x}_2)$ is given by

$$\Sigma = \begin{pmatrix} m_0 & 0 & \nu_0 & -\nu_1 \\ 0 & m_2 & \nu_1 & \nu_2 \\ \nu_0 & \nu_1 & m_0 & 0 \\ -\nu_1 & \nu_2 & 0 & m_2 \end{pmatrix} \qquad (11.119)$$

where

$$m_0 = \int_0^\infty S(\omega)\, d\omega$$

$$m_2 = \int_0^\infty \omega^2 S(\omega)\, d\omega$$

$$\nu_0 = \int_0^\infty S(\omega)\cos\tau\omega\, d\omega \qquad (11.120)$$

$$\nu_1 = \int_0^\infty \omega S(\omega)\sin\tau\omega\, d\omega$$

$$\nu_2 = \int_0^\infty \omega^2 S(\omega)\cos\tau\omega\, d\omega$$

$S(\omega)$ = spectral density function

From the covariance matrix given in Eq. (11.119), the joint probability density function $f(0, \dot{x}_1, 0, \dot{x}_2)$ in Eq. (11.118) can be written as

$$f(0, \dot{x}_1, 0, \dot{x}_2) = \frac{1}{4\pi^2 |\Sigma|^{1/2}} \exp\left\{ -\frac{1}{2|\Sigma|} M_{22}(\dot{x}_1^2 + \dot{x}_2^2) + 2M_{24}\dot{x}_1\dot{x}_2 \right\}$$

$$(11.121)$$

where

$$|\Sigma| = \{(m_0 + \nu_0)(m_2 - \nu_2) - \nu_1^2\}\{(m_0 - \nu_0)(m_2 + \nu_2) - \nu_1^2\}$$

$$M_{22} = m_2(m_0^2 - \nu_0^2) - m_0\nu_1^2$$

$$M_{24} = -\nu_2(m_0^2 - \nu_0^2) + \nu_0\nu_1^2 \qquad (11.122)$$

Next, by letting

$$\sqrt{\frac{M_{22}}{2|\Sigma|}}\, \dot{x}_1 = u \quad \text{and} \quad \sqrt{\frac{M_{22}}{2|\Sigma|}}\, \dot{x}_2 = -v \qquad (11.123)$$

the numerator of Eq. (11.118) becomes

$$\int_0^\infty \dot{x}_1 \int_{-\infty}^0 \dot{x}_2 f(0, \dot{x}_1, 0, \dot{x}_2) \, d\dot{x}_1 \, d\dot{x}_2$$

$$= \frac{|\Sigma|^{3/2}}{\pi^2 M_{22}^2} \int_0^\infty u \int_0^\infty v \exp\left\{-(u^2 + v^2) + 2\frac{M_{24}}{M_{22}} uv\right\} du \, dv$$

$$= \frac{1}{4\pi^2} \frac{1}{M_{22}^2 - M_{24}^2} \{1 + H \cot^{-1}(-H)\} \qquad (11.124)$$

where

$$H = \frac{M_{24}}{\sqrt{M_{22}^2 - M_{24}^2}}$$

Thus, the probability density function of a half period is given by

$$f(\tau) = \frac{1}{2\pi} \sqrt{\frac{m_0}{m_2}} \frac{|\Sigma|^{3/2}}{M_{22}^2 - M_{24}^2} \{1 + H \cot^{-1}(-H)\} \qquad (11.125)$$

In order to eliminate the determinant in Eq. (11.125), the following relationship may be used:

$$M_{22}^2 - M_{24}^2 = |\Sigma|(m_0^2 - v_0^2) \qquad (11.126)$$

Then, Eq. (11.125) becomes

$$f(\tau) = \frac{1}{2\pi} \sqrt{\frac{m_0}{m_2}} \frac{(M_{22} - M_{24})^{1/2}}{(m_0^2 - v_0^2)^{3/2}} \{1 + H \cot^{-1}(-H)\} \qquad (11.127)$$

By letting $\tau = T/2$, the probability density function of the period T, can be obtained from Eq. (11.127) as

$$f(T) = \tfrac{1}{2}[f(\tau)]_{\tau = T/2} \qquad (11.128)$$

The probability density function derived in Eq. (11.128) appears to be appropriate for random processes with relatively high frequencies.

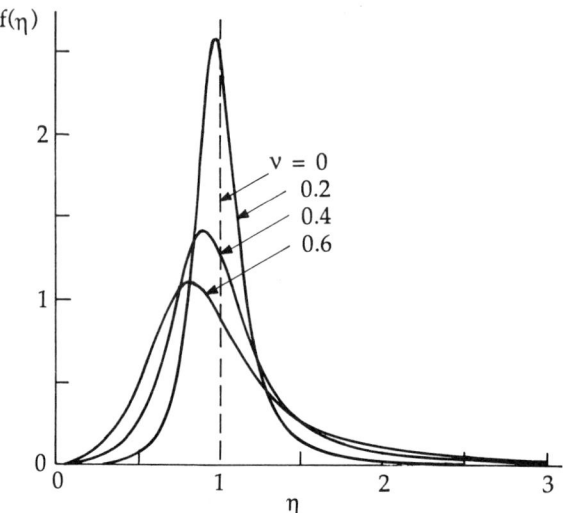

Figure 11.18 Probability density function of period (dimensionless) as a function of parameter ν (from Longuet-Higgins, 1983).

The probability density function of the periods can also be obtained from the joint distribution of amplitudes and periods developed by Longuet-Higgins (1983) (see Section 11.3.1). That is, by integrating the joint probability density function $f(\xi, \eta)$ given in Eq. (11.96) with respect to ξ, the probability density function of the periods in (dimensionless form) can be obtained as

$$f(\eta) = \left(1 + \frac{\nu^2}{4}\right) \frac{1}{2\nu\eta^2} \left\{1 + \left(1 - \frac{1}{\eta}\right)^2 \frac{1}{\nu^2}\right\}^{-3/2} \quad (11.129)$$

The probability density function $f(\eta)$ is shown in Figure 11.18 as a function of the parameter ν.

It is noted that the mean of the distribution of periods given in Eq. (11.129) is theoretically infinite, and that the variance of the distribution cannot be obtained from Eq. (11.129). In order to avoid this difficulty, Longuet-Higgins suggests the use of the average zero-crossing period as the mean period. It is given in dimensionless form as

$$\eta_{\text{ave}} = \overline{T}_0/\overline{T} = 1/\sqrt{1 + \nu^2} \quad (11.130)$$

ESTIMATION OF EXTREME AMPLITUDE AND MAXIMA

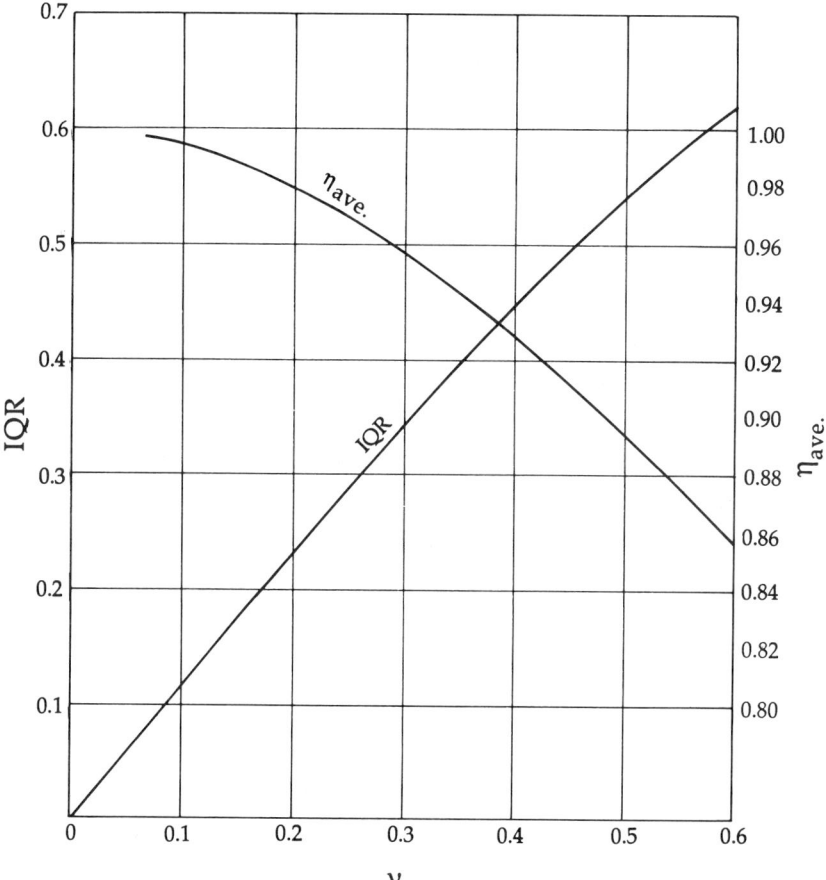

Figure 11.19 Interquartile range (IQR) and average period (dimensionless) as a function of parameter ε (from Longuet-Higgins, 1983).

The parameter ν involved in Eqs. (11.129) and (11.130) can be evaluated by Eq. (10.28) from knowledge of moments of spectral density function. Evaluation of the parameter ν from the observed data can be done by using the interquartile range (IQR), which is defined as the distance between two points $\eta_{1/4}$ and $\eta_{3/4}$ (see the definition shown in Figure 2.7) of the probability density function of η.

By applying the probability density function given in Eq. (11.129), the IQR can be evaluated as follows:

$$\text{IQR} = \eta_{3/4} - \eta_{1/4} = \frac{1}{1 - \nu\beta_3} - \frac{1}{1 - \nu\beta_1} \quad (11.131)$$

where

$$\beta_3 = \left(\frac{3}{2L} - 1\right) \bigg/ \sqrt{1 - \left(\frac{3}{2L} - 1\right)^2}$$

$$\beta_1 = \left(\frac{1}{2L} - 1\right) \bigg/ \sqrt{1 - \left(\frac{1}{2L} - 1\right)^2}$$

$$L = 1 + \frac{\nu^2}{4}$$

Figure 11.19 shows the value of IQR computed from Eq. (11.131) as a function of ν. Included also in the figure is η_{ave} given in Eq. (11.130). By obtaining the value of IQR from the cumulative distribution function constructed from observed data, the parameter ν can be evaluated from Figure 11.19, and therefrom the probability density function of period can be obtained.

11.5 ESTIMATION OF EXTREME AMPLITUDE AND MAXIMA

Let $x(t)$ be a stationary Gaussian random process with zero mean, and let m_0 be the area under the spectral density function. We will evaluate the extreme amplitude of the process that is expected to occur in a specified time T [see Figure 11.20(a)] when the process has a narrow-band spectrum. We will also evaluate the extreme value of the maxima of the process with a non-narrow-band spectrum [see Figure 11.20(b)] and compare their magnitudes.

For a Gaussian random process with a narrow-band spectrum, it was derived in Section 11.1.1 that the amplitudes of the process follow the Rayleigh probability distribution with a parameter equal to twice the variance $(2m_0)$. On the other hand, it was shown in Section 8.1.2 that the magnitude of the probable extreme value, $\overline{Y}_n$, expected n observations (or samples) can be evaluated as a solution of

$$1 - F(\bar{y}_n) = 1/n \tag{11.132}$$

where $F(\ ) =$ cumulative distribution function.

ESTIMATION OF EXTREME AMPLITUDE AND MAXIMA

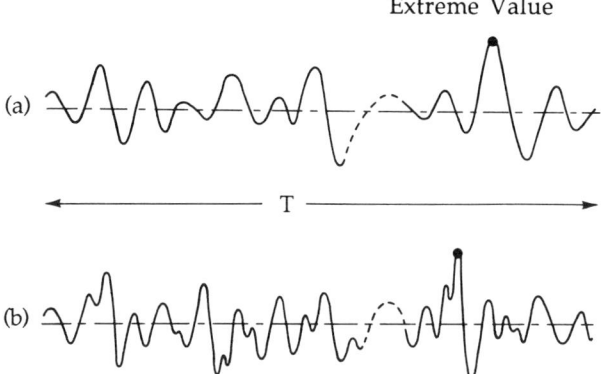

Figure 11.20 (*a*) Extreme amplitude and (*b*) extreme maxima in a specified time *T*.

Then, the probable extreme amplitude expected in n observations for the Rayleigh distribution can be obtained from the result given in Example 8.2 as

$$\bar{y}_n = \sqrt{\ln n}\,\sqrt{2m_0} \qquad (11.133)$$

Next, let the number of amplitudes, n, of a random process observed in a specified time period equal the number of zero crossings in that period, where the average number of zero crossings per unit time is given in Eq. (11.113). Assuming that the spectral analysis is carried out in second units, the average number of zero crossings in T hours becomes

$$n = \frac{T}{2\pi}(60)^2 \sqrt{m_2/m_0} \qquad (11.134)$$

Thus, from Eqs. (11.133) and (11.134), the probable extreme amplitude expected to occur in T hours can be obtained as

$$\bar{y}_n = \left\{2\ln\left(\frac{T}{2\pi}(60)^2\sqrt{m_2/m_0}\right)\right\}^{1/2} \sqrt{m_0} \qquad (11.135)$$

For a random process with a non-narrow-band spectrum, the cumulative distribution (in dimensionless form) is the integration of Eq. (11.79) with respect to ξ. It is given in Exercise 11.2 as follows:

$$F(\xi) = \frac{2}{1 + \sqrt{1 - \varepsilon^2}} \left[-\frac{1}{2}(1 - \sqrt{1 - \varepsilon^2}) + \Phi(\xi/\varepsilon) \right.$$
$$\left. - \sqrt{1 - \varepsilon^2}\, e^{-\xi^2/2} \Phi\left(\frac{\sqrt{1 - \varepsilon^2}}{\varepsilon}\xi\right) \right] \quad (11.136)$$

where ξ = dimensionless maxima = maxima/$\sqrt{m_0}$.

It is extremely difficult to solve Eq. (11.132) by using the cumulative distribution function given in Eq. (11.136). However, an approximate solution can be found for a bandwidth parameter ε of less than 0.9. That is, for $\varepsilon < 0.9$, it is possible to use the following approximation for a large number of observations (Ochi, 1973):

$$\Phi(\xi/\varepsilon) \sim 1$$

and

$$\Phi\left(\frac{\sqrt{1 - \varepsilon^2}}{\varepsilon}\xi\right) = 1 - \Phi\left(-\frac{\sqrt{1 - \varepsilon^2}}{\varepsilon}\xi\right) \sim 1 \quad (11.137)$$

Then, Eq. (11.136) becomes

$$F(\xi) = 1 - \frac{2\sqrt{1 - \varepsilon^2}}{1 + \sqrt{1 - \varepsilon^2}} \exp\{-\xi^2/2\} \quad (11.138)$$

and thereby the solution of Eq. (11.132) can be found in dimensionless form as

$$\bar{y}_n = \left\{ 2\ln\left(\frac{2\sqrt{1 - \varepsilon^2}}{1 + \sqrt{1 - \varepsilon^2}} n\right) \right\}^{1/2} \sqrt{m_0} \quad (11.139)$$

The probable extreme maxima calculated from Eq. (11.139) as a function of the bandwidth parameter ε are shown in Figure 11.21. As can be seen in the figure, the effect of ε on the extreme value is noticeable only for large ε values (say $\varepsilon > 0.6$), irrespective of the number of observations.

Next, let us express the probable extreme maxima given in Eq. (11.139), which is applicable for any non-narrow-band spectrum, in terms of time. The expected number of positive maxima per unit time was derived in Eq. (11.77). By using the present notation, we may write

$$n = \frac{1}{4\pi}\left(\frac{1 + \sqrt{1 - \varepsilon^2}}{\sqrt{1 - \varepsilon^2}}\right)\sqrt{\frac{m_2}{m_0}} \quad (11.140)$$

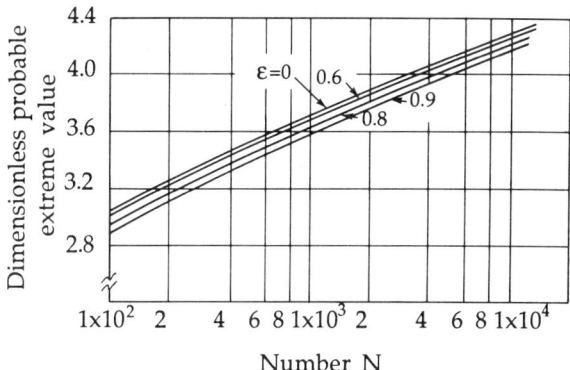

Figure 11.21 Probable extreme maxima (in dimensionless form) as a function of bandwidth parameter ε.

Thus, from Eqs. (11.139) and (11.140), the probable extreme maxima in a time period T hours is given by

$$\bar{y}_n = \left\{ 2\ln\left(\frac{T}{2\pi}(60)^2\sqrt{m_2/m_0}\right)\right\}^{1/2} \sqrt{m_0} \qquad (11.141)$$

Note that the above formula is the same as that for the extreme amplitude for the narrow-band spectrum derived in Eq. (11.135), and that $\bar{y}_n$ is no longer a function of the bandwidth parameter. Although the equation for evaluating the probable extreme values expressed in terms of time is the same irrespective of the bandwidth parameter of the spectrum, it should be remembered that the number of peaks (maxima) for a non-narrow-band spectrum is much larger than for a narrow-band spectrum during the same period of time.

EXERCISES

11.1 $x(t)$ is a Gaussian random process with the spectral density function $S(\omega)$. Let x_1 and x_2 be the values of $x(t)$ at times $t = t_1$ and $t = t_2$, respectively. Prove that

$$\text{Cov}[x_1, x_2] = \int_0^\infty S(\omega)\cos\tau\omega\, d\omega$$

and

$$\text{Cov}[x_1, \dot{x}_2] = -\int_0^\infty S(\omega)\sin\tau\omega\, d\omega$$

where

$$\tau = t_2 - t_1 \text{ and } \dot{x}_2 \text{ is the time derivative of } x_2$$

11.2 Prove that the cumulative distribution function of the dimensionless positive maxima of a Gaussian random process becomes as follows through integration of Eq. (11.79):

$$F(\xi) = \frac{2}{1 + \sqrt{1 - \varepsilon^2}} \left[-\frac{1}{2}(1 - \sqrt{1 - \varepsilon^2}) + \Phi\left(\frac{\xi}{\varepsilon}\right) \right.$$

$$\left. - \sqrt{1 - \varepsilon^2} \, e^{-\xi^2/2} \Phi\left(\frac{\sqrt{1 - \varepsilon^2}}{\varepsilon} \xi\right) \right]$$

11.3 Let $x(t)$ be a narrow-band Gaussian random process. Prove that the conditional probability density function of periods given double amplitudes (in dimensionless form $h = H/\sqrt{m_0}$) of $x(t)$ can be evaluated from Eq. (11.97) as follows:

$$f(\eta|h) = \frac{1}{2\sqrt{2\pi}} \frac{1}{\nu} \frac{1}{\Phi(h/2\nu)} \frac{h}{\eta^2} \exp\left[-\frac{h^2}{8\nu^2}\left(1 - \frac{1}{\eta}\right)^2 \right]$$

11.4 The spectral density function of a random process is given by

$$S(\omega) = \frac{A}{\omega^5} e^{-B/\omega^4}$$

where A and B are constants. Evaluate the following properties of the process:

(a) significant value, $x_{1/3}$ (amplitude)

(b) average zero-crossing period, $\overline{T}_0$

(c) mean frequency, $\overline{\omega}$

(d) modal frequency, ω_m

(e) bandwidth parameter, ε.

11.5 Let $x(t)$ be a Gaussian random process with zero mean having a narrow-band spectrum. $x(t)$ may be written as follows [see Eq. (11.84)]:

$$x(t) = \sum_{n=1}^{\infty} a_n \cos(\omega_n t + \varepsilon_n) = x_c(t) \cos \overline{\omega} t - x_s(t) \sin \overline{\omega} t$$

where $\overline{\omega}$ = mean frequency evaluated from spectrum. Prove that

$$\text{Var}[\dot{x}_c] = \text{Var}[\dot{x}_s] = \mu_2 = \int_0^\infty (\omega - \overline{\omega})^2 S(\omega) \, d\omega$$

CHAPTER 12

Statistical Analysis of Time Series Data

It was discussed in the previous chapter that the amplitude of a narrow-band Gaussian random process follows the Rayleigh probability distribution, which can be determined by a single parameter, σ^2, evaluated from the spectrum of the process. It is often convenient to obtain statistical properties of the amplitude of the process from randomly observed data (amplitudes) without carrying out spectral analysis. This can be achieved from estimation of the parameter of the distribution by applying the techniques developed in the statistical inference theory. In applying these techniques, however, a sufficiently large sample of the data is highly desirable for accurate estimation of the parameter. If the sample size is not large enough, the confidence interval for the parameter of the distribution is established for a preassigned measure of assurance.

In this section, analysis of time series data from the statistical inference viewpoint will be discussed. The principle of estimating the parameter of a distribution from an observed random sample will be outlined, and the method for establishing the confidence interval for the parameter of the probability distribution will be explained.

12.1 PRINCIPLE OF STATISTICAL ESTIMATION

Let $f(x; \theta)$ be the probability density function of a random variable X with the parameter θ. For simplicity, we consider the following case where the probability density function is defined by a single parameter. However, the principle is equally applicable when the distribution has multiparameters. The probability density function is known, and we want to estimate the value of θ from an observed random sample of size n, $(x_1, x_2, x_3, \ldots, x_n)$. Here, it is assumed that the sample elements are random and they are statistically independent.

The estimated value of θ from a random sample is called an *estimator*, and is denoted by $\hat{\theta}$. In many situations more than one estimator exists for θ; hence, the question arises as to what conditions provide the best selection of the estimator for θ.

The most commonly employed requirements for selecting the best estimator are the following:

1. The expected value of the estimator $\hat{\theta}$ should be equal to θ; $E[\hat{\theta}] = \theta$. $\hat{\theta}$, which satisfies this condition, is called an *unbiased estimator*.

2. The variance of the estimator $\hat{\theta}$ is smallest of all possible estimators; $\hat{\theta}$ is then called the *minimum variance estimator*.

The estimator that satisfies these two conditions is called the *best unbiased estimator* for the parameter θ.

There are many other properties that should be considered for selecting the best estimator such as the requirement that the properties of the estimator be consistent, efficient, and sufficient. However, this subject is beyond the scope of applied statistics. Readers who are interested in this subject may refer to Kendall and Stuart (1963), Freeman (1963), and Rohatgi (1976), among others.

The most widely considered method for finding the estimator is the *maximum likelihood method*. The principle of this method is the following. Since each element of a random sample of size n, $(x_1, x_2, \ldots, x_n)$, is statistically independent, the joint probability distribution of $x_1, x_2, \ldots, x_n$ is given by the product of $f(x_i|\theta)$, and is called the *likelihood function*, denoted by L, of the random sample. We may write

$$L(x_1, x_2, \ldots, x_n|\theta) = \prod_{i=1}^{n} f(x_i|\theta) \qquad (12.1)$$

Then, the estimator $\hat{\theta}$ is called the *maximum likelihood estimator* if $\hat{\theta}$ satisfies the condition

$$L(x_1, x_2, \ldots, x_n|\hat{\theta}) > L(x_1, x_2, \ldots, x_n|\theta') \qquad (12.2)$$

where θ' is any other possible estimator. Equation (12.2) implies that the estimator $\hat{\theta}$ is chosen such that it maximizes the likelihood function L.

In practice, the maximum likelihood estimator $\hat{\theta}$ is obtained by maximizing the logarithm of L. That is, $\hat{\theta}$ is determined as a solution of the following equation:

$$\frac{\partial}{\partial \theta} \ln L(x_1, x_2, \ldots, x_n | \theta) = 0 \tag{12.3}$$

Similarly, for a probability distribution with multiple parameters $f(x|\theta_1, \theta_2, \ldots, \theta_m)$, the maximum likelihood estimators, $\hat{\theta}_1, \hat{\theta}_2, \ldots, \hat{\theta}_m$, are determined as a simultaneous solution of the following set of m equations:

$$\frac{\partial}{\partial \theta_1} \ln L(x_1, x_2, \ldots, x_n | \theta_1, \theta_2, \ldots, \theta_m) = 0$$

$$\frac{\partial}{\partial \theta_2} \ln L(x_1, x_2, \ldots, x_n | \theta_1, \theta_2, \ldots, \theta_m) = 0 \tag{12.4}$$

$$\vdots$$

$$\frac{\partial}{\partial \theta_m} \ln L(x_1, x_2, \ldots, x_n | \theta_1, \theta_2, \ldots, \theta_m) = 0$$

The maximum likelihood estimators for parameters of probability distributions often considered in the statistical analysis of random phenomena observed in engineering and physical sciences are presented in the following example:

Example 12.1.   Rayleigh Distribution. One of the most frequently encountered statistical predictions of a random phenomenon is the evaluation of the magnitude of crest-to-trough excursions (or amplitudes) of the phenomenon. It was shown in Section 11.1 that the excursions (or amplitudes) of narrow-banded Gaussian random process with zero mean follow the Rayleigh distribution given by

$$f(x) = \frac{2x}{R} e^{-x^2/R}, \qquad 0 \leqslant x < \infty$$

The parameter R of the distribution can be estimated from a random sample $(x_1, x_2, \ldots, x_n)$ of the crest-to-trough excursions of a narrow-band

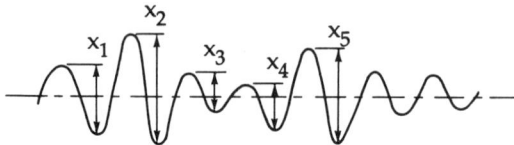

Figure 12.1 Random sample of crest-to-trough excursions.

Gaussian random process, an example of which is shown in Figure 12.1. The likelihood function can be obtained as

$$L = \prod_{i=1}^{n} \frac{2x_i}{R} \exp\left\{-\frac{x_i^2}{R}\right\} = \left(\frac{2}{R}\right)^n \left(\prod_{i=1}^{n} x_i\right) \exp\left\{-\frac{\sum_{i=1}^{n} x_i^2}{R}\right\}$$

Taking the logarithm and differentiating with respect to R yields

$$\frac{d}{dR} \ln L = -\frac{n}{R} + \sum_{i=1}^{n} x_i^2 / R^2$$

The above equation is then set equal to zero so that the maximum likelihood estimator $\hat{R}$ is estimated to be

$$\hat{R} = \frac{1}{n} \sum_{i=1}^{n} x_i^2$$

In evaluating the excursions (or amplitudes), $\sqrt{\hat{R}}$ is used. Hence, the square root of $\hat{R}$ obtained above is often called the *root-mean-square (rms) value*. The estimator can be used as the parameter of the Rayleigh distribution if the sample size is sufficiently large. ∎

Example 12.2. **Normal Distribution.** Estimation of two parameters (mean and variance) is necessary for the Normal probability distribution given by

$$f(x) = \frac{1}{\sqrt{2\pi}\,\sigma} e^{-(x-\mu)^2/2\sigma^2}$$

PRINCIPLE OF STATISTICAL ESTIMATION

We may write the likelihood function of the distribution as

$$L = \prod_{i=1}^{n} \frac{1}{\sqrt{2\pi}\,\sigma} e^{-(x_i-\mu)^2/2\sigma^2} = \left(\frac{1}{2\pi\sigma^2}\right)^{n/2} e^{-(1/2\sigma^2)\sum_{i=1}^{n}(x_i-\mu)^2}$$

and therefrom evaluate

$$\frac{\partial}{\partial \mu} \ln L = \frac{1}{\sigma^2} \sum_{i=1}^{n} (x_i - \mu)$$

$$\frac{\partial}{\partial \sigma^2} \ln L = -\frac{n}{2}\frac{1}{\sigma^2} + \frac{1}{2\sigma^4} \sum_{i=1}^{n} (x_i - \mu)^2$$

By equating the derivatives to zero and by solving the equations simultaneously, we can obtain the following maximum likelihood estimators:

$$\hat{\mu} = \frac{1}{n} \sum_{i=1}^{n} x_i$$

$$\hat{\sigma}^2 = \frac{1}{n} \sum_{i=1}^{n} (x_i - \hat{\mu})^2$$

It is noted that the estimator $\hat{\mu}$ is unbiased, but $\hat{\sigma}^2$ is not. The expected value of $\hat{\sigma}^2$ is equal to $(n-1)\sigma^2/n$, and thereby $\hat{\sigma}^2 = \{1/(n-1)\} \Sigma(x_i - \hat{\mu})^2$ should be used as the unbiased estimator for σ^2 for small sample size. ∎

In estimating the parameter of the Rayleigh distribution which is applicable for the crest-to-trough excursions (or amplitudes) of a narrow-band Gaussian process, it is often necessary to estimate the variance of the Gaussian process from a random sample of deviations from the mean-line $(x_1, x_2, \ldots, x_n)$ as shown in Figure 12.2. Then, the parameter R is estimated by applying the relationship $R = 8\sigma^2$ given in connection with Eq. (11.25). Since the deviation x is taken from the mean-line, the variance can be estimated by $\hat{\sigma}^2 = (1/n)\sum_{i=1}^{n} x_i^2$. $\hat{\sigma}$ is also called the *root-mean-square* (*rms*) *value* (see rms value defined in Example 12.1). Therefore, extreme care must be given to the definition of the rms value in estimating the parameter of the distribution associated with crest-to-trough excursions (or amplitudes) of a narrow-band random process.

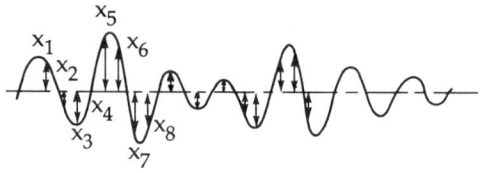

Figure 12.2 Random sample of deviations from the mean value.

Table 12.1

Estimation of Various Statistics from Crest-to-Trough Excursions of a Narrow-Band Random Process from Random Sample

	Crest - to - trough excursions	Deviations from the mean
Random sample		
Distribution of x	Rayleigh distribution with parameter R	Normal distribution with $\mu = 0$, σ^2
Estimator	$\hat{R} = \dfrac{1}{n} \sum_{i=1}^{n} x_i^2$	$\hat{\sigma}^2 = \dfrac{1}{n} \sum_{i=1}^{n} x_i$
RMS - Value	$\sqrt{\hat{R}}$	$\hat{\sigma}$
Estimation of crest - to - trough excursions		
Significant value	$\sqrt{2}\,\sqrt{\hat{R}}$	$4\hat{\sigma}$
Probable extreme value	$\sqrt{\ln n}\,\sqrt{\hat{R}}$	$2\sqrt{2\ln n}\,\hat{\sigma}$

PRINCIPLE OF STATISTICAL ESTIMATION

Table 12.1 summarizes the formulas for estimating significant values as well as probable extreme values of crest-to-trough excursions of a narrow-band random process from observed random sample.

Example 12.3. **Uniform Distribution.**

$$f(x|\theta) = 1/\theta, \quad \text{for } 0 \leq x \leq \theta$$

The likelihood function of the uniform distribution is simply given by $L = 1/\theta^n$, which is either an ever-increasing or decreasing function of θ. Hence, it is not possible to find $\hat{\theta}$ by differentiating the likelihood function L. However, L can be maximized by selecting θ as small as possible. This implies that $\hat{\theta}$ is associated with the largest x. Hence, the maximum likelihood estimator for this case is given by the largest observed x value in the sample. It is noted that the estimator thusly evaluated is biased. That is, $E[\max x_i] = n\theta/(n + 1)$. ∎

Example 12.4. **Weibull Distribution.**

$$f(x) = c\lambda^c x^{c-1} e^{-(\lambda x)^c}, \quad 0 \leq x < \infty$$

The likelihood function becomes

$$L = (c\lambda^c)^n \left(\prod_{i=1}^{n} x_i^{c-1} \right) \exp\left\{ -\lambda^c \sum_{i=1}^{n} x_i^c \right\}$$

Then, as the solution of the two equations $\partial/\partial \lambda \ln L = 0$ and $\partial/\partial c \ln L = 0$, we have the following maximum likelihood estimators for c and λ:

$$\hat{c} = \left[\left(\sum_{i=1}^{n} x_i^{\hat{c}} \ln x_i \right) \left(\sum_{i=1}^{n} x_i^{\hat{c}} \right)^{-1} - \frac{1}{n} \sum_{i=1}^{n} \ln x_i \right]^{-1}$$

$$\hat{\lambda} = \left(\frac{1}{n} \sum_{i=1}^{n} x_i^{\hat{c}} \right)^{-1/\hat{c}}$$

The estimator $\hat{c}$ can be evaluated from the first equation by the iteration method by providing an arbitrary assigned starting value. On estimating the $\hat{c}$ value, the estimator $\hat{\lambda}$ can be determined from the second equation. ∎

Example 12.5. **Gamma Distribution.**

$$f(x) = \frac{1}{\Gamma(m)} \lambda^m x^{m-1} e^{-\lambda x}, \qquad 0 \leq x < \infty$$

The likelihood function is given by

$$L = \frac{1}{\{\Gamma(m)\}^n} \lambda^{mn} \left(\prod_{i=1}^{n} x_i \right)^{m-1} e^{-\lambda \sum_{i=1}^{n} x_i}$$

From $\partial/\partial \lambda \ln L = 0$ and $\partial/\partial m \ln L = 0$, the following two solutions can be derived:

$$\hat{\lambda} = \hat{m} \bigg/ \left(\sum_{i=1}^{n} x_i/n \right) = \hat{m}/\bar{x}$$

$$\psi(\hat{m}) = \Gamma'(\hat{m})/\Gamma(m) = \ln \hat{\lambda} + \frac{1}{n} \sum_{i=1}^{n} \ln x_i$$

where

$$\bar{x} = \text{sample mean}$$

$$\psi(\hat{m}) = \text{psi (or digamma) function.}$$

By eliminating $\hat{\lambda}$ from the two equations, we have

$$\psi(\hat{m}) - \ln \hat{m} = \frac{1}{n} \sum_{i=1}^{n} \ln x_i - \ln \bar{x}$$

The estimator $\hat{m}$ can be evaluated from the above equation, and then $\hat{\lambda}$ can be obtained. ∎

Example 12.6. **Asymptotic Type I Extreme Value Distribution.** The probability density function of the asymptotic Type I extreme value distribution is given in Eq.(8.42). For simplicity, we may write the density function for a random variable X instead of ^1Y_n. That is,

$$f(x) = \alpha \exp\{-\alpha(x-u) - e^{-\alpha(x-u)}\}$$

Then, the likelihood function can be written as

$$L = \alpha^n \exp\left\{-\sum_{i=1}^{n}\alpha(x_i - u) - \sum_{i=1}^{n} e^{-\alpha(x_i - u)}\right\}$$

From $\partial/\partial u \ln L = 0$ and $\partial/\partial \alpha \ln L = 0$, we can derive the following two solutions:

$$\frac{1}{n}\sum_{i=1}^{n} e^{-\hat{\alpha}(x_i - \hat{u})} = 1$$

$$\bar{x} = \frac{1}{\hat{\alpha}} + \hat{u} + \frac{1}{n}\sum_{i=1}^{n} x_i e^{-\hat{\alpha}(x_i - \hat{u})} - \frac{\hat{u}}{n}\sum_{i=1}^{n} e^{-\hat{\alpha}(x_i - \hat{u})}$$

By eliminating $\hat{u}$ from the above two equations, we have

$$\bar{x} = \frac{1}{\hat{\alpha}} + \left(\sum_{i=1}^{n} x_i e^{-\hat{\alpha}x_i}\right)\left(\sum_{i=1}^{n} e^{-\hat{\alpha}x_i}\right)^{-1}$$

Thus, $\hat{\alpha}$ can be evaluated by iteration method. Then, the parameter $\hat{u}$ can be obtained from the first equation associated with the solution $\hat{\alpha}$ and $\hat{u}$, which can be written as

$$\hat{u} = -\frac{1}{\hat{\alpha}} \ln\left(\frac{1}{n}\sum_{i=1}^{n} e^{-\hat{\alpha}x_i}\right) \qquad \blacksquare$$

12.2 CONFIDENCE INTERVALS

As shown in the previous section, the estimator $\hat{\theta}$ depends on the sample size n. Obviously the larger the sample size the more accurate estimation of the unknown parameter can be made. Therefore, it is highly desirable that some measure of assurance in the estimation be reflected in the estimator θ. One way to provide this measure of assurance is to evaluate upper and lower limits of the estimator $\hat{\theta}$. In other words, we estimate an interval, called the *confidence interval*, in which the true value of the unknown parameter θ lies with a measure of assurance $1 - \alpha$, called the *confidence coefficient*. Here, we may interpret α as the probability of committing a possible error in the estimation.

It is noted that the confidence interval of the parameter of a probability distribution $f(x|\theta)$ is evaluated through its estimator $\hat{\theta}$, which is a random variable, and it has its own probability density function different from $f(x|\theta)$. Hence, in establishing the confidence interval, we must first find the probability density function of the estimator $\hat{\theta}$. We will next explain the procedure for deriving the confidence interval for some distributions often considered in the statistical analysis of random processes.

12.2.1 Confidence Interval for the Rayleigh Distribution Parameter

Consider a random sample of size n, $(x_1, x_2, \ldots, x_n)$, obtained from an observation of the crest-to-trough excursions of a narrow-band Gaussian random process as shown in Figure 12.1. The excursions obey the Rayleigh probability distribution with parameter R, and it is shown in Example 12.1 that the likelihood estimator of the parameter is given by

$$\hat{R} = \frac{1}{n}\sum x_i^2$$

We may first obtain the probability density function of the random variable $Y_i = X_i^2$, where X_i follows the Rayleigh distribution. The probability density function of Y_i can be obtained from Example 7.3. That is,

$$f(y_i) = \frac{1}{R}e^{-y_i/R} \qquad (12.5)$$

The density function $f(y_i)$ is that of the exponential distribution, and its characteristic function can be written as

$$\phi_{y_i}(t) = \int_0^\infty e^{ity_i}\frac{1}{R}e^{-y_i/R}\,dy_i = (1 - iRt)^{-1} \qquad (12.6)$$

Note that y_i are statistically independent; hence, following the property of the characteristic function given in Section 4.2, the characteristic function of $\hat{R}$ becomes

$$\phi_{\hat{R}}(t) = \left(1 - \frac{iRt}{n}\right)^{-n} \qquad (12.7)$$

which is the characteristic function for the gamma probability distribution (see Section 6.2.1), and thereby it is found that $\hat{R}$ follows the gamma probability law given by

$$f(\hat{R}) = \frac{(n/R)^n}{\Gamma(n)}\hat{R}^{n-1}e^{-(n/R)\hat{R}}, \qquad 0 \leq \hat{R} < \infty \qquad (12.8)$$

CONFIDENCE INTERVALS

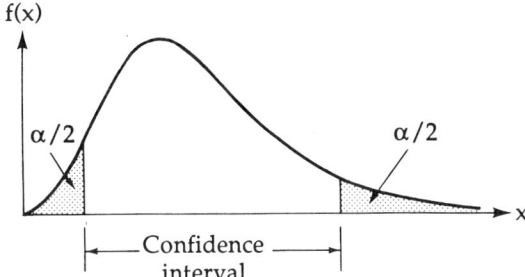

Figure 12.3 Definition of confidence interval for the confidence coefficient $1 - \alpha$.

Thus, the confidence interval of $\hat{R}$ can be obtained from the above equation. However, it may be more convenient to transform the gamma distribution into the χ^2 distribution. That is, following the result derived in Example 6.6, it can be shown that if the random variable $\hat{R}$ has the gamma distribution with parameter n and n/R as obtained in Eq. (12.8), then the random variable $2(n/R)\hat{R}$ obeys the χ^2 distribution with $2n$ degrees of freedom.

In determining the confidence interval of a distribution, it is a common practice to choose the location of the confidence interval such that the possibility of committing an error may occur at the lower and upper tail portions of the distribution with equal probability. Hence, for the confidence coefficient $1 - \alpha$, the probability of committing an error is $\alpha/2$ at both ends of the distribution as shown in Figure 12.3. Thus, for the present problem the required confidence interval should be determined from the following probability of the random variable $2(n/R)\hat{R}$:

$$\Pr\left\{\chi^2_{(2n)}\left(\frac{\alpha}{2}\right) < 2\frac{n}{R}\hat{R} < \chi^2_{(2n)}\left(1 - \frac{\alpha}{2}\right)\right\} = 1 - \alpha \qquad (12.9)$$

where

$\chi^2_{(2n)}\left(\frac{\alpha}{2}\right), \chi^2_{(2n)}\left(1 - \frac{\alpha}{2}\right)$ = lower and upper values of the χ^2 distribution with $2n$ degrees of freedom for the confidence coefficient $1 - \alpha$.

Thus, we can obtain the confidence interval of the parameter R with the confidence coefficient $1 - \alpha$ as follows:

$$\frac{2n\hat{R}}{\chi^2_{(2n)}(1 - \alpha/2)} < R < \frac{2n\hat{R}}{\chi^2_{(2n)}(\alpha/2)} \qquad (12.10)$$

The confidence interval given in Eq. (12.10) can be applied to any number of samples of size n. However, the confidence interval can be estimated from the normal distribution if the sample size is large, $n \geq 30$. This is because the χ^2 distribution is asymptotically equal to the normal distribution as shown in Example 6.7. Since the mean and variance of the random variable $2(n/R)\hat{R}$ are $2n$ and $4n$, respectively, it can be obtained that the random variable $\{2(n/R)\hat{R} - 2n\}/\sqrt{4n}$ is approximately normal with zero mean and unit variance for $n \geq 30$. This yields the following probability for the confidence coefficient $1 - \alpha$:

$$\Pr\left\{-U\left(\frac{\alpha}{2}\right) < \frac{2(n/R)\hat{R} - 2n}{\sqrt{4n}} < U\left(\frac{\alpha}{2}\right)\right\} = 1 - \alpha \qquad (12.11)$$

where

$-U\left(\dfrac{\alpha}{2}\right), U\left(\dfrac{\alpha}{2}\right)$ = lower and upper values of the standardized normal distribution for the confidence coefficient $1 - \alpha$ (see Figure 12.4).

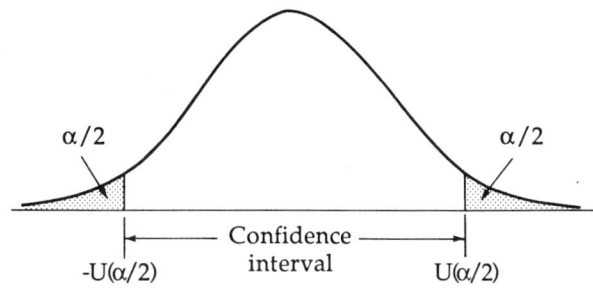

Figure 12.4 Confidence interval of the standardized normal distribution for the confidence coefficient $1 - \alpha$.

Thus, we have the following confidence interval for the parameter R with the confidence coefficient $1 - \alpha$ when the sample size $n \geqslant 30$:

$$\frac{\hat{R}}{1 + [U(\alpha/2)/\sqrt{n}]} < R < \frac{\hat{R}}{1 - [U(\alpha/2)/\sqrt{n}]} \qquad (12.12)$$

12.2.2 Confidence Interval for the Variance of the Normal Distribution

It was stated in connection with Figure 12.2 that one way to estimate the parameter of the Rayleigh distribution associated with the crest-to-trough excursions (or amplitude) of a narrow-band Gaussian process is through the estimation of the variance of the process that is normally distributed. The estimator in this case is given by

$$\hat{\sigma}^2 = \frac{1}{n} \sum_{i=1}^{n} x_i^2 \qquad (12.13)$$

where

x_i = element of a random sample obtained from deviations from the mean

It is given in Theorem 6.3 that $(1/\sigma^2)\sum_{i=1}^{n} x_i^2$ has a χ^2 distribution with n degrees of freedom if the mean value is equal to zero. Hence, it can be obtained that $n\hat{\sigma}^2/\sigma^2$ has a χ^2 distribution with n degrees of freedom. This yields the following probability of the random variable $n\hat{\sigma}^2/\sigma^2$ for a confidence coefficient $1 - \alpha$:

$$\Pr\left\{\chi^2_{(n)}(\alpha/2) < \frac{n\hat{\sigma}^2}{\sigma^2} < \chi^2_{(n)}(1 - \alpha/2)\right\} = 1 - \alpha \qquad (12.14)$$

Hence, the confidence interval of σ^2 with the confidence coefficient $1 - \alpha$ can be obtained as

$$\frac{n\hat{\sigma}^2}{\chi^2_{(n)}(1 - \alpha/2)} < \sigma^2 < \frac{n\hat{\sigma}^2}{\chi^2_{(n)}(\alpha/2)} \qquad (12.15)$$

Based on the estimation of the variance σ^2, the confidence interval of the parameter R of the Rayleigh distribution applicable from crest-to-trough excursion of the narrow-band Gaussian random process can be obtained as follows:

$$\frac{8n\hat{\sigma}^2}{\chi^2_{(n)}(1 - \alpha/2)} < R < \frac{8n\hat{\sigma}^2}{\chi^2_{(n)}(\alpha/2)} \qquad (12.16)$$

Two confidence intervals of the parameter R of the Rayleigh distribution have been derived: one is based on the estimator $\hat{R}$ and the other is based on the estimator $\hat{\sigma}^2$. Figure 12.5 shows a comparison of the confidence intervals evaluated by the two methods for a random sample taken from observation of ocean wave heights. Included also in the figure is the R value evaluated through spectral analysis, which may be considered as the population value (true value). As can be seen in the figure, the confidence intervals estimated by the $\hat{\sigma}^2$ method are much narrower than those estimated by the $\hat{R}$ method for the same confidence coefficient. This can be

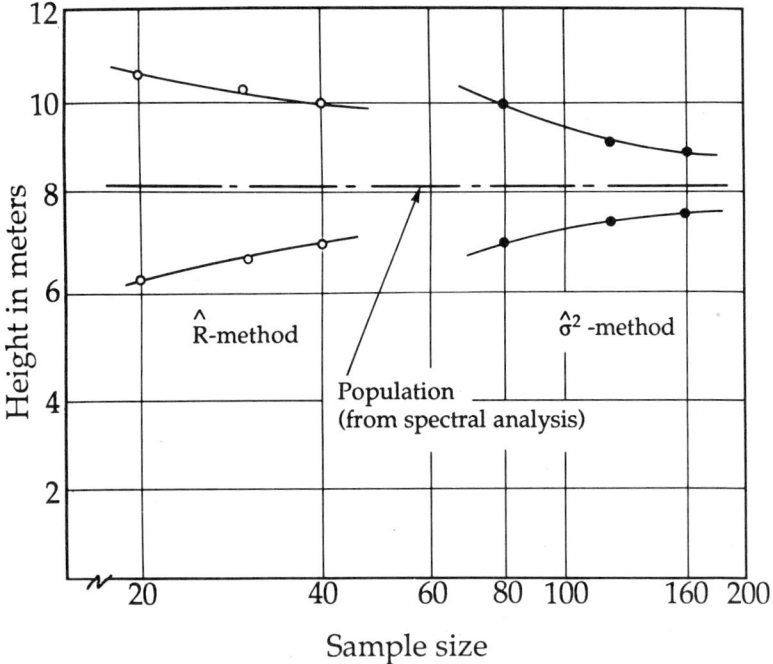

Figure 12.5 Comparison of confidence intervals obtained by $\hat{R}$ method and $\hat{\sigma}^2$ method as a function of sample size.

CONFIDENCE INTERVALS

attributed to the fact that considerably more samples can be taken in the former than in the latter method during the same observation time.

Next, let us consider two sets of independently observed data, both of which are from populations having normal distributions with zero means but with variances that may or may not be the same. We wish to know which is the case.

In order to carry out statistical inference on this problem, let us write the sample variances as s_1^2 and s_2^2, with sample size n_1 and n_2, and let the population variances be σ_1^2 and σ_2^2, respectively. Then, we have from Theorem 6.3 that $n_1 s_1^2/\sigma^2$ and $n_2 s_2^2/\sigma^2$ are distributed following the χ^2 distribution with n_1 and n_2 degrees of freedom, respectively. Hence, the problem at issue is equivalent to examining the identity of two χ^2 distributions. For this we may use the following theorem regarding the F distribution:

Theorem 12.1. Let X_1 and X_2 be the two independent random variables having χ^2 distributions with r_1 and r_2 degrees of freedom, respectively. Then, the random variable

$$F = \frac{X_1/r_1}{X_2/r_2}$$

has the F distribution whose probability density function is given by

$$f(F) = \frac{(n_1/n_2)^{n_1/2}}{B(n_1/2, n_2/2)} \frac{F^{(n_1/2)-1}}{[1 + (n_1/n_2)F]^{(n_1+n_2)/2}}, \quad 0 \leq F < \infty$$

(12.17)

where $B(\ )$ = beta function.

Proof. Consider the joint probability density function of two independent random variables X_1 and X_2. From the definition of the χ^2 distribution we have

$$f(x_1, x_2) = \frac{1}{\Gamma(r_1/2)\Gamma(r_2/2)2^{(r_1+r_2)/2}} x_1^{(r_1/2)-1} x_2^{(r_2/2)-1} e^{-(x_1+x_2)/2},$$

$$0 \leq x_1 < \infty, 0 \leq x_2 < \infty \quad (12.18)$$

Let $F = (X_1/r_1)/(X_2/r_2)$ and $U = X_2$. Then, we have $X_1 = (r_1/r_2)FU$ and $X_2 = U$. By applying the technique of changing the random variables from (X_1, X_2) to (F, U), the joint probability density function of F and U becomes

$$f(F, u) = \frac{1}{\Gamma(r_1/2)\Gamma(r_2/2)2^{(r_1+r_2)/2}} \left(\frac{r_1}{r_2}\right)^{r_1/2} F^{(r_1/2)-1} u^{(r_1+r_2)/2-1}$$

$$\times \exp\left\{-\frac{u}{2}\left(\frac{r_1}{r_2}F + 1\right)\right\} \qquad (12.19)$$

The marginal probability density function $f(F)$ can then be derived by integrating $f(F, u)$ with respect to u.

By applying Theorem 12.1, the problem concerning the equality of variance of two normal distributions with zero means can be solved following the procedure developed in statistical tests. A brief description of the statistical test is given below. The reader who is interested in details of the test may refer to Bury (1975) and Kendall and Stuart (1963), among others.

For the present problem, we wish to test the hypothesis that $\sigma_1^2/\sigma_2^2 = 1$ against the alternative $\sigma_1^2/\sigma_2^2 \neq 1$. The region where the hypothesis is rejected is called the *critical region*, and the probability of committing a possible error by rejecting the hypothesis when it is true is called the *level of significance*, α. Now, we have shown that both $n_1 s_1^2/\sigma_1^2$ and $n_2 s_2^2/\sigma_2^2$ follow χ^2 distributions with n_1 and n_2 degrees of freedom, respectively; hence, from Theorem 12.1, it can be said that $F = (s_1^2/\sigma_1^2)/(s_2^2/\sigma_2^2)$ follows the F distribution with (n_1, n_2) degree of freedom. Under the hypothesis $\sigma_1^2/\sigma_2^2 = 1$, we have $F = s_1^2/s_2^2$. By choosing $F_1(\alpha/2)$ and $F_2(\alpha/2)$ as the lower and upper bounds, respectively, we may accept the hypothesis if

$$Pr\{F_1(\alpha/2) < s_1^2/s_2^2 < F_2(\alpha/2)\} = 1 - \alpha \qquad (12.20)$$

Inversely, we may reject the hypothesis with a level of significance α if

$$s_1^2/s_2^2 < F_1(\alpha/2) \quad \text{or} \quad s_1^2/s_2^2 > F_2(\alpha/2) \qquad (12.21)$$

Another example of application of the F distribution will be shown in Chapter 17 in connection with the Poisson random process.

12.2.3 Confidence Interval for the Parameter of the Poisson Distribution

The maximum likelihood estimator for the parameter μ of the Poisson distribution is given by the sample mean $\hat{\mu} = (1/n)\sum_{i=1}^{n} x_i$ (Exercise 12.1). In order to determine the confidence interval of the parameter μ, we may

CONFIDENCE INTERVALS

first consider the probability distribution of $\Sigma x_i = n\hat{\mu}$, and assume that $n\mu$ as well as $n\hat{\mu}$ are large. It can be seen from the characteristic function of the Poisson distribution that Σx_i also has a Poisson distribution with parameter $n\mu$. Since we assume that $n\hat{\mu}$ is large, the random variable Σx_i is approximately normally distributed with mean $n\mu$ and variance $n\mu$. This implies that $(\Sigma x_i - n\mu)/\sqrt{n\mu} = (\hat{\mu} - \mu)/\sqrt{\mu/n}$ has a standardized normal distribution. Thus, as was the case shown in Eq. (12.11), we have the following probability with the confidence coefficient $1 - \alpha$:

$$Pr\left\{-U\left(\frac{\alpha}{2}\right) < \frac{\hat{\mu} - \mu}{\sqrt{\mu/n}} < U\left(\frac{\alpha}{2}\right)\right\} = 1 - \alpha \qquad (12.22)$$

Then, the confidence interval of μ can be found from the solution of the following formula:

$$\left|\frac{\bar{\mu} - \mu}{\sqrt{\mu/n}}\right| < U\left(\frac{\alpha}{2}\right)$$

which is equivalent to

$$n\mu^2 - \left\{2n\hat{\mu} + U^2\left(\frac{\alpha}{2}\right)\right\}\mu + n\hat{\mu}^2 < 0 \qquad (12.23)$$

As the solution of μ, we have

$$\hat{\mu} + \frac{U^2(\alpha/2)}{2n} - U\left(\frac{\alpha}{2}\right)\sqrt{\frac{\hat{\mu}}{n} + \frac{U^2(\alpha/2)}{4n^2}}$$

$$< \mu < \hat{\mu} + \frac{U^2(\alpha/2)}{2n} + U\left(\frac{\alpha}{2}\right)\sqrt{\frac{\hat{\mu}}{n} + \frac{U^2(\alpha/2)}{4n^2}} \qquad (12.24)$$

Since $n\mu$ is assumed to be large, we may neglect $(1/2n)U^2(\alpha/2)$ in comparison to $\hat{\mu}$. Thus, we can derive the confidence interval of the parameter μ of the Poisson distribution with confidence coefficient $1 - \alpha$ as follows:

$$\hat{\mu} - U\left(\frac{\alpha}{2}\right)\sqrt{\frac{\hat{\mu}}{n}} < \mu < \hat{\mu} + U\left(\frac{\alpha}{2}\right)\sqrt{\frac{\hat{\mu}}{n}} \qquad (12.25)$$

EXERCISES

12.1 Show that the maximum likelihood estimator of the Poisson distribution $p(x) = \mu^x e^{-\mu}/x!$ is given by the sample mean $\hat{\mu} = (1/n)\sum_{i=1}^{n} x_i$.

12.2 Let X be a random variable representing the lifetime of a mechanical system. It is assumed that X follows the exponential distribution $f(x) = \lambda e^{-\lambda x}$, where $0 \leq x \leq \infty$. Show that the maximum likelihood estimator of the parameter is given by $\hat{\lambda} = (1/n)\sum_{i=1}^{n} x_i$, and prove that $2n\lambda/\hat{\lambda}$ follows the χ^2 distribution with $2n$ degrees of freedom.

12.3 In testing the quality of the production of a merchandise, a sample of size n, where n is fairly large, is randomly drawn. It is found that k in n samples are defective. Evaluate the confidence interval of the fraction, p, of the merchandise that is defective.
[Hint] The estimator $\hat{p} = k/n = (1/n)\sum_{i=1}^{n} x_i$ approximately follows a normal distribution with mean p and variance $p(1-p)/n$.

12.4 Let $(x_1, x_2, \ldots, x_n)$ be a random sample of size n obtained from a population which obeys a log-normal distribution defined in Eq. (6.10). Show that the maximum likelihood estimators of the parameter μ and σ^2 of the log-normal distribution are given by

$$\hat{\mu} = \frac{1}{n} \sum_{i=1}^{n} \ln x_i \quad \text{and} \quad \sigma^2 = \frac{1}{n} \sum_{i=1}^{n} (\ln x_i - \hat{\mu})^2$$

12.5 It is known that the failure of a mechanical system follows an exponential distribution with a parameter indicating the rate of failure per unit time. The average value of the system's lifetime obtained from a sample of size 10 is 2550 hours. Evaluate the confidence interval of the probability that the system will operate longer than 2000 hours without failure.
[Hint] The probability of n occurrences of failure in time T follows a Poisson distribution with the mean value νT, where ν is the failure rate per unit time. By letting $n = 0$, the probability that the system will operate for time T without failure is $e^{-\nu T}$.

CHAPTER 13

Wiener–Lévy and Markov Processes

There are two extremely important random processes for formulating stochastic models for many problems in engineering and the physical sciences. These are the Wiener–Lévy process and the Poisson process. The former deals with a continuous state space, while the latter pertains to a discrete state space. However, both processes possess the property of independent increments in a continuous time space, and hence are Markov processes.

The Markov process concept has received considerable attention as a stochastic solution of physical phenomena because of its distinctive characteristics. An example will be shown in Chapter 15 in applying the Markov process concept to a random vibration system with strong nonlinear characteristics.

In this chapter, we discuss the Wiener–Lévy process and the Markov process in detail. The Poisson process will be discussed separately in Chapter 17 as a counting stochastic process.

13.1 WIENER–LÉVY PROCESS

The Wiener–Lévy process is known as the Brownian motion process since it gives a rigorous mathematical explanation of physical Brownian motion. *Brownian motion* is a ceaseless random fluctuating motion of a microscopic particle suspended in a fluid or gas. The phenomenon was found and studied by the botanist Robert Brown in connection with the erratic motion of pollen grains suspended in fluids, although a similar phenomenon was observed earlier by a physician Jan Ingenhousz. A quantitative theory of the Brownian motion was first presented by Einstein based on kinematic theory and statistical mechanics. Then, later a more rigorous mathematical explanation was developed by Wiener and Lévy based on stochastic process theory, called the Wiener–Lévy process.

The Wiener–Lévy process can be applied to many problems in diverse areas in science such as quantum mechanics, statistical tests, diffusion phenomena, and economics. It can be expressed as the integral of a white noise process, and it is a limiting form of the random walk. Hence, it may be well to outline the random walk theory prior to discussing the Wiener–Lévy process.

13.1.1 *Random Walk*

Consider a simplified version of the random movement of a particle in one dimension, namely movement along a straight line. The particle moves a fixed distance $+\Delta x$ or $-\Delta x$ with equal probability of positive or negative direction at each step as shown in Figure 13.1.

Let us consider the total displacement at time t at the end of n movements in the time interval 0 to t. It can be expressed as a sum of each movement Δx_i, which is a sequence of independent, identically distributed random variables. That is,

$$Y(t) = \sum_{i=1}^{n} \Delta X_i \qquad (13.1)$$

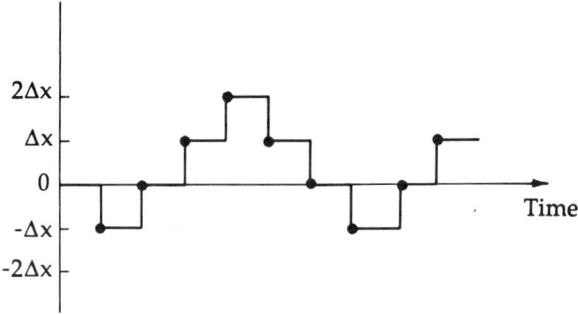

Figure 13.1 Time history of particle movement.

We first derive the probability function of $Y(t)$ for unit displacement at each step, and then later modify the probability function for an arbitrary displacement Δx. Since the movement at each step obeys the Bernoulli probability law, its characteristic function can be written as

$$\phi_x(u) = \tfrac{1}{2}(e^{iu} + e^{-iu}) = \cos u \qquad (13.2)$$

and thereby, the characteristic function of Y becomes

$$\phi_y(u) = (\cos u)^n \qquad (13.3)$$

By applying Theorem 4.5, the probability function applicable for the random variable Y can be derived as follows:

$$p(Y = k) = Pr\{\text{total distance is } k \text{ after } n \text{ steps}\}$$

$$= \frac{1}{2\pi} \int_{-\pi}^{\pi} (\cos u)^n e^{-iku} \, du$$

$$= \frac{1}{2\pi} \frac{2\pi}{2^n(n+1)} \frac{1}{B[(n+k+2)/2, (n-k+2)/2]}$$

$$= \frac{n!}{2^n} \frac{1}{[(n+k)/2]! [(n-k)/2]!} \qquad (13.4)$$

where $B(\ )$ = beta function. If n is an even number (or odd number), then k is chosen to be the largest even number (or odd number) that satisfies $|k| \leq n$.

Next, let us consider the probability when n is large. For this, $(\cos t)^n$ in Eq. (13.4) may be written as

$$\lim_{n \to \infty} (\cos u)^n = \lim_{n \to \infty} \left(1 - \frac{u^2}{2!} + \frac{u^4}{4!} + \cdots \right)^n$$

$$= \exp\left\{-n\left(\frac{u^2}{2!} - \frac{u^4}{4!} + \cdots\right)\right\} \qquad (13.5)$$

Then, Eq. (13.4) becomes

$$p(Y = k) = \frac{1}{2\pi} \int_{-\pi}^{\pi} \exp\left\{-iku - n\left(\frac{u^2}{2!} - \frac{u^4}{4!} + \cdots\right)\right\} du \quad (13.6)$$

By letting $\sqrt{n}\, u = v$, and by writing $p(Y = k)$ as the probability density function $f(Y = k)$, we have for large n

$$f(Y = k) = \frac{1}{2\pi\sqrt{n}} \int_{-\infty}^{\infty} e^{-i(k/\sqrt{n})v - (v^2/2)} dv$$

$$= \frac{1}{\sqrt{2\pi n}} e^{-k^2/2n} \quad (13.7)$$

This is a normal distribution with the mean zero and variance n.

If the particle moves a distance Δx in each step (instead of unity), then the variance of the normal distribution will be $n(\Delta x)^2$. Since the particle moves n steps in time t, we may write $n = t/\Delta t$, where Δt = time interval between successive movements. Then, the variance can be written as $(\Delta x)^2(t/\Delta t)$, and thereby the probability density function of Y can be written as

$$f(y) = \frac{1}{\sqrt{2\pi}\, \Delta x \sqrt{t/\Delta t}} e^{-y^2/[2(\Delta x)^2(t/\Delta t)]} \quad (13.8)$$

By letting $\Delta t \to 0$ and $\Delta x \to 0$ in such a way that $(\Delta x)^2/\Delta t$ is a constant σ^2, the variance in Eq. (13.8) becomes $\sigma^2 t$. Here, $(\Delta x)^2/\Delta t$ represents the mean square displacement of the particle per unit time. It can also be proved that $Y(t)$ is a random process with stationary independent increments. For example, for $t_1 < t_2 < t_3 < t_4$ the increments $\{f_y(t_2) - f_y(t_1)\}$ and $\{f_y(t_3) - f_y(t_4)\}$ are statistically independent. Here, $f_y(t_i)$ stands for the probability density function $f(y; t_i)$.

Furthermore, the increment $\{f_y(t_2) - f_y(t_1)\}$ is independent of $f_y(t_1)$; and thereby we have

$$E\big[\{f_y(t_2) - f_y(t_1)\}f_y(t_1)\big] = E\big[f_y(t_1)f_y(t_2)\big] - E\big[\{f_y(y_1)\}^2\big] = 0$$

Hence

$$E\big[f_y(t_1)f_y(t_2)\big] = E\big[\{f_y(t_1)\}^2\big] = \sigma^2 t_1$$

WIENER–LÉVY PROCESS

which results in the autocorrelation function

$$R_{yy}(t_1, t_2) = \sigma^2 t_1, \quad t_1 < t_2$$

Under these condition the random process $y(t)$ satisfies the requirement of the Wiener–Lévy process. Therefore, we can conclude that the limiting form of the random walk as $n \to \infty$ (or $\Delta t \to 0$) is the Wiener–Lévy process.

13.1.2 Wiener–Lévy Process

The definition of the Wiener–Lévy process is given in Section 9.3.2. That is, a stochastic process $x(t)$ is called a Wiener–Lévy process if $x(t)$ has stationary independent increments and every increment is normally distributed. Furthermore, $x(0) = 0$ and its mean value function is zero. Various properties associated with the Wiener–Lévy process are summarized in the following:

1. The variance of the increment $x(t_2) - x(t_1)$ for $t_1 < t_2$ is proportional to $t_2 - t_1$. That is,

$$\text{Var}[x(t_2) - x(t_1)] = \sigma^2(t_2 - t_1)$$

where $t_1 < t_2$ and σ^2 is a parameter.

Proof. Since the increment of the Wiener–Lévy process is a stationary, independent process, the property follows from the result given in Example 9.3.

2. The Wiener–Lévy process is a Gaussian random process.

Proof. Consider n increments of the process defined as $y_1 = x(t_1)$, $y_2 = x(t_2) - x(t_1)$, $y_3 = x(t_3) - x(t_2), \ldots, y_n = x(t_n) - x(t_{n-1})$, where $t_1 < t_2 < t_3 < \cdots < t_n$. Since $x(t)$ is a Wiener–Lévy process, $y_1, y_2, \ldots, y_n$ are independent and normally distributed and thereby they are jointly normally distributed. Then, we may write $x(t_1), x(t_2), \ldots$ as a linear combination of $y_1, y_2, y_3, \ldots$. That is, $x(t_1) = y_1, x(t_2) = y_1 + y_2, \ldots, x(t_n) = y_1 + y_2 + \cdots + y_n$. Here, $x(t_1), x(t_2), \ldots$ are jointly normally distributed, and thereby $x(t)$ is a normal (Gaussian) random process.

3. $\text{Cov}[x(t_1), x(t_2)] = \text{Var}[x(t_1)]$ for $t_1 < t_2$.

The proof is given in Example 9.4. It is noted that $x(t)$ is not a covariance stationary process.

4. The autocorrelation function of the Wiener–Lévy process is given by $R(t_1, t_2) = \sigma^2 t_1$ for $t_1 < t_2$.

   The proof follows from Item (3) since $E[x(t)] = 0$.

5. The Wiener–Lévy process is a Markov process.

   Since the Wiener–Lévy process has the property of independent increments, it is a Markov process. The proof will be given in Section 13.2.1.

6. For a Wiener–Lévy process with $E[x(t)] = 0$ and $\text{Cov}[x(t_1), x(t_2)] = \sigma^2 t_1$ for $t_1 < t_2$, the integrated process $z(t) = \int x(t)\, dt$ has $E[z(t)] = 0$ and $\text{Var}[z(t)] = \sigma^2 t^3 / 3$.

   The proof is given in Example 10.12.

7. The Wiener–Lévy process is not differentiable in the mean square.

   The proof is given in Example 10.14. As shown in the example, however, the derivative process is a Gaussian white noise. Inversely speaking, we can state that the integrated Gaussian white noise is a Wiener–Lévy process.

8. The conditional probability distribution of the Wiener–Lévy process $f\{x(t)|x(t_0) = x_0\}$, $t_0 < t$, satisfies the following differential equation known as the *diffusion equation*:

$$\frac{\partial}{\partial t} f\{x(t)|x_0\} = \frac{\sigma^2}{2} \frac{\partial^2}{\partial x^2} f\{x(t)|x_0\} \tag{13.9}$$

Proof. Since the Wiener–Lévy process is a Gaussian random process, the joint probability distribution $f\{x(t), x(t_0)\}$ is a bivariate normal distribution. Let us assume that $E[x] = 0$. Then, from the result given in Example 3.13, the conditional distribution $f\{x(t)|x(t_0) = x_0\}$ is a normal distribution with the following conditional mean and variance:

$$E[x(t)|x(t_0) = x_0] = \rho \sqrt{\frac{\text{Var}[x(t)]}{\text{Var}[x(t_0)]}}\, x_0$$

$$\text{Var}[x(t)|x(t_0)] = (1 - \rho^2)\, \text{Var}[x(t)]$$

We may write ρ, Var$[x(t)]$ and Var$[x(t_0)]$ in terms of the autocorrelation function and apply the property given in Item (4). Then, we have

$$E[x(t)|x(t_0) = x_0] = \frac{R_{xx}(t_0, t)}{R_{xx}(t_0, t_0)} x_0 = \frac{\sigma^2 t_0}{\sigma^2 t_0} x_0 = x_0$$

$$\text{Var}[x(t)|x(t_0) = x_0] = R_{xx}(t, t) - \frac{R_{xx}^2(t_0, t)}{R_{xx}(t_0, t_0)}$$

$$= \sigma^2 t - \sigma^2 t_0 = \sigma^2(t - t_0)$$

Thus, the conditional probability density function becomes

$$f\{x(t)|x_0\} = \frac{1}{\sqrt{2\pi}\,\sigma\sqrt{t - t_0}} e^{-(x - x_0)^2/2\sigma^2(t - t_0)} \tag{13.10}$$

By carrying out partial differentiation on the above probability density function, we can derive the desired result.

13.2 MARKOV PROCESS

13.2.1 *Chapman–Kolmogorov Equation*

The definition of a Markov random process is given in Section 9.2.4. However, we may repeat the definition here for the convenience of further discussion on this subject. That is, a stochastic process $x(t)$ is said to be a Markov process if it satisfies the following conditional probability:

$$Pr\{x(t_n) \leq x_n | x(t_{n-1}) = x_{n-1}, \ldots, x(t_2) = x_2, x(t_1) = x_1\}$$

$$= Pr\{x(t_n) \leq x_n | x(t_{n-1}) = x_{n-1}\}$$

where $t_1 < t_2 < \cdots < t_{n-1} < t_n$.

We may simply write the above conditional probability as

$$Pr\{x(t_n) \leq x_n | \text{all } x(t), t \leq t_{n-1}\} = Pr\{x(t_n) \leq x_n | x(t_{n-1}) = x_{n-1}\}$$

$$\tag{13.11}$$

where $t_1 < t_2 < \cdots < t_{n-1} < t_n$.

We define the Markov process with respect to the latest time t_n in the above. We may also define the Markov process with respect to the earliest time t_1 as follows:

$$Pr\{x(t_1) \leq x_1 | \text{all } x(t), t \geq t_2\} = Pr\{x(t_1) \leq x_1 | x(t_2) = x_2\} \quad (13.12)$$

It was also defined in Chapter 9 that a random process $x(t)$ is called a Markov chain if the set of states consists of discrete values, while $x(t)$ is called a diffusion process if the set consists of continuous values.

Theorem 13.1. If the increment $x(t_n) - x(t_{n-1})$ of a random process $x(t)$ is independent of all $x(t)$ for $t \leq t_{n-1}$, then the random process $x(t)$ is a Markov process.

Proof. Let us consider the following conditional probability of the increment $x(t_n) - x(t_{n-1})$ given all $x(t)$ for $t \leq t_{n-1}$:

$$Pr\{x(t_n) - x(t_{n-1}) \leq x_n - x_{n-1} | x(t_{n-1}) = x_{n-1},$$
$$\ldots, x(t_2) = x_2, x(t_1) = x_1\}$$
$$= Pr\{x(t_n) \leq x_n | x(t_{n-1}) = x_{n-1}, \ldots, x(t_2) = x_2, x(t_1) = x_1\}$$
$$(13.13)$$

From the independent increment condition, the conditional probability given on the left-hand side of Eq. (13.13) can be written as

$$Pr\{x(t_n) - x(t_{n-1}) \leq x_n - x_{n-1} | x(t_{n-1}) = x_{n-1}, \ldots, x(t_1) = x_1\}$$
$$= Pr\{x(t_n) \leq x_n | x(t_{n-1}) = x_{n-1}\} \quad (13.14)$$

Thus, from Eqs. (13.13) and (13.14), it can be derived that an independent increment process is a Markov process.

Based on Theorem 13.1, the Poisson process and the Wiener–Lévy process are Markov processes since they have an independent increment property.

In the following, we may write a random process $x(t)$ whose state is x_j at time t_j as $x_j(t_j)$, for brevity. Let us consider the conditional probability density function $f\{x(t)|x_0(t_0)\}$ of a Markov process. $f\{x(t)|x_0(t_0)\}$ is the probability density function of a random process $x(t)$, given that the state

MARKOV PROCESS

Figure 13.2 Schematic representation of Chapman–Kolmogorov equation given in Eq. (13.18).

was x_0 at time t_0, and it is called the *transition probability density function* of a Markov process.

In passing from a state x_0 to the present state x in time $(t - t_0)$, the process passes through some state x_j at time t_j as shown in Figure 13.2. Under this situation, we consider the joint probability density function of $x(t)$ and $x_0(t_0)$. It can be written as

$$f\{x(t), x_0(t_0)\} = \int f\{x(t), x_0(t_0), x_j(t_j)\} \, dx_j$$

$$= \int f\{x(t)|x_0(t_0), x_j(t_j)\} f\{x_0(t_0), x_j(t_j)\} \, dx_j \quad (13.15)$$

Since $x(t)$ is a Markov process, we can write for $t_0 < t_j < t$,

$$f\{x(t)|x_0(t_0), x_j(t_j)\} = f\{x(t)|x_j(t_j)\} \quad (13.16)$$

Hence, the joint probability density function becomes,

$$f\{x(t), x_0(t_0)\} = \int f\{x(t)|x_j(t_j)\} f\{x_0(t_0), x_j(t_j)\} \, dx_j \quad (13.17)$$

By dividing both sides by $f\{x_0(t_0)\}$, the conditional probability density function $f\{x(t)|x_0(t_0)\}$ can be obtained as

$$f\{x(t)|x_0(t_0)\} = \int f\{x(t)|x_j(t_j)\} f\{x_j(t_j)|x_0(t_0)\} \, dx_j \quad (13.18)$$

Equation (13.18) is called the *Chapman–Kolmogorov equation* for the transition probability density $f\{x(t)|x_0(t_0)\}$ of a Markov process, and it provides the fundamental property of the process. The solution of the Chapman–Kolmogorov equation, although difficult to obtain in practice, yields the probability density function applicable for a Markov process (see Section 13.3 for practical solution).

For a Markov chain, we may define the transition probability function as $p\{x(t)|x_0(t_0)\}$, and the Chapman–Kolmogorov equation may be written as follows:

$$p\{x(t)|x_0(t_0)\} = \sum_j p\{x(t)|x_j(t_j)\} p\{x_j(t_j)|x_0(t_0)\} \quad (13.19)$$

Next, we may express the Chapman–Kolmogorov equation given in Eq. (13.18) in terms of the transition time interval. That is, by writing $t_j = t_0 + \tau$, and $t = t_0 + \tau + \Delta\tau$, Eq. (13.18) becomes,

$$f\{x(t_0 + \tau + \Delta\tau)|x_0(t_0)\} = \int f\{x(t_0 + \tau + \Delta\tau)|x_j(t_0 + \tau)\}$$

$$\times f\{x_j(t_0 + \tau)|x_0(t_0)\}\, dx_j \quad (13.20)$$

Let us assume that the Markov process is homogeneous as defined in Eq. (9.9), then the transition probability density function is invariant with respect to a time shift. Hence, we may subtract t_0 from all transition probability density functions in Eq. (13.20), and furthermore, we may subtract τ from the first transition probability density function on the right side of the equation. Then, the Chapman–Kolmogorov equation becomes

$$f\{x(\tau + \Delta\tau)|x_0(0)\} = \int f\{x(\Delta\tau)|x_j(0)\} f\{x_j(\tau)|x_0(0)\}\, dx_j \quad (13.21)$$

in which all transition probability density functions are expressed in terms of the time intervals.

In applying the Chapman–Kolmogorov equation given in Eq. (13.21) it is often assumed that the time interval τ is sufficiently long that the initial state $x_0(0)$ becomes immaterial. This results in the conditional probability density function $f\{x(\tau + \Delta\tau)|x_0(0)\}$ can be substituted by $f\{x(\tau + \Delta\tau)\}$. An example of this simplification will be shown in Section 13.3.

13.2.2 Two-State Markov Chain and Markov Process

A Markov chain or Markov process whose state consists of two discrete elements is a simple mathematical model in stochastic processes, but the application of its concept is extremely useful for many practical problems. As an example, in this section we may consider the two discrete states to represent weather conditions: rain (wet) and no rain (dry). For brevity, let us assume that the stochastic process considered here is homogeneous with respect to time and let the two states be denoted by 0 and 1. Then, there are four possible transitions of states: (0 0), (0 1), (1 0), and (1 1).

For the two-state Markov chain, it is assumed that we have information concerning the probability that the present state is 0 (or 1) followed by a state 0 (or 1) after a one-step transition. In general, p_{ij} (where $i, j = 0, 1$) is called the *one-step transition probability* that represents the probability of changing its state at each step. It can be presented in matrix form called the *transition probability matrix* as follows:

$$\mathsf{P} = \begin{pmatrix} p_{00} & p_{01} \\ p_{10} & p_{11} \end{pmatrix} \tag{13.22}$$

The probabilities of states after n steps are given in the following theorem:

Theorem 13.2. Let the transition probability matrix be given by

$$\mathsf{P} = \begin{pmatrix} p_{00} & p_{01} \\ p_{10} & p_{11} \end{pmatrix}$$

Then, the n-step transition probability matrix, denoted by $\mathsf{P}(n)$ can be obtained by

$$\mathsf{P}(n) = \begin{pmatrix} p_{00}(n) & p_{01}(n) \\ p_{10}(n) & p_{11}(n) \end{pmatrix} = \mathsf{P}^n$$

where

$$p_{ij}(n) = n\text{-step transition probability}$$
$$= Pr\{\text{state after } n \text{ steps} = j \,|\, \text{initial state} = i\}$$
$$i, j = 0, 1.$$

Proof. By applying the Chapman–Kolmogorov equation given in Eq. (13.19), the two-step transition probabilities can be obtained as follows:

$$p_{00}(2) = p_{00}p_{00} + p_{01}p_{10}$$
$$p_{01}(2) = p_{00}p_{01} + p_{01}p_{11}$$
$$p_{10}(2) = p_{10}p_{00} + p_{11}p_{10}$$
$$p_{11}(2) = p_{10}p_{01} + p_{11}p_{11}$$

This implies that the two-step transition probability matrix can be evaluated by

$$P(2) = P^2$$

For the three-step transition probability $p_{00}(3)$, for example, the Chapman–Kolmogorov equation yields

$$p_{00}(3) = p_{00}p_{00}p_{00} + p_{01}p_{10}p_{00} + p_{00}p_{01}p_{10} + p_{01}p_{11}p_{10}$$

which is equal to the element p_{00} of the matrix P^3. Thus, we have

$$P(3) = P(2)P = P^3$$

This relationship can be extended to the n-step transition probability matrix and thereby the theorem can be proved.

Let us examine the n-step transition probabilities in detail. For this, we define $P(n)$ as the probability row vector indicating the process is in state 0 or 1 after n steps. That is,

$$P(n) = (p_0(n), p_1(n)) \tag{13.23a}$$

In particular, the initial probability may be written as

$$P(0) = (p_0(0), p_1(0)) \tag{13.23b}$$

Here, the elements of the row vector $P(n)$ can be expressed by the following recurrence relations:

$$p_0(n) = p_0(n-1)p_{00} + p_1(n-1)p_{10}$$
$$p_1(n) = p_0(n-1)p_{01} + p_1(n-1)p_{11} \tag{13.24}$$

The above equation can be simply written as

$$P(n) = P(n-1)P \tag{13.25}$$

and by iteration we have

$$P(n) = P(0)P^n \tag{13.26}$$

Equation (13.26) implies that the n-step transition probabilities can be evaluated from knowledge of the initial probability $P(0)$ and the one-step transition probability matrix P.

MARKOV PROCESS

In order to evaluate $\mathbf{P}^n$ in practice, application of the characteristic equation of the matrix $\mathbf{P}$ simplifies the computation to a great extent. That is, from the solution of the following characteristic equation

$$|\mathbf{P} - \lambda I| = 0 \tag{13.27}$$

we obtain two solutions λ_1 and λ_2. Next choose a matrix $\mathbf{Q}$ such that its first and second columns, q_1 and q_2, respectively, satisfy the following relationships:

$$\mathbf{P}q_1 = \lambda_1 q_1 \quad \text{and} \quad \mathbf{P}q_2 = \lambda_2 q_2 \tag{13.28}$$

Then, we can express the transition probability matrix as

$$\mathbf{P} = \mathbf{Q} \begin{pmatrix} \lambda_1 & 0 \\ 0 & \lambda_2 \end{pmatrix} \mathbf{Q}^{-1} \tag{13.29}$$

and thereby $\mathbf{P}^n$ can be obtained as follows:

$$\mathbf{P}^n = \mathbf{Q} \begin{pmatrix} \lambda_1^n & 0 \\ 0 & \lambda_2^n \end{pmatrix} \mathbf{Q}^{-1} \tag{13.30}$$

For the one-step transition probability matrix $\mathbf{P}$ given in Eq. (13.22), we have $\lambda_1 = 1$ and $\lambda_2 = p_{00} + p_{11} - 1$ as the solution of the characteristic equation, and thereby $\mathbf{P}$ can be expressed as

$$\mathbf{P} = Q \begin{pmatrix} 1 & 0 \\ 0 & p_{00} + p_{11} - 1 \end{pmatrix} \mathbf{Q}^{-1} \tag{13.31}$$

where

$$Q = \begin{pmatrix} 1 & p_{01} \\ 1 & -p_{10} \end{pmatrix}$$

$$Q^{-1} = \frac{1}{p_{01} + p_{10}} \begin{pmatrix} p_{10} & p_{01} \\ 1 & -1 \end{pmatrix}$$

and $p_{00} + p_{11} - 1 < 1$ unless $p_{00} + p_{11} = 0$ or 2. Then, we can evaluate the transition probability matrix after n steps $\mathbf{P}(n)$ as

$$\mathbf{P}(n) = \mathbf{P}^n = \frac{1}{p_{01} + p_{10}} \begin{pmatrix} 1 & p_{01} \\ 1 & -p_{10} \end{pmatrix} \begin{pmatrix} 1 & 0 \\ 0 & (p_{00} + p_{11} - 1)^n \end{pmatrix} \begin{pmatrix} p_{10} & p_{01} \\ 1 & 1 \end{pmatrix}$$

$$= \frac{1}{p_{01} + p_{10}} \begin{pmatrix} p_{10} & p_{01} \\ p_{10} & p_{01} \end{pmatrix} + \frac{(p_{00} + p_{11} - 1)^n}{p_{01} + p_{10}} \begin{pmatrix} p_{01} & -p_{01} \\ -p_{10} & p_{10} \end{pmatrix} \quad (13.32)$$

It is noted that the first term in Eq. (13.32) is constant, while the second term is a transient term and tends to become zero as n increases, since $p_{00} + p_{11} - 1 < 1$. By letting $p_{00} + p_{11} - 1 = \delta$, we may write the above equation as

$$\mathbf{P}(n) = \mathbf{P}^n = \frac{1}{p_{01} + p_{10}} \begin{pmatrix} p_{10} + p_{01}\delta^n & p_{01}(1 - \delta^n) \\ p_{10}(1 - \delta^n) & p_{01} + p_{10}\delta^n \end{pmatrix} \quad (13.33)$$

From Eq. (13.32) or (13.33) we obtain the limiting state for $n \to \infty$ as

$$\lim_{n \to \infty} p_{00}(n) = \lim_{n \to \infty} p_{10}(n) = \frac{p_{10}}{p_{01} + p_{10}} = \frac{1 - p_{11}}{2 - (p_{00} + p_{11})}$$

$$\lim_{n \to \infty} p_{01}(n) = \lim_{n \to \infty} p_{11}(n) = \frac{p_{01}}{p_{01} + p_{10}} = \frac{1 - p_{00}}{2 - (p_{00} + p_{11})} \quad (13.34)$$

Example 13.1. Rainfall data obtained in City A show that the probability of a dry day followed by a dry day is 0.725, while the probability of a wet day followed by a wet day is 0.640. Hence, we have the following one-step transition probabilities:

$$\mathbf{P} = \begin{pmatrix} 0.725 & 0.275 \\ 0.360 & 0.640 \end{pmatrix}$$

From this information we can estimate the long-term transition probabilities by Eq. (13.34) as follows:

$$\mathbf{P}^n = \begin{pmatrix} 0.567 & 0.433 \\ 0.567 & 0.433 \end{pmatrix}$$

Let the probability of rain tomorrow be 0.60. That is p(0) = (0.40, 0.60). Then, the probability of rain 2 days hence can be evaluated by applying Eqs. (13.26) and (13.33). That is, we have

$$P(2) = \begin{pmatrix} 0.625 & 0.375 \\ 0.491 & 0.509 \end{pmatrix}$$

and thereby

$$p(2) = p(0)P(2) = (0.545 \quad 0.455)$$

Thus, the probability of rain is 0.455. ∎

Additional useful theorems concerning the two-state Markov chains are given in following without proof. Readers who are interested in details on these subjects may refer to Bhat (1972) and Cox and Miller (1965), among others.

Theorem 13.3. Let the transition probability matrix P be given by

$$P = \begin{pmatrix} p_{00} & p_{01} \\ p_{10} & p_{11} \end{pmatrix}$$

with $|p_{00} + p_{11} - 1| = \delta < 1$, and let $\mu_{ij}(n)$, where $i, j = 0, 1$, be the expected number of visits to state j in n steps having initially started from state i. Then, the matrix of μ_{ij} is given as follows:

$$\begin{pmatrix} \mu_{00} & \mu_{01} \\ \mu_{10} & \mu_{11} \end{pmatrix}$$

$$= \frac{1}{p_{01} + p_{10}} \begin{pmatrix} np_{10} + \dfrac{p_{00}p_{01}(1 - \delta^n)}{p_{01} + p_{10}} & np_{01} - \dfrac{p_{00}p_{01}(1 - \delta^n)}{p_{01} + p_{10}} \\ np_{10} - \dfrac{p_{10}\delta(1 - \delta^n)}{p_{01} + p_{10}} & np_{01} + \dfrac{p_{10}\delta(1 - \delta^n)}{p_{01} + p_{10}} \end{pmatrix}$$

(13.35)

Theorem 13.4. Let τ_j ($j = 0, 1$) be the number of steps staying in state j before changing state for a Markov process having the transition probability matrix given in Theorem 13.3. Then, the mean and variance of τ_j are given as follows:

$$E[\tau_0] = p_{00}/(1 - p_{00})$$

$$E[\tau_1] = p_{11}/(1 - p_{11})$$

$$\text{Var}[\tau_0] = p_{00}/(1 - p_{00})^2$$

$$\text{Var}[\tau_1] = p_{11}/(1 - p_{11})^2 \qquad (13.36)$$

Example 13.2. For the rainfall data for City A given in Example 13.1, the expected number of wet days in 1 month, when the first day is rain, can be evaluated by computing μ_{11} given in Eq. (13.35). It can be obtained as 13.3. The mean and variance of continuous wet days can be evaluated from Eq. (13.36). They are

$$E[\tau_1] = 1.78 \quad \text{and} \quad \text{Var}[\tau_1] = 4.94 \qquad \blacksquare$$

Next, let us consider the two-state Markov process $x(t)$ that takes states 0 and 1 at random time intervals. Suppose the process $x(t)$ is in state 0 at time t, it may move on to state 1 at time $t + \Delta t$ or it may stay in state 0. If we assume that the probability of changing state from 0 to 1 in time interval Δt is $\lambda \Delta t$ (where $\lambda > 0$), then the probability of sustaining state 0 in Δt becomes $1 - \lambda \Delta t$. These are denoted as p_{01} and p_{00}, respectively. A similar expression $\mu \Delta t$ is assumed when the initial state is 1 at time t. We may write the transition probabilities in time interval Δt as follows:

$$\begin{pmatrix} p_{00}(\Delta t) & p_{01}(\Delta t) \\ p_{10}(\Delta t) & p_{11}(\Delta t) \end{pmatrix} = \begin{pmatrix} 1 - \lambda \Delta t & \lambda \Delta t \\ \mu \Delta t & 1 - \mu \Delta t \end{pmatrix} \qquad (13.37)$$

Since $X(t)$ is a Markov process homogeneous in time, the probability that the process remains in state 0 during time interval $t + \Delta t$ can be written, by applying the Chapman–Kolmogorov equation, as

$$p_{00}(t + \Delta t) = p_{00}(t) p_{00}(\Delta t) + p_{01}(t) p_{10}(\Delta t) \qquad (13.38)$$

Then, by inserting $p_{00}(\Delta t)$ and $p_{10}(\Delta t)$ given in Eq.(13.37) and by letting

Δt be small, the following linear differential equation can be obtained:

$$\lim_{\Delta t \to 0} \frac{p_{00}(t + \Delta t) - p_{00}(t)}{\Delta t} = \frac{d}{dt} p_{00}(t)$$

$$= -\lambda p_{00}(t) + \mu p_{01}(t) = \mu - (\lambda + \mu) p_{00}(t) \quad (13.39)$$

As the solution of this differential equation we may write

$$p_{00}(t) = \frac{\mu}{\lambda + \mu} + k e^{-(\lambda + \mu) t} \quad (13.40)$$

where k = constant to be determined from the initial condition.

Let the initial condition $x(0) = 0$ and thereby $p_{00}(0) = 1$. This yields the constant k to be $\lambda/(\lambda + \mu)$. Thus, the transition probability $p_{00}(t)$ can be obtained as

$$p_{00}(t) = \frac{\mu}{\lambda + \mu} + \frac{\lambda}{\lambda + \mu} e^{-(\lambda + \mu) t} \quad (13.41)$$

From the relationship $p_{00}(t) + p_{01}(t) = 1$, $p_{01}(t)$ can be obtained as

$$p_{01}(t) = \frac{\lambda}{\lambda + \mu} - \frac{\lambda}{\lambda + \mu} e^{-(\lambda + \mu) t} \quad (13.42)$$

By applying the Chapman–Kolmogorov equation to $p_{11}(t + \Delta t)$ and by carrying out similar analysis procedure, we can derive

$$p_{11}(t) = \frac{\lambda}{\lambda + \mu} + \frac{\mu}{\lambda + \mu} e^{-(\lambda + \mu) t}$$

and

$$p_{10}(t) = \frac{\mu}{\lambda + \mu} - \frac{\mu}{\lambda + \mu} e^{-(\lambda + \mu) t} \quad (13.43)$$

The limiting value of the transition probabilities for $t \to \infty$ can be easily obtained from Eqs. (13.41) through (13.43) as

$$p_{00} = p_{10} = \frac{\mu}{\lambda + \mu}$$

$$p_{01} = p_{11} = \frac{\lambda}{\lambda + \mu} \quad (13.44)$$

Two useful theorems on the two-state Markov process equivalent to those stated in Theorems 13.3 and 13.4 are given in the following without proof.

Theorem 13.5. The transition probability matrix $P(t)$ of a Markov process is given by

$$P(t) = \begin{pmatrix} p_{00}(t) & p_{01}(t) \\ p_{10}(t) & p_{11}(t) \end{pmatrix}$$

Let $\mu_{ij}(t)$, where $i, j = 0, 1$, be the expected length of time that the process spends in state j during a time interval $(0, t)$ having initially started from $x(0) = i$. Then, the matrix of $\mu_{ij}(t)$ can be obtained as follows:

$$\begin{pmatrix} \mu_{00}(t) & \mu_{01}(t) \\ \mu_{10}(t) & \mu_{11}(t) \end{pmatrix} = \frac{1}{\lambda + \mu} \begin{pmatrix} \mu t + \dfrac{\lambda}{\lambda + \mu}\{1 - \alpha(t)\} & \lambda t - \dfrac{\lambda}{\lambda + \mu}\{1 - \alpha(t)\} \\ \mu t - \dfrac{\mu}{\lambda + \mu}\{1 - \alpha(t)\} & \lambda t + \dfrac{\mu}{\lambda + \mu}\{1 - \alpha(t)\} \end{pmatrix},$$

where
$$\alpha(t) = \exp\{-(\lambda + \mu)t\}. \tag{13.45}$$

Theorem 13.6. Consider a Markov process with the transition probability matrix given in Theorem 13.5. Let τ_j ($j = 0, 1$) be the length of time the process spends in state j before changing state, and let us assume that τ_0 and τ_1 are statistically independent and obey the following exponential distributions, respectively:

$$f(\tau_0) = \lambda e^{-\lambda \tau_0}$$

$$f(\tau_1) = \mu e^{-\mu \tau_1}$$

Then, the mean and variance of τ_j are given as follows:

$$E[\tau_0] = 1/\lambda, \quad \text{Var}[\tau_0] = 1/\lambda^2$$

$$E[\tau_1] = 1/\mu, \quad \text{Var}[\tau_1] = 1/\mu^2 \tag{13.46}$$

Example 13.3. Rainfall data in City B show that the time duration of no rain (dry) and rain (wet) are considered to be independent random variables and both obey the exponential distribution with means of 40 hours and 8 hours, respectively. Suppose it is initially wet starting at noon on January 1. The probability that it would be wet at 2 PM on January 3 (50 hours later) can be evaluated by p_{11} in Eq. (13.43) with $\lambda = 1/40 = 0.025$ and $\mu = 1/8 = 0.125$ and $t = 50$. That is, $p_{11} = 0.167$.

The expected length of time it would be wet during a time period of 50 hours beginning at noon on January 1 can be evaluated by μ_{11} given in Eq. (13.45). That is $\mu_{11} = 13.9$ hours. ∎

13.3 FOKKER–PLANCK EQUATION

13.3.1 *Derivation of Fokker–Planck Equation*

As stated in the previous section, the probability density function applicable for a Markov process can be obtained through finding the transition probability density function of the Chapman–Kolmogorov equation, and by letting the transition time interval be sufficiently long. Unfortunately, the Chapman–Kolmogorov equation is an integral equation and hence its solution is not readily obtainable. For the transition probability density of a Markov process, however, there exists an equivalent differential equation known as the Fokker–Planck equation; and thereby the solution of the Chapman–Kolmogorov equation is equivalent to that of the Fokker–Planck equation. The derivation of the Fokker–Planck equation given by Wang and Uhlenbeck (1945) is outlined below:

We first generalize the Chapman–Kolmogorov equation to a multidimensional case. That is, we consider a random vector $\mathbf{x}(t)$ where elements are random processes $x_1(t), x_2(t), \ldots, x_n(t)$. Dimensions of the elements $x_1(t), x_2(t)$, etc. may not be the same; instead, they may be displacement, velocity, acceleration, etc., of a random phenomenon.

For convenience, we consider the transition probability density function at time $t + \Delta t$ with a state $\mathbf{x}'(t + \Delta t)$ given that the state was $\mathbf{x}_0(0)$ at time $t = 0$ as illustrated in Figure 13.3. Then, the Chapman–Kolmogorov equation may be written as

$$f\{\mathbf{x}'(t + \Delta t)|\mathbf{x}_0(0)\} = \int \cdots \int f\{\mathbf{x}'(t + \Delta t)|\mathbf{x}(t)\}$$

$$\times f\{\mathbf{x}(t)|\mathbf{x}_0(0)\}\, dx_1\, dx_2\, \cdots\, dx_n \quad (13.47)$$

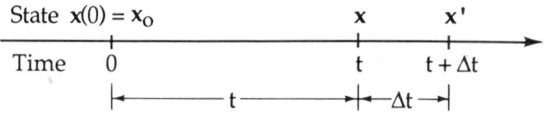

Figure 13.3 Schematic representation of Chapman–Kolmogorov equation for multidimensional random processes **x** given in Eq. (13.47).

where $\mathbf{x}'(t + \Delta t)$ implies that at time $(t + \Delta t)$, each element of the vector $\mathbf{x}(t)$ takes the value $x_1(t + \Delta t) = x'_1, x_2(t = \Delta t) = x'_2, \ldots, x_n(t + \Delta t) = x'_n$.

Let us consider an arbitrary scalar function of $\mathbf{x}'(t + \Delta t)$, denoted by $R\{\mathbf{x}'(t + \Delta t)\}$, and assume that

$$\lim_{x'_j \to \pm\infty} R\{\mathbf{x}'(t + \Delta t)\} = 0$$

$$\lim_{x'_j \to \pm\infty} \frac{\partial^n}{\partial x'^n_j} R\{\mathbf{x}'(t + \Delta t)\} = 0 \qquad (13.48)$$

By multiplying Eq. (13.47) by $R\{\mathbf{x}'(t + \Delta t)\}$, and by integrating with respect to $x'_1, x'_2, \ldots, x'_n$, we have

$$\int_{-\infty}^{\infty} \cdots \int_{-\infty}^{\infty} R\{\mathbf{x}'(t + \Delta t)\} \cdot f\{\mathbf{x}'(t + \Delta t)|\mathbf{x}_0(0)\}\, dx'_1\, dx'_2 \cdots dx'_n$$
$$= \int_{-\infty}^{\infty} \cdots \int_{-\infty}^{\infty} dx_1\, dx_2 \cdots dx_n \int_{-\infty}^{\infty} \cdots \int_{-\infty}^{\infty} R\{\mathbf{x}'(t + \Delta t)\}$$
$$\times f\{\mathbf{x}'(t + \Delta t)|\mathbf{x}(t)\} f\{\mathbf{x}(t)|\mathbf{x}_0(0)\}\, dx'_1\, dx'_2 \cdots dx'_n \qquad (13.49)$$

We may expand $R\{\mathbf{x}'(t + \Delta t)\}$ in a Taylor series in the neighborhood of **x**. That is,

$$R\{\mathbf{x}'(t + \Delta t)\} = R\{\mathbf{x}(t)\} + \sum_i (x'_i - x_i)\frac{\partial}{\partial x_i} R\{\mathbf{x}(t)\}$$
$$+ \frac{1}{2!}\sum_i\sum_j (x'_i - x_i)(x'_j - x_j)\frac{\partial^2}{\partial x_i\,\partial x_j} R\{\mathbf{x}(t)\}$$
$$+ \frac{1}{3!}\sum_i\sum_j\sum_k (x'_i - x_i)(x'_j - x_j)(x'_k - x_k)$$
$$\times \frac{\partial^3}{\partial x_i\,\partial x_j\,\partial x_k} R\{\mathbf{x}(t)\} + \cdots \qquad (13.50)$$

FOKKER–PLANCK EQUATION

By substituting Eq. (13.50) on the right-hand side (RHS) of Eq. (13.49), we have

RHS of Eq. (13.49)

$$= \int_{-\infty}^{\infty} \cdots \int_{-\infty}^{\infty} dx_1 \, dx_2 \cdots dx_n$$

$$\times \Bigg[R\{\mathbf{x}(t)\} \cdot f\{\mathbf{x}(t)|\mathbf{x}_0(0)\}$$

$$\times \int_{-\infty}^{\infty} \cdots \int_{-\infty}^{\infty} f\{\mathbf{x}'(t + \Delta t)|\mathbf{x}(t)\} \, dx'_1 \, dx'_2 \cdots dx'_n$$

$$+ \sum_i \frac{\partial}{\partial x_i} R\{\mathbf{x}(t)\} \cdot f\{\mathbf{x}(t)|\mathbf{x}_0(0)\}$$

$$\times \int_{-\infty}^{\infty} \cdots \int_{-\infty}^{\infty} (x'_i - x_i) f\{\mathbf{x}'(t + \Delta t)|\mathbf{x}(t)\} \, dx'_1 \, dx'_2 \cdots dx'_n$$

$$+ \frac{1}{2!} \sum_i \sum_j \frac{\partial^2}{\partial x_i \, \partial x_j} R\{\mathbf{x}(t)\} \cdot f\{\mathbf{x}(t)|\mathbf{x}_0(0)\}$$

$$\times \int_{-\infty}^{\infty} \cdots \int_{-\infty}^{\infty} (x'_i - x_i)(x'_j - x_j)$$

$$\times f\{\mathbf{x}'(t + \Delta t)|\mathbf{x}(t)\} \, dx'_1 \, dx'_2 \cdots dx'_n$$

$$+ \frac{1}{3!} \sum_i \sum_j \sum_k \frac{\partial^3}{\partial x_i \, \partial x_j \, \partial x_k} R\{\mathbf{x}(t)\} \cdot f\{\mathbf{x}(t)|\mathbf{x}_0(0)\}$$

$$\times \int_{-\infty}^{\infty} \cdots \int_{-\infty}^{\infty} (x'_i - x_i)(x'_j - x_j)(x'_k - x_k)$$

$$\times f\{\mathbf{x}'(t + \Delta t)|\mathbf{x}(t)\} \, dx'_1 \, dx'_2 \cdots dx'_n + \cdots \Bigg] \quad (13.51)$$

Note that

$$\int_{-\infty}^{\infty} \cdots \int_{-\infty}^{\infty} f\{\mathbf{x}'(t + \Delta t)|\mathbf{x}(t)\} \, dx'_1 \, dx'_2 \cdots dx'_n = 1 \quad (13.52)$$

We may use the following notation:

$$a_i\{\mathbf{x}(t)\} = \int_{-\infty}^{\infty} \cdots \int_{-\infty}^{\infty} (x_i' - x_i) f\{\mathbf{x}'(t + \Delta t) | \mathbf{x}(t)\} \, dx_1' \, dx_2' \cdots dx_n'$$

$$b_{ij}\{\mathbf{x}(t)\} = \int_{-\infty}^{\infty} \cdots \int_{-\infty}^{\infty} (x_i' - x_i)(x_j' - x_j)$$

$$\times f\{\mathbf{x}'(t + \Delta t) | \mathbf{x}(t)\} \, dx_1' \, dx_2' \cdots dx_n'$$

$$c_{ijk}\{\mathbf{x}(t)\} = \int_{-\infty}^{\infty} \cdots \int_{-\infty}^{\infty} (x_i' - x_i)(x_j' - x_j)(x_k' - x_k)$$

$$\times f\{\mathbf{x}'(t + \Delta t) | \mathbf{x}(t)\} \, dx_1 \, dx_2 \cdots dx_n \quad (13.53)$$

Then, Eq. (13.51), which represents the right-hand side of Eq. (13.49), becomes

RHS of Eq. (13.49)

$$= \int_{-\infty}^{\infty} \cdots \int_{-\infty}^{\infty} \Bigg[R\{\mathbf{x}(t)\} \cdot f\{\mathbf{x}(t) | \mathbf{x}_0(0)\} + \sum_i a_i\{\mathbf{x}(t)\}$$

$$\times \frac{\partial}{\partial x_i} R\{\mathbf{x}(t)\} f\{\mathbf{x}(t) | \mathbf{x}_0(0)\} + \frac{1}{2!} \sum_i \sum_j b_{ij}\{\mathbf{x}(t)\}$$

$$\times \frac{\partial^2}{\partial x_i \, \partial x_j} R\{\mathbf{x}(t)\} \cdot f\{\mathbf{x}(t) | \mathbf{x}_0(0)\}$$

$$+ \frac{1}{3!} \sum_i \sum_j \sum_k c_{ijk}\{\mathbf{x}(t)\} \frac{\partial^3}{\partial x_i \, \partial x_j \, \partial x_k} R\{\mathbf{x}(t)\}$$

$$\times f\{\mathbf{x}(t) | \mathbf{x}_0(0)\} + \cdots \Bigg] dx_1 \, dx_2 \cdots dx_n \quad (13.54)$$

By replacing $x'_1, x'_2, \ldots, x'_n$ on the left-hand side of Eq. (13.49) by $x_1, x_2, \ldots, x_n$, and by combining Eqs. (13.49) and (13.54), we have

$$\int_{-\infty}^{\infty} \cdots \int_{-\infty}^{\infty} \Bigg[R\{\mathbf{x}(t)\}(f\{\mathbf{x}(t+\Delta t)|\mathbf{x}_0(0)\} - f\{\mathbf{x}(t)|\mathbf{x}_0(0)\})$$

$$- \sum_i a_i\{\mathbf{x}(t)\} \cdot \frac{\partial}{\partial x_i} R\{\mathbf{x}(t)\} \cdot f\{\mathbf{x}(t)|\mathbf{x}_0(0)\}$$

$$- \frac{1}{2!} \sum_i \sum_j b_{ij}\{\mathbf{x}(t)\} \frac{\partial^2}{\partial x_i \, \partial x_j} R\{\mathbf{x}(t)\} \cdot f\{\mathbf{x}(t)|\mathbf{x}_0(0)\}$$

$$- \frac{1}{3!} \sum_i \sum_j \sum_k c_{ijk}\{\mathbf{x}(t)\} \frac{\partial^3}{\partial x_i \, \partial x_j \, \partial x_k} R\{\mathbf{x}(t)\}$$

$$\times f\{\mathbf{x}(t)|\mathbf{x}_0(0)\} \cdots \Bigg] dx_1 \, dx_2 \, \cdots \, dx_n = 0 \qquad (13.55)$$

By carrying out the integration by parts for each term and by taking into consideration the assumptions given in Eq. (13.48), we have

$$\int_{-\infty}^{\infty} \cdots \int_{-\infty}^{\infty} a_i\{\mathbf{x}(t)\} \frac{\partial}{\partial x_i} R\{\mathbf{x}(t)\} \cdot f\{\mathbf{x}(t)|\mathbf{x}_0(0)\} \, dx_1 \, dx_2 \, \cdots \, dx_n$$

$$= - \int_{-\infty}^{\infty} \cdots \int_{-\infty}^{\infty} R\{\mathbf{x}(t)\}$$

$$\times \frac{\partial}{\partial x_i} \Big(a_i\{\mathbf{x}(t)\} f\{\mathbf{x}(t)|\mathbf{x}_0(0)\} \Big) dx_1 \, dx_2 \, \cdots \, dx_n$$

$$\int_{-\infty}^{\infty} \cdots \int_{-\infty}^{\infty} b_{ij}\{\mathbf{x}(t)\} \frac{\partial^2}{\partial x_i \, \partial x_j} R\{\mathbf{x}(t)\} f\{\mathbf{x}(t)|\mathbf{x}_0(0)\} \, dx_1 \, dx_2 \, \cdots \, dx_n$$

$$= \int_{-\infty}^{\infty} \cdots \int_{-\infty}^{\infty} R\{\mathbf{x}(t)\}$$

$$\times \frac{\partial^2}{\partial x_i \, \partial x_j} \Big(b_{ij}\{\mathbf{x}(t)\} f\{\mathbf{x}(t)|\mathbf{x}_0(0)\} \Big) dx_1 \, dx_2 \, \cdots \, dx_n$$

$$\int_{-\infty}^{\infty} \cdots \int_{-\infty}^{\infty} c_{ijk}\{\mathbf{x}(t)\} \frac{\partial^3}{\partial x_i \, \partial x_j \, \partial x_k} R\{\mathbf{x}(t)\}$$

$$\times f\{\mathbf{x}(t)|\mathbf{x}_0(0)\} \, dx_1 \, dx_2 \cdots dx_n$$

$$= -\int_{-\infty}^{\infty} \cdots \int_{-\infty}^{\infty} R\{\mathbf{x}(t)\}$$

$$\times \frac{\partial^3}{\partial x_i \, \partial x_j \, \partial x_k} \left(c_{ijk}\{\mathbf{x}(t)\} f\{\mathbf{x}(t)|\mathbf{x}_0(0)\} \right) dx_1 \, dx_2 \cdots dx_n$$

(13.56)

Then, Eq. (13.55) becomes

$$\int_{-\infty}^{\infty} \cdots \int_{-\infty}^{\infty} R\{\mathbf{x}(t)\}$$

$$= \left[f\{\mathbf{x}(t + \Delta t)|\mathbf{x}_0(0)\} - f\{\mathbf{x}(t)|\mathbf{x}_0(0)\} \right.$$

$$+ \sum_i \frac{\partial}{\partial x_i} \left(a_i\{\mathbf{x}(t)\} \cdot f\{\mathbf{x}(t)|\mathbf{x}_0(0)\} \right)$$

$$- \frac{1}{2!} \sum_i \sum_j \frac{\partial^2}{\partial x_i \, \partial x_j} \left(b_{ij}\{\mathbf{x}(t)\} \cdot f\{\mathbf{x}(t)|\mathbf{x}_0(0)\} \right)$$

$$+ \frac{1}{3!} \sum_i \sum_j \sum_k \frac{\partial^3}{\partial x_i \, \partial x_j \, \partial x_k} \left(c_{ijk}\{\mathbf{x}(t)\} f\{\mathbf{x}(t)|\mathbf{x}_0(0)\} \right) \cdots \left. \right]$$

$$\times dx_1 \, dx_2 \cdots dx_n = 0 \quad (13.57)$$

Since $R\{\mathbf{x}(t)\}$ is an arbitrary function, the integrand must be equal to zero. Furthermore, if the integrand is divided by Δt, where $\Delta t \to 0$, the following formula can be derived:

$$\frac{\partial}{\partial t} f\{\mathbf{x}(t)|\mathbf{x}_0(0)\} + \sum_i \frac{\partial}{\partial x_i} \left(A_i\{\mathbf{x}(t)\} \cdot f\{\mathbf{x}(t)|\mathbf{x}_0(0)\} \right)$$

$$- \frac{1}{2!} \sum_i \sum_j \frac{\partial^2}{\partial x_i \partial x_j} \left(B_{ij}\{\mathbf{x}(t)\} \cdot f\{\mathbf{x}(t)|\mathbf{x}_0(0)\} \right)$$

$$+ \frac{1}{3!} \sum_i \sum_j \sum_k \frac{\partial^3}{\partial x_i \partial x_j \partial x_k} \left(C_{ijk}\{\mathbf{x}(t)\} \cdot f\{\mathbf{x}(t)|\mathbf{x}_0(0)\} \right) \cdots = 0$$

(13.58)

where $f\{\mathbf{x}(t)|\mathbf{x}_0(t)\}$ = transition probability density function

$$A_i\{\mathbf{x}(t)\} = \lim_{\Delta t \to 0} \frac{1}{\Delta t} a_i\{\mathbf{x}(t)\}$$

$$B_{ij}\{\mathbf{x}(t)\} = \lim_{\Delta t \to 0} \frac{1}{\Delta t} b_{ij}\{\mathbf{x}(t)\}$$

$$C_{ijk}\{\mathbf{x}(t)\} = \lim_{\Delta t \to 0} \frac{1}{\Delta t} c_{ijk}\{\mathbf{x}(t)\}, \cdots$$

It is noted that from the definition of $a_i\{\mathbf{x}(t)\}, b_{ij}\{\mathbf{x}(t)\}, c_{ijk}\{\mathbf{x}(t)\}$ given in Eq. (13.53), the coefficients, $A_i\{\mathbf{x}(t)\}, B_{ij}\{\mathbf{x}(t)\}, C_{ijk}\{\mathbf{x}(t)\}$ can be simply expressed as expected values in the following manner:

$$A_i\{\mathbf{x}(t)\} = \lim_{\Delta t \to 0} \frac{1}{\Delta t} E[\Delta x_i | \mathbf{x}]$$

$$B_{ij}\{\mathbf{x}(t)\} = \lim_{\Delta t \to 0} \frac{1}{\Delta t} E[(\Delta x_i)(\Delta x_j) | \mathbf{x}]$$

$$C_{ijk}\{\mathbf{x}(t)\} = \lim_{\Delta t \to 0} \frac{1}{\Delta t} E[(\Delta x_i)(\Delta x_j)(\Delta x_j) | \mathbf{x}]$$

where

$$\Delta x_i = x_i(t + \Delta t) - x_i(t), \cdots \quad (13.59)$$

Equation (13.58) is called the *Fokker–Planck equation* or the *Kolmogorov forward equation*, the solution of which is the transition probability density function $f\{\mathbf{x}(t)|\mathbf{x}_0(0)\}$ of the Markov process $\mathbf{x}(t)$. If we assume that the process is stationary, then the first term of Eq. (13.58) becomes zero. We may also assume that a sufficiently long time has passed since the initial transient disturbance so that the conditional probability density function $f\{\mathbf{x}(t)|\mathbf{x}_0(t)\}$ can be substituted by $f\{\mathbf{x}(t)\}$. Under these assumptions, the Fokker–Planck equation can be written as

$$\sum_i \frac{\partial}{\partial x_i}[A_i(\mathbf{x}) \cdot f(\mathbf{x})] - \frac{1}{2!}\sum_i \sum_j \frac{\partial^2}{\partial x_i \, \partial x_j}[B_{ij}(\mathbf{x}) \cdot f(\mathbf{x})]$$

$$+ \frac{1}{3!}\sum_i \sum_j \sum_k \frac{\partial^3}{\partial x_i \, \partial x_j \, \partial x_k}[C_{ijk}(\mathbf{x}) \cdot f(\mathbf{x})] + \cdots = 0 \quad (13.60)$$

For a one-dimensional Markov process, the Fokker–Planck equation given in Eq. (13.60) becomes

$$\frac{\partial}{\partial x}[A(x)f(x)] - \frac{1}{2!}\frac{\partial^2}{\partial x^2}[B(x)f(x)] + \frac{1}{3!}\frac{\partial^3}{\partial x^3}[C(x)f(x)] \cdots = 0$$

$$(13.61)$$

where

$$A(x) = \lim_{\Delta t \to 0} \frac{1}{\Delta t} E[\Delta x | x]$$

$$B(x) = \lim_{\Delta t \to 0} \frac{1}{\Delta t} E[(\Delta x)^2 | x]$$

$$C(x) = \lim_{\Delta t \to 0} \frac{1}{\Delta t} E[(\Delta x)^3 | x], \ldots, \text{and}$$

$$\Delta x = x(t + \Delta t) - x(t).$$

Next, let us consider a Markov chain for a system with a time–state relationship shown in Figure 13.4. We may write the transition probability as

$$p_{ij}(t + \Delta t) = \Pr\{x(t + \Delta t) = j | x(0) = i\} \quad (13.62)$$

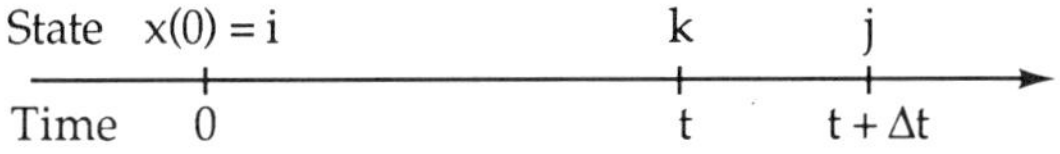

FOKKER–PLANCK EQUATION

```
State  x(0) = i              k        j
       +─────────────────────+────────+──→
Time   0                     t      t + Δt
```

Figure 13.4 Schematic representation of Chapman–Kolmogorov equation given in Eq. (13.63).

Then, the Chapman–Kolmogorov equation can be written as

$$p_{ij}(t + \Delta t) = \sum_{k} p_{ik}(t) p_{kj}(\Delta t) \tag{13.63}$$

By subtracting p_{ij} from both sides of Eq. (13.63) and by dividing by time Δt, we have

$$\frac{1}{\Delta t}\{p_{ij}(t + \Delta t) - p_{ij}(t)\}$$

$$= \frac{1}{\Delta t}\left\{\sum_{k \neq j} p_{ik}(t) p_{kj}(\Delta t) + p_{ij}(\Delta t) p_{jj}(\Delta t) - p_{ij}(t)\right\}$$

$$= \sum_{k \neq j} p_{ik}(t) \frac{1}{\Delta t} p_{kj}(\Delta t) - p_{ij}(t) \frac{1}{\Delta t}\{1 - p_{jj}(\Delta t)\} \tag{13.64}$$

Let $\Delta t \to 0$, and we may write

$$\lim_{\Delta t \to 0} \frac{1}{\Delta t}\{1 - p_{jj}(\Delta t)\} = \lambda_j(t)$$

$$\lim_{\Delta t \to 0} \frac{1}{\Delta t} p_{kj}(\Delta t) = \lambda_{kj}(t) \tag{13.65}$$

Here, the functions $\lambda_j(t)$ and $\lambda_{kj}(t)$ become constants λ_j and λ_{kj}, respectively, for a homogeneous Markov chain. Then, Eq. (13.64) can be written as

$$\frac{d}{dt} p_{ij}(t) = \sum_{k \neq j} \lambda_{kj} p_{ik}(t) - \lambda_j(t) p_{ij}(t) \tag{13.66}$$

This differential equation referred to a fixed state i and is known as *Kolmogorov forward equation* applicable for a Markov chain.

13.3.2 *Application to Nonlinear Vibration Systems*

As an example of application of the Fokker–Planck equation to practical problems, we present the solution of a nonlinear vibration system having a Gaussian random process as an excitation.

The Markov process approach was first applied to the solution of a nonlinear system by Andronov et al. in the 1930s, and significant progress in application to physical and engineering problems was made during the succeeding 20 years. Among others, Wang and Uhlenbeck (1945), Chang and Kazda (1959), Ariaratnam (1960), Lyon (1960a), Crandall (1962), and Caughey (1959, 1963a) made important contributions in this area.

Let us write the equation of a nonlinear vibration system as

$$\ddot{y} + g(\dot{y}) + h(y) = f_x(t) \tag{13.67}$$

where

$g(\dot{y})$ = damping per unit mass (or moment of inertia) including nonlinear terms

$h(y)$ = restoring per unit mass (or moment of inertia) including nonlinear terms

$f_x(t)$ = excitation per unit mass (or moment of inertia).

It is assumed that the input random process $x(t)$ is a stationary Gaussian random process with mean value zero and having a white noise spectrum A. Let the elements of the output random vector $\mathbf{y}(t)$ be the displacement, y, and velocity $\dot{y}$, and we may write $y = y_1$ and $\dot{y} = y_2$. Then, Eq. (13.67) can be expressed in the following form of two first order equations:

$$\dot{y}_1 = y_2$$

$$\dot{y}_2 = -g(y_2) - h(y_1) + f_x(t) \tag{13.68}$$

The coefficients A and B of the Fokker–Planck equation shown in Eqs. (13.58) and (13.59) can be evaluated for the present problem as follows:

$$A_1 = \lim_{\Delta t \to 0} \frac{1}{\Delta t} E[\Delta y_1] = \lim_{\Delta t \to 0} \frac{1}{\Delta t} E[\dot{y}_1 \Delta t] = \dot{y}_1 = \dot{y}$$

$$A_2 = \lim_{\Delta t \to 0} \frac{1}{\Delta t} E[\Delta y_2] = -g(y_2) - h(y_1)$$
$$= -g(\dot{y}) - h(y)$$

$$B_{11} = \lim_{\Delta t \to 0} \frac{1}{\Delta t} E\left[(\Delta y_1)^2\right] = \lim_{\Delta t \to 0} \frac{1}{\Delta t} E\left[(\dot{y} \Delta t)^2\right] = 0$$

$$B_{12} = B_{21} = \lim_{\Delta t \to 0} \frac{1}{\Delta t} E[(\Delta y_1)(\Delta y_2)] = 0$$

$$B_{22} = \lim_{\Delta t \to 0} \frac{1}{\Delta t} E\left[(\Delta y_2)^2\right] = \lim_{\Delta t \to 0} \frac{1}{\Delta t} E\left[(\dot{y}_2 \Delta t)^2\right]$$

$$= \lim_{\Delta t \to 0} \frac{1}{\Delta t} E\bigg[\{-g(y_2) - h(y_1)\}^2 (\Delta t)^2$$

$$- 2\{g(y_2) + h(y_1)\} \Delta t \int_t^{t+\Delta t} f_x(u)\, du$$

$$+ \int_t^{t+\Delta t} \int_t^{t+\Delta t} f_x(u) f_x(v)\, du\, dv\bigg]$$

$$= \lim_{\Delta t \to 0} \frac{1}{\Delta t} E\left[\int_t^{t+\Delta t} \int_t^{t+\Delta t} f_x(u) f_x(v)\, du\, dv\right] = \pi A, \quad (13.69)$$

where A = constant spectral density function of the excitation.

It is noted with respect to B_{22} that $E[f_x(u) f_x(v)]$ is equal to $I(u)\delta(v - u)$, where $I(u)$ is the intensity function of the shot noise and $\delta(v - u)$ is the delta function. We assume that $I(u)$ is a constant I and the shot noise is a weakly stationary random process. Then $I\delta(\tau)$, where $\tau = v - u$, is essentially an autocorrelation function and hence following the definition given in Eq. (10.34) the spectral density function is equal to I/π. Inversely, by letting the spectral density function of the vibration system be a constant A, $E[f_x(u)f_x(v)]$ becomes $\pi A\,\delta(\tau)$, and this results in $B_{22} = \pi A$.

From the results obtained in Eq. (13.69), the Fokker–Planck equation for the nonlinear system given in Eq. (13.67) becomes

$$-\dot{y}\frac{\partial}{\partial y}f(y,\dot{y}) + \frac{\partial}{\partial \dot{y}}\left(\{g(\dot{y}) + h(y)\}f(y,\dot{y})\right) + \frac{\pi A}{2}\frac{\partial^2}{\partial \dot{y}^2}f(y,\dot{y}) = 0$$

(13.70)

where

$f(y,\dot{y}) =$ joint probability density function of displacement and velocity.

Although an analytical solution to Eq. (13.70) in closed form is difficult to derive, it can be obtained if the damping of the system is linear. That is, let the linear damping be $\alpha\dot{y}$ and the nonlinear restoring be $h(y)$. Then, the Fokker–Planck equation becomes

$$-\dot{y}\frac{\partial}{\partial y}f(y,\dot{y}) + \frac{\partial}{\partial \dot{y}}\left(\{\alpha\dot{y} + h(y)\}f(y,\dot{y})\right) + \frac{\pi A}{2}\frac{\partial^2}{\partial \dot{y}^2}f(y,\dot{y}) = 0$$

(13.71)

The above equation may be written as

$$\left(\alpha\frac{\partial}{\partial \dot{y}} - \frac{\partial}{\partial y}\right)\left[\dot{y}f(y,\dot{y}) + \frac{\pi A}{2\alpha}\frac{\partial}{\partial \dot{y}}f(y,\dot{y})\right]$$

$$+ \frac{\partial}{\partial \dot{y}}\left[h(y)f(y,\dot{y}) + \frac{\pi A}{2\alpha}\frac{\partial}{\partial y}f(y,\dot{y})\right] = 0 \quad (13.72)$$

Then, the solution is given by the following pair of equations:

$$\dot{y}f(y,\dot{y}) + \frac{\pi A}{2\alpha}\frac{\partial}{\partial \dot{y}}f(y,\dot{y}) = 0$$

$$h(y)f(y,\dot{y}) + \frac{\pi A}{2\alpha}\frac{\partial}{\partial y}f(y,\dot{y}) = 0 \quad (13.73)$$

FOKKER–PLANCK EQUATION

Assuming that y and $\dot{y}$ are statistically uncorrelated, we can derive the probability density function $f(\dot{y})$ from the first equation of (13.73) as

$$f(\dot{y}) = (\text{constant})e^{-(\alpha/\pi A)\dot{y}^2} \tag{13.74}$$

and from the second equation, we have the probability density function of $f(y)$. That is

$$f(y) = (\text{constant})\exp\left\{-\frac{2\alpha}{\pi A}\int_0^y h(y)\,dy\right\} \tag{13.75}$$

where the constants are determined such that $f(y)$ and $f(\dot{y})$ satisfy the property of the probability density function. Further discussion on this subject will be given in Chapter 15.

CHAPTER 14

Linear System and Stochastic Prediction

In this chapter the concept of the stochastic process is applied to predict the response of a physical vibratory system to a random excitation under the condition that the system has linear response characteristics. Methods for evaluating the frequency characteristics of the system are discussed, and the relationship between the input and output of the system is presented; in particular, the relationship between the input and output autocorrelation functions and spectral density functions.

14.1 LINEAR SYSTEM AND UNIT IMPULSE RESPONSE

14.1.1 *Linear System*

Consider a vibratory system subject to a random excitation $x(t)$ illustrated in Figure 14.1. If the excitation is applied irregularly, the system will vibrate randomly, and the magnitude as well as the frequency components of the response, $y(t)$, may not be the same as those of the input, $x(t)$. The behavior of the response depends entirely on the system's individual response characteristics as will be discussed in Section 14.2.

Let us express the response (output) of a system, $y(t)$, to an excitation (input) $x(t)$ as

$$L[x(t)] = y(t) \qquad (14.1)$$

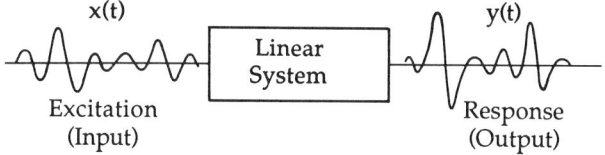

Figure 14.1 Schematic representation of a linear system.

The system is called *time invariant* if the output follows the same time shift that takes place in the input. That is,

$$L[x(t+\tau)] = y(t+\tau) \tag{14.2}$$

The system is called *linear* if (1) the magnitude of the output follows that of the input, namely,

$$L[ax(t)] = ay(t) \tag{14.3}$$

and (2) the system satisfies an additive property; that is,

$$L[x_1(t) + x_2(t)] = y_1(t) + y_2(t) \tag{14.4}$$

In summary, we may define the time-invariant linear system as follows:

Definition 14.1. The output of a vibratory system, $y(t)$, to an input, $x(t)$, is expressed as $L[x(t)] = y(t)$. The system is defined as a *time-invariant linear system* if it satisfies the following conditions:

$$L[x(t+\tau)] = y(t+\tau)$$

$$L[a_1 x_1(t) + a_2 x_2(t)] = a_1 y_1(t) + a_2 y_2(t)$$

Stochastic analysis of oscillating phenomena excited by random input observed in physics and the engineering fields is most commonly carried out under the linearity assumption. For some vibratory systems, however, the assumption of linearity may be questionable, and it is desirable to check whether or not the system is linear. In this case, one method of examining the degree of linearity of the system is to evaluate the coherency function discussed in Section 14.3.3.

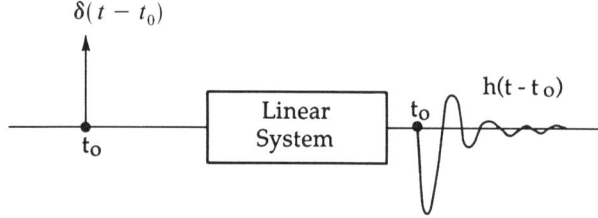

Figure 14.2 Definition of impulse response function $h(t)$.

Example 14.1. The input of a time invariant linear system is given by $e^{i\omega t}$. We may write the output as

$$y(t + t_0) = L[e^{i\omega(t+t_0)}] = e^{i\omega t_0}L[e^{i\omega t}] = e^{i\omega t_0}y(t)$$

By letting $t = 0$, we have

$$y(t_0) = e^{i\omega t_0}y(0)$$

Since t_0 is arbitrary, we may write t_0 as t. Hence, we can write

$$y(t) = e^{i\omega t}y(0)$$

This implies that the output of the system is also given by $e^{i\omega t}$. ∎

14.1.2 Impulse Response Function

We first consider the response of a linear system under a particular situation as shown in Figure 14.2. That is, let the input be an impulse at time t_0 represented by the unit impulse $\delta(t - t_0)$ and let the response of the system be $h(t - t_0)$. This response (output) is called the *impulse response function* of the system representing the unique response characteristics of the system.

Next, consider a random vibratory input $x(t)$, and obtain the response of the system $y(t)$ from information of the impulse response function. For this, the input $x(t)$ is divided into many impulses of varying intensity as shown in Figure 14.3. The impulse function at time t_k for an input of strength $x(t_k)\Delta t_k$ becomes $x(t_k) \cdot \delta(t - t_k) \cdot \Delta t_k$. Then, the impulse response function can be written as $x(t_k) \cdot h(t - t_k)\Delta t_k$. Since the response

LINEAR SYSTEM AND UNIT IMPULSE RESPONSE

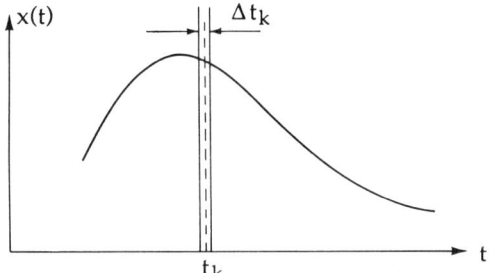

Figure 14.3 Unit impulse function $x(t_k)\delta(t - t_k)\Delta t_k$ of an arbitrary input $x(t)$.

of the system is the accumulation of the impulse responses, we have

$$\text{Output } y(t) = \sum_k x(t_k) \cdot h(t - t_k) \Delta t_k \qquad (14.5)$$

By letting $\Delta t_k \to 0$, we can write Eq. (14.5) in the form of integration. That is,

$$y(t) = \lim_{\Delta t_k \to 0} \sum_k x(t_k) \cdot h(t - t_k) \Delta t_k$$

$$= \int_{-\infty}^{\infty} x(\tau) h(t - \tau) \, d\tau = x(t) * h(t) \qquad (14.6)$$

Since we consider the steady-state random process of a time-invariant linear system, we may write the above equation as

$$y(t) = \int_{-\infty}^{\infty} x(t - \tau) h(\tau) \, d\tau \qquad (14.6a)$$

Equation (14.6) shows that the output $y(t)$ of a linear system can be evaluated as the convolution integral of the input and frequency response function.

The relationship derived in Eq. (14.6) is in the time domain. If we consider the input–output relationship in the frequency domain, then by the property of the Fourier transform (see Appendix A), the convolution integral in the time domain can be expressed as the product of Fourier transforms in the frequency domain. That is,

$$Y(\omega) = X(\omega) H(\omega) \qquad (14.7)$$

where $X(\omega)$, $Y(\omega)$, and $H(\omega)$ are the Fourier transform of $x(t)$, $y(t)$, and

$h(t)$, respectively. For instance, $H(\omega)$ is given by

$$H(\omega) = \int_{-\infty}^{\infty} h(t) e^{-i\omega t} \, dt \tag{14.8}$$

Here, the function $H(\omega)$ is called the *frequency response function* (or *transfer function*). The squared value, $|H(\omega)|^2$, is referred to the *response amplitude operator* in the engineering field. From Eqs. (14.6) and (14.7), we have the following theorem:

Theorem 14.1. The output $y(t)$ of a linear system to an arbitrary random input $x(t)$ is given by the following convolution integral in the time domain:

$$y(t) = x(t) * h(t)$$

where $h(t)$ is the impulse response function. In the frequency domain, the relationship can be expressed as

$$Y(\omega) = X(\omega) H(\omega)$$

where $X(\omega)$, $Y(\omega)$, and $H(\omega)$ are the Fourier transform of $x(t)$, $y(t)$, and $h(t)$, respectively.

Example 14.2. The input of a linear system is given by $ke^{i\omega t}$, where k is a constant. Then, the output can be written as

$$Y(\omega) = ke^{i\omega t} H(\omega)$$

By letting $\omega = 0$, the input is a constant k, and the output becomes

$$Y(0) = kH(0)$$

where $Y(0)$ implies a response to static input, and $H(0)$ represents the response characteristics of the system to a static input. ∎

Example 14.3. Let us consider the response of a system to a unit step function, $u(t)$, defined as $u(t) = 1$ for $t > 0$ and $u(t) = 0$ otherwise (see Appendix C). Since the system is linear, the response can be written as

$$y(t) = \int_{-\infty}^{\infty} u(t - \tau) h(\tau) \, d\tau$$

LINEAR SYSTEM AND UNIT IMPULSE RESPONSE

where $h(t)$ is the frequency response function, and

$$u(t - \tau) = \begin{cases} 1 & \text{for } t > \tau \\ 0 & \text{for } t < \tau \end{cases}$$

Hence, we have

$$y(t) = \int_{-\infty}^{\infty} 1 \cdot h(\tau) \, d\tau$$

For a physically realizable system, we may write

$$y(t) = \int_{0}^{\infty} h(\tau) \, d\tau$$

It is noted that $y(\infty) = \int_{0}^{\infty} h(\tau) \, d\tau = H(0)$. ∎

14.1.3 Input and Output Mean Levels

If the mean value of the input of a linear system is not zero but is a constant x_0 as shown in Figure 14.4, then question arises as to the mean value of the output.

The mean value of the output may be written from Eq. (14.6a) as

$$E[y(t)] = \int_{-\infty}^{\infty} E[x(t - \tau)] \cdot h(\tau) \, d\tau \tag{14.9}$$

Since we consider a steady-state random process and a time-invariant linear system, we can write Eq. (14.9) as

$$E[y(t)] = E[x(t)] \int_{-\infty}^{\infty} h(\tau) \, d\tau \tag{14.10}$$

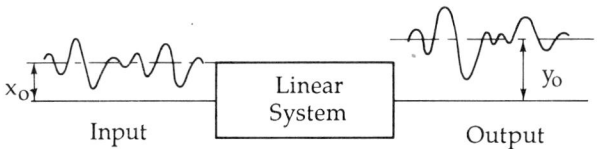

Figure 14.4 Input and output mean levels of a linear system.

On the other hand, the frequency response function is defined in Eq. (14.8) as

$$H(\omega) = \int_{-\infty}^{\infty} h(\tau) e^{-i\omega\tau} d\tau$$

By letting $\omega = 0$, we have

$$H(0) = \int_{-\infty}^{\infty} h(\tau) d\tau \qquad (14.11)$$

and hence, Eq. (14.10) can be written as

$$E[y(t)] = E[x(t)] \cdot H(0) \qquad (14.12)$$

The mean value of the output, therefore, can be evaluated from knowledge of $H(0)$, which represents the response of the system to a static (nonvibratory) input.

14.2 EVALUATION OF IMPULSE AND FREQUENCY RESPONSE FUNCTIONS

It was shown in the previous section that if we know either the impulse response function, $h(t)$, of a linear system in the time domain, or the frequency response function, $H(\omega)$, then it is possible to obtain the time history of the response of the system to an arbitrarily given input $x(t)$. Subsequently, we want to know how to obtain either the impulse response function $h(t)$ or the frequency response function $H(\omega)$ of the system. These can be evaluated either theoretically or experimentally as will be discussed in the following.

14.2.1 Theoretical Approach

The frequency response function, $H(\omega)$, of a linear system can be simply obtained as a solution of differential equation of a vibratory system excited by a harmonic force (or moment) with a frequency ω. It may be written as the real part of $Xe^{i\omega t}$, where X is the excitation amplitude.

For a system having a single degree of freedom, the differential equation equation is given by

$$m\ddot{y} + c\dot{y} + ky = \text{Re}\{Xe^{i\omega t}\} \qquad (14.13)$$

EVALUATION OF IMPULSE AND FREQUENCY RESPONSE FUNCTIONS 373

where the first term represents the inertia force (or moment), the second and third terms refer to the damping and restoring forces (or moments), respectively, of the system, and $\text{Re}\{Xe^{i\omega t}\}$ is the real part of $Xe^{i\omega t}$. Equation (14.13) can also be written in terms of the natural frequency of undamped oscillation of the system, denoted by ω_0, as

$$\ddot{y} + 2\zeta\omega_0\dot{y} + \omega_0^2 y = \text{Re}\{X_0 e^{i\omega t}\} \tag{14.14}$$

where

$\omega_0 = \sqrt{k/m}$ = natural frequency of undamped oscillation

$\zeta = c/(2m\omega_0)$ = damping factor

$2m\omega_0$ = critical damping

$X_0 = X/m$ = complex amplitude of the excitation per unit mass (or moment of inertia).

As a general solution of Eq. (14.14), we consider (1) the general solution of the homogeneous equation (free oscillation), and (2) the particular solution of the nonhomogenous equation. The former yields a damped free oscillation with its natural frequency given in the following form:

$$y(t) = Ke^{-\zeta\omega_0 t} e^{\pm i\sqrt{1-\zeta^2}\,\omega_0 t} \tag{14.15}$$

where K is determined from the initial condition. As can be seen in Eq. (14.15), the oscillation damps out with time and it is not a steady-state oscillation; hence, this is not the solution we are seeking.

For the particular solution, let us write

$$y(t) = \text{Re}\{Y_0 e^{i\omega t}\} \tag{14.16}$$

Then, we can derive

$$Y_0 = \frac{X_0}{(\omega_0^2 - \omega^2) + i(2\zeta\omega_0\omega)} \tag{14.17}$$

Hence, the frequency response function as defined in Eq. (14.8) can be obtained as

$$H(\omega) = \frac{1}{(\omega_0^2 - \omega^2) + i(2\zeta\omega_0\omega)} \tag{14.18}$$

The absolute value of $H(\omega)$ becomes

$$|H(\omega)| = \frac{1}{\sqrt{(\omega_0^2 - \omega^2)^2 + (2\zeta\omega_0\omega)^2}} \tag{14.19}$$

Next, let us consider an input that is an impulse represented by the unit impulse function $\delta(t)$. The differential equation of the oscillation is given by

$$\ddot{y} + 2\zeta\omega_0\dot{y} + \omega_0^2 y = \frac{1}{m}\delta(t) \tag{14.20}$$

From the definition of the unit impulse function, $\delta(t) = 0$ everywhere except $t = 0$. This implies that the oscillation induced by $\delta(t)$ is essentially a free damped oscillation that is given in Eq. (14.15). Hence, we may write the response as

$$y(t) = e^{-\zeta\omega_0 t}\left(A\cos\sqrt{1 - \zeta^2}\,\omega_0 t + iB\sin\sqrt{1 - \zeta^2}\,\omega_0 t\right) \tag{14.21}$$

As an initial condition, let $y(t) = 0$ at $t = 0$. Then, we have $A = 0$ and $y(t)$ becomes

$$Y(t) = iBe^{-\zeta\omega_0 t}\sin\sqrt{1 - \zeta^2}\,\omega_0 t \tag{14.22}$$

On the other hand, from the concept of momentum associated with the impulse, we have for the unit mass system,

$$\dot{y}(0) = \int_0^\infty \delta(t)\,dt = 1 \tag{14.23}$$

From Eqs. (14.22) and (14.23), the constant B can be determined as $B = 1/(i\sqrt{1 - \zeta^2}\,\omega_0)$. Since the response is that excited by the impulse $\delta(t)$, we may write the response $y(t)$ as the impulse response function $h(t)$ defined in the previous section. Thus, we have

$$h(t) = \frac{1}{\sqrt{1 - \zeta^2}\,\omega_0}e^{-\zeta\omega_0 t}\sin\sqrt{1 - \zeta^2}\,\omega_0 t \tag{14.24}$$

It is noted that $\sqrt{1 - \zeta^2}\,\omega_0$ in the above equation is equal to the frequency of the damped vibrating system. It can be proved that the frequency response function $H(\omega)$ derived in Eq. (14.18) is the Fourier transform of $h(t)$ obtained in Eq. (14.24). The proof is left for the reader's exercise.

14.2.2 Experimental Approach

The frequency response function, $H(\omega)$, of a linear system may be obtained by carrying out a series of experiments in which the response of a system to regular sinusoidal excitation of various frequencies are measured. Let $x(t) = a_i \cos \omega_i t$ be a sinusoidal excitation to the system and let the response of the system to this excitation be $y(t) = r_i \cos(\omega_i t + \phi)$. Then, the ratio r_i/a_i is the frequency response of the system to the input excitation for the frequency ω_i. By carrying out experiments for various frequencies, the frequency response function $H(\omega)$ can be obtained by connecting points r_i/a_i for various frequencies (see Figure 14.5).

Evaluation of the impulse response function, $h(t)$, of the system can also be made through experiments by generating an impulse whose shape is similar to that of the delta function $\delta(t)$, though sometimes it may not be easy to do. For example, for evaluating impulse response functions of ships and offshore structures, an impulse wave is produced by generating a wave train in the experimental tank, the frequency of which decreases linearly with time from the highest to the lowest frequency in such a way that the fast moving (low-frequency) waves catch up the slower (high-frequency)

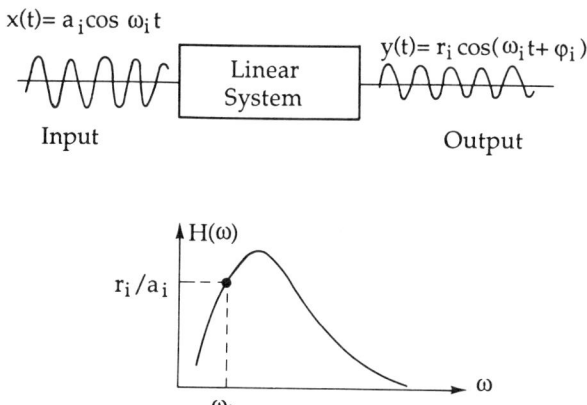

Figure 14.5 Evaluation of frequency response function $H(\omega)$ through experimental approach.

waves to coalesce at some point in space and time so as to produce an extremely large single wave that can be considered as a unit impulse.

An impulse response, called the weighting function, associated with impulsive wave can be theoretically obtained as follows (Davis and Zarnick 1964):

$$w(at) = a\left[\cos\frac{\pi(at)^2}{2}\left\{\frac{1}{2} + \frac{t}{|t|}C(at)\right\} + \sin\frac{\pi(at)^2}{2}\left\{\frac{1}{2} + \frac{t}{|t|}S(at)\right\}\right]$$

(14.25)

where

$$a = \sqrt{g/(2\pi x)}$$

x = distance from wave generating source

$$C(at) = \int_0^{at} \cos\frac{\pi z^2}{2}\, dz$$

$$S(at) = \int_0^{at} \sin\frac{\pi z^2}{2}\, dz.$$

Figure 14.6 shows Eq. (14.25) in dimensionless form. If we generate a wave train with constant amplitude and linearly decreasing frequency (equivalent to the wave train shown in Figure 14.6 with the time scale reversed), a very large wave results at a distance x from the wave-generating source.

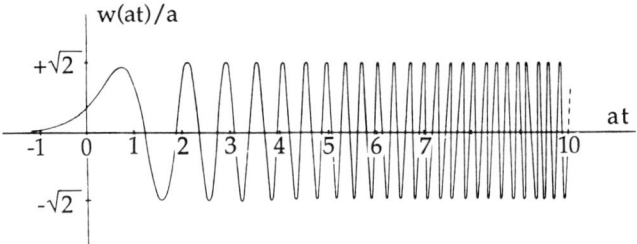

Figure 14.6 Initial wave profile that yields coalescence of waves at a point distance x given in Eq. (14.25) (from Davis and Zarnick, 1964).

14.3 INPUT AND OUTPUT SPECTRAL RELATIONSHIP

14.3.1 *Autocorrelations and Spectral Density Functions*

Let us evaluate the autocorrelation function of the output of a time-invariant linear system, denoted by $R_{yy}(\tau)$, from the knowledge of the autocorrelation function of the input, $R_{xx}(\tau)$, and the impulse response function $h(t)$. By definition, the output autocorrelation function may be written as

$$R_{yy}(\tau) = \lim_{T \to \infty} \frac{1}{2T} \int_{-T}^{T} y(t) \cdot y(t + \tau) \, dt \tag{14.26}$$

On the other hand, Theorem 14.1 states that the output of a linear system can be expressed by a convolution integral of the input $x(t)$ and the impulse response function $h(t)$. That is,

$$y(t) = \int_{-\infty}^{\infty} x(t - \alpha) \cdot h(\alpha) \, d\alpha$$

and

$$y(t + \tau) = \int_{-\infty}^{\infty} x(t + \tau - \beta) \cdot h(\beta) \, d\beta \tag{14.27}$$

Then, Eq. (14.26) becomes

$$R_{yy}(\tau) = \lim_{T \to \infty} \frac{1}{2T} \int_{-T}^{T} \left\{ \int_{-\infty}^{\infty} x(t - \alpha) h(\alpha) \, d\alpha \right.$$

$$\left. \times \int_{-\infty}^{\infty} x(t + \tau - \beta) h(\beta) \, d\beta \right\} dt$$

By letting $t - \alpha = u$, we have

$$R_{yy}(\tau) = \lim_{T \to \infty} \frac{1}{2T} \int_{-T-\alpha}^{T-\alpha} \int_{-\infty}^{\infty} \int_{-\infty}^{\infty} h(\alpha) h(\beta)$$

$$\times x(u) x(u + \tau + \alpha - \beta) \, d\alpha \, d\beta \, du$$

$$= \int_{-\infty}^{\infty} \int_{-\infty}^{\infty} h(\alpha) h(\beta) \cdot R_{xx}(\tau + \alpha - \beta) \, d\alpha \, d\beta \tag{14.28}$$

The above equation gives the relationship between the input and output autocorrelation functions. In particular, if the mean of the output is zero, then the variance of the output can be evaluated by

$$R_{yy}(0) = \int_{-\infty}^{\infty} \int_{-\infty}^{\infty} h(\alpha) h(\beta) R_{xx}(\alpha - \beta) \, d\alpha \, d\beta \quad (14.29)$$

Next, by applying the Wiener–Khintchine's theorem, the spectral density function of the output can be obtained from Eq. (14.28). That is, by letting $\tau + \alpha - \beta = \lambda$ in Eq. (14.28), the output spectral density function can be written as

$$S_{yy}(\omega) = \frac{1}{\pi} \int_{-\infty}^{\infty} R_{yy}(\tau) e^{-i\omega\tau} \, d\tau$$

$$= \frac{1}{\pi} \int_{-\infty}^{\infty} h(\alpha) e^{i\omega\alpha} \, d\alpha \int_{-\infty}^{\infty} h(\beta) e^{-i\omega\beta} \, d\beta \int_{-\infty}^{\infty} R_{xx}(\lambda) e^{-i\omega\lambda} \, d\lambda$$

$$= H^*(\omega) H(\omega) S_{xx}(\omega) = |H(\omega)|^2 S_{xx}(\omega) \quad (14.30)$$

It should be noted that the relationship obtained in Eq. (14.30) can also be derived directly from the definition of the output spectral density function. That is, the output spectral density function can be written from Definition 10.2 as follows:

$$S_{yy}(\omega) = \lim_{T \to \infty} \frac{1}{2\pi T} |Y(\omega)|^2 \quad (14.31)$$

Then, by Theorem 14.1, we can write

$$S_{yy}(\omega) = \lim_{T \to \infty} \frac{1}{2\pi T} |X(\omega)|^2 |H(\omega)|^2 = S_{xx}(\omega) |H(\omega)|^2 \quad (14.32)$$

The relationship derived above leads to the following theorem:

Theorem 14.2. The output spectral density function, $S_{yy}(\omega)$, of a time-invariant linear system is equal to the product of the input spectral density function, $S_{xx}(\omega)$, and the square of the frequency response function, $H(\omega)$. That is,

$$S_{yy}(\omega) = S_{xx}(\omega) |H(\omega)|^2$$

INPUT AND OUTPUT SPECTRAL RELATIONSHIP

Theorem 14.2 is extremely useful and is most commonly applied for evaluating the stochastic properties of linear systems appearing in the engineering and physical sciences. The quantity $|H(\omega)|^2$ is often called the *response amplitude operator* of the system in the engineering field.

Example 14.4. The input of a linear system has a white noise spectrum, A, and the impulse response function is given by

$$h(t) = e^{-\alpha t}, \quad \text{for } t \geq 0$$

Then, the frequency response function becomes

$$H(\omega) = \int_0^\infty e^{-\alpha t} \cdot e^{-i\omega t} \, dt = \frac{1}{\alpha + i\omega}$$

By applying Theorem 14.2, the output spectral density function can be obtained by

$$S_{yy}(\omega) = A|H(\omega)|^2 = A/(\alpha^2 + \omega^2) \qquad \blacksquare$$

14.3.2 Response of a System of Dual Inputs

Let us consider the response of a linear system that is subject to two inputs $x_1(t)$ and $x_2(t)$ as shown in Figure 14.7. Here, $x_1(t)$ and $x_2(t)$ are not necessarily uncorrelated. Let the impulse response functions of the system for these two inputs be $h_1(t)$ and $h_2(t)$, respectively. We may write the response $y(t)$ as

$$y(t) = \int_{-\infty}^\infty x_1(t - \lambda) h_1(\lambda) \, d\lambda + \int_{-\infty}^\infty x_2(t - \lambda) h_2(\lambda) \, d\lambda \quad (14.33)$$

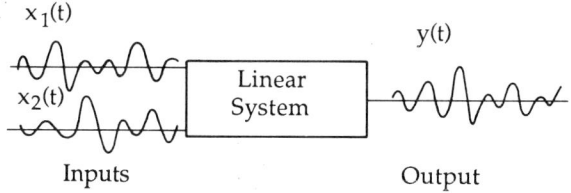

Figure 14.7 Dual inputs and output of a linear system.

Following the definition given in Eq. (10.4), the autocorrelation function of the response can be obtained as

$$R_{yy}(\tau) = \lim_{T \to \infty} \frac{1}{2T} \int_{-T}^{T} y(t) y(t+\tau) \, dt$$

$$= \lim_{T \to \infty} \frac{1}{2T} \int_{-T}^{T} \left\{ \int_{-\infty}^{\infty} x_1(t-\lambda) h_1(\lambda) \, d\lambda \right.$$

$$\left. + \int_{-\infty}^{\infty} x_2(t-\lambda) h_2(\lambda) \, d\lambda \right\}$$

$$\times \left\{ \int_{-\infty}^{\infty} x_1(t+\tau-\mu) h_1(\mu) \, d\mu \right.$$

$$\left. + \int_{-\infty}^{\infty} x_2(t+\tau-\mu) h_2(\mu) \, d\mu \right\} dt$$

$$= \lim_{T \to \infty} \frac{1}{2T} \left[\int_{-T}^{T} \left\{ \int_{-\infty}^{\infty} x_1(t-\lambda) h_1(\lambda) \, d\lambda \right. \right.$$

$$\left. \times \int_{-\infty}^{\infty} x_1(t+\tau-\mu) h_1(\mu) \, d\mu \right\} dt$$

$$+ \int_{-T}^{T} \left\{ \int_{-\infty}^{\infty} x_2(t-\lambda) h_2(\lambda) \, d\lambda \int_{-\infty}^{\infty} x_1(t+\tau-\mu) h_1(\mu) \, d\mu \right\} dt$$

$$+ \int_{-T}^{T} \left\{ \int_{-\infty}^{\infty} x_1(t-\lambda) h_1(\lambda) \, d\lambda \int_{-\infty}^{\infty} x_2(t+\tau-\mu) h_2(\mu) \, d\mu \right\} dt$$

$$\left. + \int_{-T}^{T} \left\{ \int_{-\infty}^{\infty} x_2(t-\lambda) h_2(\lambda) \, d\lambda \int_{-\infty}^{\infty} x_2(t+\tau-\mu) h_2(\mu) \, d\mu \right\} dt \right]$$

$$= \int_{-\infty}^{\infty} \int_{-\infty}^{\infty} \{ R_{11}(\lambda + \tau - \mu) h_1(\lambda) h_1(\mu)$$

$$+ R_{21}(\lambda + \tau - \mu) h_2(\lambda) h_1(\mu)$$

$$+ R_{12}(\lambda + \tau - \mu) h_1(\lambda) h_2(\mu)$$

$$+ R_{22}(\lambda + \tau - \mu) h_2(\lambda) h_2(\mu) \} \, d\lambda \, d\mu \qquad (14.34)$$

where

$$R_{11}(\tau) = \text{autocorrelation function of } x_1(t)$$
$$R_{22}(\tau) = \text{autocorrelation function of } x_2(t)$$
$$R_{12}(\tau) = \text{cross-correlation function of } x_1(t) \text{ and } x_2(t)$$
$$R_{21}(\tau) = \text{cross-correlation function of } x_2(t) \text{ and } x_1(t).$$

Then, the spectral density function of the response, $S_{yy}(\omega)$, can be evaluated by applying Theorem 10.2. For example, the spectral density function for the first term of Eq. (14.34) can be obtained as follows:

$$\frac{1}{\pi}\int_{-\infty}^{\infty}\left\{\int_{-\infty}^{\infty}\int_{-\infty}^{\infty} R_{11}(\lambda+\tau-\mu)h_1(\lambda)h_1(\mu)e^{-i\omega\tau}\,d\lambda\,d\mu\right\}d\tau$$

$$= \frac{1}{\pi}\int_{-\infty}^{\infty} R_{11}(\lambda+\tau-\mu)e^{-i\omega(\tau+\lambda-\mu)}\,d\tau$$

$$\times \int_{-\infty}^{\infty} h_1(\lambda)e^{i\omega\lambda}\,d\lambda \int_{-\infty}^{\infty} h_1(\mu)e^{-i\omega\mu}\,d\mu$$

$$= S_{11}(\omega)H_1^*(\omega)H_1(\omega) = S_{11}(\omega)|H_1(\omega)|^2 \qquad (14.35)$$

where

$$S_{11}(\omega) = \text{spectral density of } x_1(t)$$
$$H_1(\omega) = \text{frequency response function for } x_1(t)$$
$$H_1^*(\omega) = \text{conjugate function of } H_1(\omega).$$

Similarly, the spectral density function for the remaining terms of Eq. (14.34) can be obtained as $S_{21}(\omega)H_2^*(\omega)H_1(\omega)$, $S_{12}(\omega)H_1^*(\omega)H_2(\omega)$, and $S_{22}(\omega)|H_2(\omega)|^2$, where $S_{21}(\omega)$ and $S_{12}(\omega)$ are cross-spectral density functions defined in Eq. (10.45). Thus, the output spectral density function $S_{yy}(\omega)$ becomes

$$S_{yy}(\omega) = S_{11}(\omega)|H_1(\omega)|^2 + S_{21}(\omega)H_2^*(\omega)H_1(\omega)$$
$$+ S_{12}(\omega)H_1^*(\omega)H_2(\omega) + S_{22}(\omega)|H_2(\omega)|^2 \qquad (14.36)$$

If the inputs $x_1(t)$ and $x_2(t)$ are uncorrelated, then the second and third terms in the above equation are zero.

14.3.3 Coherency Function

The discussion so far pertains to the stochastic relationship between the input and output of a system assuming that the system is linear. It is often necessary, however, to examine whether or not the system can be considered linear. The linearity of a system can be clarified by carrying out the auto- and cross-spectral analysis of the system as follows:

Let us first consider the cross-correlation function between input and output of a linear system having an impulse response function $h(t)$. That is,

$$R_{xy}(\tau) = \lim_{T \to \infty} \frac{1}{2T} \int_{-T}^{T} x(t) y(t+\tau) \, dt$$

$$= \lim_{T \to \infty} \frac{1}{2T} \int_{-T}^{T} x(t) \left\{ \int_{-\infty}^{\infty} x(t+\tau-\mu) h(\mu) \, d\mu \right\} dt$$

$$= \int_{-\infty}^{\infty} \left\{ \lim_{T \to \infty} \frac{1}{2T} \int_{-T}^{T} x(t) x(t+\tau-\mu) \, dt \right\} h(\mu) \, d\mu$$

$$= \int_{-\infty}^{\infty} R_{xx}(\tau-\mu) h(\mu) \, d\mu \qquad (14.37)$$

By applying the Wiener–Khintchine theorem, the cross-spectral density function can be written as

$$S_{xy}(\omega) = \frac{1}{\pi} \int_{-\infty}^{\infty} \left\{ \int_{-\infty}^{\infty} R_{xx}(\tau-\mu) h(\mu) \, d\mu \right\} e^{-i\omega\tau} \, d\tau$$

$$= \int_{-\infty}^{\infty} \left\{ \frac{1}{\pi} \int_{-\infty}^{\infty} R_{xx}(\tau-\mu) e^{-i\omega(\tau-\mu)} \, d\tau \right\} h(\mu) e^{-i\omega\mu} \, d\mu$$

$$= S_{xx}(\omega) H(\omega) \qquad (14.38)$$

Similarly, we can derive

$$S_{xy}(-\omega) = S_{xx}(\omega) H^*(\omega) \qquad (14.39)$$

INPUT AND OUTPUT SPECTRAL RELATIONSHIP

From Eqs. (14.38) and (14.39), we have

$$|H(\omega)|^2 = \frac{S_{xy}(\omega)S_{xy}(-\omega)}{\{S_{xx}(\omega)\}^2} = \frac{S_{xy}(\omega)S_{xy}^*(\omega)}{\{S_{xx}(\omega)\}^2}$$

$$= \frac{\{C_{xy}(\omega)\}^2 + \{Q_{xy}(\omega)\}^2}{\{S_{xx}(\omega)\}^2} \qquad (14.40)$$

where

$C_{xy}(\omega)$ = cospectrum of $x(t)$ and $y(t)$

$Q_{xy}(\omega)$ = quadrature spectrum of $x(t)$ and $y(t)$.

On the other hand, if the system is linear, then $|H(\omega)|^2$ can be written as follows from Theorem 14.2:

$$|H(\omega)|^2 = \frac{S_{yy}(\omega)}{S_{xx}(\omega)} \qquad (14.41)$$

Thus, if the system is linear, it must satisfy both Eqs. (14.40) and (14.41). This results in

$$\{C_{xy}(\omega)\}^2 + \{Q_{xy}(\omega)\}^2 = S_{xx}(\omega)S_{yy}(\omega) \qquad (14.42)$$

If the system is not linear, the relationship given in Eq. (14.42) does not hold. Hence, we may develop a measure of the linearity of a system based on Eq. (14.42) which is called the coherency function $\gamma(\omega)$ defined in Eq. (10.56) of Chapter 10. We may repeat the definition here:

$$\gamma(\omega) = \frac{\{C_{xy}(\omega)\}^2 + \{Q_{xy}(\omega)\}^2}{S_{xx}(\omega)S_{yy}(\omega)}, \qquad 0 \leq \gamma(\omega) \leq 1$$

It should be noted that some define the coherency function as the square root of $\gamma(\omega)$ given in the above equation.

An example of the coherency function is shown in Figure 14.8 as a function of ω. It is clear from the definition of $\gamma(\omega)$ that $\gamma(\omega) = 1$ implies that the system is linear. In reality, however, $\gamma(\omega)$ is not usually equal to 1 even for a linear system. It is often observed that $\gamma(\omega)$ is on the order of

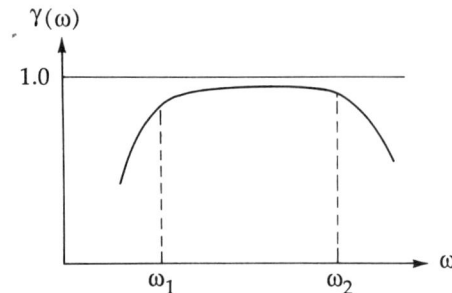

Figure 14.8 Coherency function $\gamma(\omega)$.

0.90 or greater within a certain range of frequencies, ω_1 and ω_2, as indicated in the figure.

The reader may recall the following definition of the correlation coefficient between two random variables introduced in Chapter 3:

$$\rho = \frac{\text{Cov}[x, y]}{\sqrt{\text{Var}[x]\,\text{Var}[y]}}$$

The correlation coefficient $\rho = 1$ implies that two random variables are completely correlated, while $\rho = 0$ implies they are uncorrelated. The coherency function $\gamma(\omega)$ defined in this section manifests the same characteristics as the correlation coefficient, the significant difference between them being the coherency function is a function of frequency ω.

EXERCISES

14.1 For a linear system

$$\ddot{y} + 2\zeta\omega_0 \dot{y} + \omega_0^2 y = \text{Re}\{X_0 e^{i\omega t}\}$$

the frequency response function, $H(\omega)$, is given by [see Eq. (14.18)]

$$H(\omega) = \frac{1}{(\omega^2 - \omega_0^2) + i(2\zeta\omega_0\omega)}$$

On the other hand, the impulse response function, $h(t)$, of the system becomes [see Eq. (14.24)]

$$h(t) = \frac{1}{\sqrt{1-\zeta^2}\,\omega_0} e^{-\zeta\omega_0 t} \sin\sqrt{1-\zeta^2}\,\omega_0 t$$

Show that $H(\omega)$ is the Fourier transform of $h(t)$.

14.2 Let the input of a linear system be $t\mathsf{u}(t)$, where $\mathsf{u}(t)$ is the unit step function. Let $r(t)$ be the response of the system to the input $t\mathsf{u}(t)$. Prove that the response of this system for an input $f(t)$ is given by

$$\int_{-\infty}^{\infty} r(t-\lambda)\ddot{f}(\lambda)\,d\lambda$$

where $\ddot{f}(t)$ is the acceleration of $f(t)$.

14.3 Let us assume that the acceleration spectrum of a randomly vibrating system is given by

$$S_x(f) = \begin{cases} 1\,\text{g}^2\text{-sec} & \text{for } 0.1 \leqslant f \leqslant 0.3 \\ 0 & \text{otherwise} \end{cases}$$

Obtain the number of vibrations expected in 1 hour.

14.4 The impulse response function, $h(t)$, of a linear system is given by

$$h(t) = \begin{cases} e^{-\alpha t} & \text{for } t > 0 \\ 0 & \text{for } t < 0 \end{cases}$$

Let the input $x(t)$ be a stationary random process with zero mean and its autocorrelation function is given by

$$R_x(\tau) = c\delta(\tau)$$

where c is a constant, and $\delta(\tau)$ is a delta function. Obtain the spectral density function and autocorrelation function of the output $y(t)$.

14.5 Consider the response (vertical motion) of a system induced by a random excitation. The input spectrum is obtained from measurements of acceleration and is given as a function of frequency f in Hz as follows:

$$S_{\ddot{x}}(f) = \begin{cases} a \text{ g}^2\text{-sec} & \text{for } 0.1 \leq f \leq 0.2 \\ 0 & \text{otherwise} \end{cases}$$

The frequency response function of the vertical motion of the system obtained through experiments is given as

$$H(\omega) = \begin{cases} b & \text{for } 0.5 \leq \omega \leq 1.5 \\ 0 & \text{otherwise} \end{cases}$$

Obtain the response spectrum of the vertical motion of this structure as a function of frequency ω in rps.

CHAPTER 15

Nonlinear Systems and Stochastic Prediction

15.1 NONLINEAR SYSTEMS

In the previous chapter a method for predicting responses of a system to a random input was developed assuming that the system was linear. Although most random phenomena observed in engineering and the physical sciences can be considered linear, there are many random phenomena for which linear stochastic prediction techniques can no longer be applied. As an example, motions of a moored buoy in a seaway are significantly different from those of a freely floating buoy because of the restrictions induced by mooring lines. Restrictions are, in general, imposed on the damping force (or moment) and/or the restoring force (or moment) of the system.

Let us consider a nonlinear vibratory system with a single degree of freedom for which the equation of motion is given as follows:

$$m\ddot{y} + c\dot{y} + ky + g_0(y, \dot{y}) = f_{0x}(t) \tag{15.1}$$

where $g(y, \dot{y})$ represents the nonlinear term that, in general, is a function of displacement and/or velocity. For brevity, let us divide each term of Eq. (15.1) by the coefficient m of the inertia term, which yields the following equation:

$$\ddot{y} + \alpha\dot{y} + \omega_0^2 y + g(y, \dot{y}) = f_x(t) \tag{15.2}$$

where

α = linear damping

ω_0 = natural frequency of undamped oscillation

$g(y, \dot{y})$ = nonlinear force (or moment) per unit mass (or moment of inertia)

$f_x(t)$ = excitation force (or moment) per unit mass (or moment of inertia)

In many instances, the nonlinear damping force (or moment) can be expressed in a quadratic form of the velocity $\beta \dot{y}|\dot{y}|$, while nonlinear restoring force (or moment) can be expressed in higher order terms of the displacement, ry^3, for example.

The excitation $f_x(t)$ in Eq. (15.2) is a random process; hence, it is extremely difficult to obtain an exact solution $y(t)$. All solutions presented in the following sections, therefore, are all approximate solutions. It should be noted, furthermore, that the solution $y(t)$ itself for a random input $f_x(t)$ is not the ultimate goal in stochastic analysis of a nonlinear system; instead, more relevant information is the probability distribution of the amplitude of the response. To elaborate, let us consider a nonlinear system with the input $x(t)$ shown in Figure 15.1. Here, $x(t)$ is assumed to be a Gaussian random process. Because of the nonlinear characteristics, the output (response) $y(t)$ is usually no longer a Gaussian random process; hence, the statistical characteristics of its amplitude cannot be evaluated through the probability density functions presented in Chapter 11 unless the system

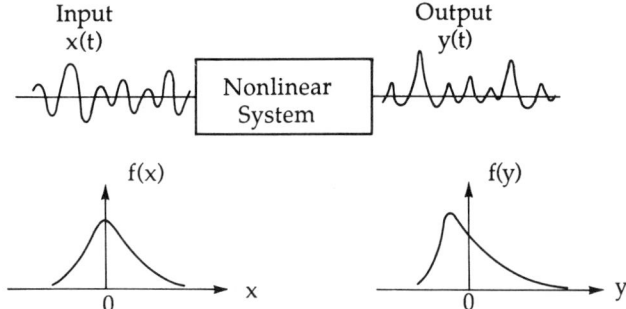

Figure 15.1 Schematic representation of input and output of a nonlinear system.

nonlinearity is weak. In principle, it is highly desirable to obtain the probability distribution of the output $y(t)$ first, and then derive the probability distribution of the amplitude of the output. Methods pertaining to the former are presented in the following sections.

The statistical properties of the response $y(t)$ of a nonlinear system to a random input can be evaluated by applying the following approaches:

- Equivalent linearization technique,
- Perturbation technique, and
- Markov process approach through the solution of the Fokker–Planck equation.
- Bispectral analysis.

15.2 EQUIVALENT LINEARIZATION TECHNIQUE

The equivalent linearization technique that has been most commonly applied to obtain the solution of a nonlinear system is that originally developed by Kryloff and Bogoliuboff (1947). The principle of this method is to express the equation of motion of a nonlinear system in linear form and consider an equivalent balance of energy during one cycle such that work per cycle of the two systems (nonlinear and equivalent linear system) is the same.

In the Kryloff–Bogoliuboff method, the excitation to the nonlinear system is considered to be sinusoidal. For a nonlinear system with random excitation, a method to evaluate the root-mean-square (rms) value of the response has been developed through the equivalent linearization technique that considers minimization of the error associated with a linear transformation (Lyon 1960b; Crandall 1962, 1964; and Caughey 1963b, among others).

Let us write the equation of motion of a linear system that is equivalent to that of the nonlinear system given in Eq. (15.2) as

$$\ddot{y} + \alpha_e \dot{y} + \omega_e^2 y = f_x(t) + \varepsilon(t) \tag{15.3}$$

where $\varepsilon(t)$ is the error associated with the linearization, and it is a random process. The coefficients α_e and ω_e^2 are determined such that the error, $\varepsilon(t)$, is as small as possible. One way to minimize the error is to minimize the mean squared value $E[\varepsilon^2]$, where $E[\ ]$ stands for the expected value of the random variable. From Eqs. (15.2) and (15.3), we can write the error as

$$\varepsilon = (\alpha_e - \alpha)\dot{y} + (\omega_e^2 - \omega_0^2)y - g(y, \dot{y}) \tag{15.4}$$

Then the expected value of ε^2 becomes

$$E[\varepsilon^2] = (\alpha_e - \alpha)^2 E[\dot{y}^2] + (\omega_e^2 - \omega_0^2)^2 E[y^2] + E[g^2(y, \dot{y})]$$
$$- 2(\alpha_e - \alpha) E[\dot{y} g(y, \dot{y})] - 2(\omega_e^2 - \omega_0^2) E[y g(y, \dot{y})]$$
$$+ 2(\alpha_e - \alpha)(\omega_e^2 - \omega_0^2) E[y\dot{y}] \qquad (15.5)$$

In order to determine whether or not minimization of $E[\varepsilon^2]$ is possible, the first and second derivatives of $E[\varepsilon^2]$ are evaluated with respect to parameters α_e and ω_e^2. These are,

$$\frac{\partial}{\partial \alpha_e} E[\varepsilon^2] = 2\{(\alpha_e - \alpha) E[\dot{y}^2] + (\omega_e^2 - \omega_0^2) E[y\dot{y}] - E[\dot{y} g(y, \dot{y})]\}$$

$$\frac{\partial}{\partial \omega_e^2} E[\varepsilon^2] = 2\{(\omega_e^2 - \omega_0^2) E[y^2] + (\alpha_e - \alpha) E[y\dot{y}] - E[y g(y, \dot{y})]\}$$

$$\frac{\partial^2}{\partial \alpha_e^2} E[\varepsilon^2] = 2 E[\dot{y}^2] > 0$$

$$\frac{\partial^2}{\partial (\omega_e^2)^2} E[\varepsilon^2] = 2[y^2] > 0 \qquad (15.6)$$

Thus, it is possible to minimize $E[\varepsilon^2]$ by letting

$$\frac{\partial}{\partial \alpha_e} E[\varepsilon^2] = 0 \quad \text{and} \quad \frac{\partial}{\partial \omega_e^2} E[\varepsilon^2] = 0 \qquad (15.7)$$

That is,

$$(\alpha_e - \alpha) E[\dot{y}^2] + (\omega_e^2 - \omega_0^2) E[y\dot{y}] - E[\dot{y} g(y, \dot{y})] = 0$$
$$(\omega_e^2 - \omega_0^2) E[y^2] + (\alpha_e - \alpha) E[y\dot{y}] - E[y g(y, \dot{y})] = 0 \qquad (15.8)$$

In the case that nonlinearities in the damping and restoring forces (or moments) are separable, we may write

$$g(y, \dot{y}) = g_1(y) + g_2(\dot{y}) \qquad (15.9)$$

We may assume that the excitation $f_x(t)$ is a Gaussian random process. Since Eq. (15.3) is a linear equation, the response $y(t)$ is also Gaussian and thereby we have $E[y\dot{y}] = 0$. Thus, we can derive α_e and ω_e^2 from Eq. (15.8) as follows:

$$\alpha_e = \alpha + \frac{E[\dot{y} g_2(\dot{y})]}{E[\dot{y}^2]}$$

$$\omega_e^2 = \omega_0^2 + \frac{E[y g_1(y)]}{E[y^2]} \tag{15.10}$$

If the system has nonlinearities only in the damping term, then $\omega_e = \omega_0$ and thereby the first equation of Eq. (15.10) is sufficient to obtain the response $y(t)$. Similarly, for a system with nonlinearities only in the restoring term, the second equation of Eq. (15.10) is sufficient to evaluate the response.

As an example, let us consider a system with nonlinearities only in the damping term, which is given by $g_2(\dot{y}) = \beta \dot{y}|\dot{y}|$. The equation of the system may be written as

$$\ddot{y} + \alpha \dot{y} + \beta \dot{y}|\dot{y}| + \omega_0^2 y = f_x(t) \tag{15.11}$$

Here, the excitation $f_x(t)$ is a Gaussian random process with mean value zero. Then, the probability distribution of the velocity $\dot{y}$ is also normally distributed with mean value zero. By writing its unknown variance as $\sigma_{\dot{y}}^2$, we have

$$E[\dot{y}^2] = \sigma_{\dot{y}}^2$$

$$E[\dot{y} g(\dot{y})] = \int_{-\infty}^{\infty} \dot{y}^2 |\dot{y}| \frac{1}{\sqrt{2\pi}\,\sigma_{\dot{y}}} e^{-\dot{y}^2/2\sigma_{\dot{y}}^2} d\dot{y} = \frac{2\sqrt{2}}{\sqrt{\pi}} \beta \sigma_{\dot{y}}^3 \tag{15.12}$$

Hence, from Eq. (15.10), we can obtain an equivalent damping coefficient as

$$\alpha_e = \alpha + \frac{2\sqrt{2}}{\sqrt{\pi}} \beta \sigma_{\dot{y}} \tag{15.13}$$

On the other hand, the square of the frequency response function for the equivalent linear system, denoted by $|H_e(\omega)|^2$, can be written from

Eq. (14.19) as follows:

$$|H_e(\omega)|^2 = \frac{1}{\left(\omega_0^2 - \omega^2\right)^2 + (\alpha_e\omega)^2} \qquad (15.14)$$

Since $|H_e(\omega)|^2$ is a function of α_e, we may expand it into a Taylor series in the neighborhood of α using the relationship given in Eq. (15.13). By letting $|H(\omega)|^2$ be equal to $|H_e(\omega)|^2$ with $\alpha_e = \alpha$, we have

$$|H_e(\omega)|^2 = |H(\omega)|^2 + \frac{2\sqrt{2}}{\sqrt{\pi}}\beta\sigma_{\dot{y}}\frac{\partial}{\partial\alpha}|H(\omega)|^2 \qquad (15.15)$$

Here, the partial derivative of $|H(\omega)|^2$ in the second term can be written as

$$\frac{\partial}{\partial\alpha}|H(\omega)|^2 = -2\alpha\omega^2|H(\omega)|^4$$

$$= 2|H(\omega)|^2 \,\mathrm{Re}\!\left\{\frac{i\omega}{\left(\omega_0^2 - \omega^2\right) - i\alpha\omega}\right\}$$

$$= 2|H(\omega)|^2 \,\mathrm{Re}\{i\omega H^*(\omega)\} \qquad (15.16)$$

Thus, we can derive

$$|H_e(\omega)|^2 = |H(\omega)|^2\left[1 - \frac{4\sqrt{2}}{\sqrt{\pi}}\alpha\beta\sigma_{\dot{y}}\omega^2|H(\omega)|^2\right]$$

$$= |H(\omega)|^2\left[1 + \frac{4\sqrt{2}}{\sqrt{\pi}}\beta\sigma_{\dot{y}}\,\mathrm{Re}\{i\omega H^*(\omega)\}\right] \qquad (15.17)$$

By applying Theorem 14.2, which gives the relationship between input and output spectral density functions of a linear system, we can write

$$S_{yy}(\omega) = S_{xx}(\omega)|H(\omega)|^2\left[1 - \frac{4\sqrt{2}}{\sqrt{\pi}}\alpha\beta\sigma_{\dot{y}}\omega^2|H(\omega)|^2\right]$$

$$= S_{xx}(\omega)|H(\omega)|^2\left[1 + \frac{4\sqrt{2}}{\sqrt{\pi}}\beta\sigma_{\dot{y}}\,\mathrm{Re}\{i\omega H^*(\omega)\}\right] \qquad (15.18)$$

EQUIVALENT LINEARIZATION TECHNIQUE 393

The second term in Eq. (15.18) gives the effect of nonlinear damping on the response spectrum. The unknown response standard deviation $\sigma_{\dot{y}}$ involved in the above equation can be determined by applying the iteration method. That is,

1. Assign a value to $\sigma_{\dot{y}}$ and evaluate $S_{yy}(\omega)$ by Eq. (15.18) for a given input spectrum $S_{xx}(\omega)$.
2. Evaluate the second moment of $S_{yy}(\omega)$, which is equal to the variance of the response velocity $\sigma_{\dot{y}}^2$ (see Section 10.3.2). Compare $\sigma_{\dot{y}}$ thus evaluated with the assigned value.
3. Repeat the procedure until the evaluated variance $\sigma_{\dot{y}}$ agrees with the assigned value.

Next, let us consider the case in which nonlinearities exist only in the restoring term, which may be written by

$$\ddot{y} + \alpha \dot{y} + \omega_0^2(y + ry^3) = f_x(t) \qquad (15.19)$$

A nonlinear system of this type is called a *Duffing system*. In this case, we can evaluate ω_e^2 from Eq. (15.10) as

$$\omega_e^2 = \omega_0^2 + 3r\omega_0^2\sigma_y^2 \qquad (15.20)$$

and we may write $|H_e(\omega)|^2$ as follows:

$$|H_e(\omega)|^2 = \frac{1}{(\omega_e^2 - \omega^2) + (\alpha\omega)^2} \qquad (15.21)$$

By expanding $|H_e(\omega)|^2$ in a Taylor series in the neighborhood of ω_0^2, we have from Eq. (15.20)

$$|H_e(\omega)|^2 = |H(\omega)|^2 + 3r\omega_0^2\sigma_y^2 \frac{\partial}{\partial(\omega_0^2)}|H(\omega)|^2 \qquad (15.22)$$

It can easily be verified that

$$\frac{\partial}{\partial(\omega_0^2)}|H(\omega)|^2 = -2(\omega_0^2 - \omega^2)|H(\omega)|^4 = -2\operatorname{Re}\{H(\omega)\}|H(\omega)|^2$$

$$(15.23)$$

Thus, we have

$$|H_e(\omega)|^2 = |H(\omega)|^2\left[1 - 6\omega_0^2 r\sigma_y^2(\omega_0^2 - \omega^2)|H(\omega)|^2\right]$$
$$= |H(\omega)|^2\left[1 - 6\omega_0^2 r\sigma_y^2 \operatorname{Re}\{H(\omega)\}\right] \quad (15.24)$$

and hence the response spectrum $S_{yy}(\omega)$ of the Duffing system for an input spectrum $S_{xx}(\omega)$ can be given by

$$S_{yy}(\omega) = S_{xx}(\omega)|H(\omega)|^2\left[1 - 6\omega_0^2 r\sigma_y^2(\omega_0^2 - \omega^2)|H(\omega)|^2\right]$$
$$= S_{xx}(\omega)|H(\omega)|^2\left[1 - 6\omega_0^2 r\sigma_y^2 \operatorname{Re}\{H(\omega)\}\right] \quad (15.25)$$

The unknown variance of the response σ_y^2 can be evaluated through an iteration procedure similar to that discussed in regard to a system with a nonlinear damping term.

Example 15.1. Let us assume that the input spectrum of a Duffing system is a white noise spectrum given by $S_{xx}(\omega) = A$ for $0 \leq \omega < \infty$. In order to evaluate the variance of the response σ_y^2, we may write the response spectrum as follows by applying $|H_e(\omega)|^2$ given in Eq. (15.22):

$$S_{yy}(\omega) = A\left\{|H(\omega)|^2 + 3r\omega_0^2\sigma_y^2 \frac{\partial}{\partial(\omega_0^2)}|H(\omega)|^2\right\}$$

The variance σ_y^2 can be evaluated by integrating $S_{yy}(\omega)$ with respect to ω. Since integration of $|H(\omega)|^2$ becomes

$$\int_0^\infty |H(\omega)|^2 d\omega = \int_0^\infty \frac{d\omega}{(\omega_0^2 - \omega^2)^2 + (\alpha\omega)^2} = \frac{\pi}{2\omega_0^2\alpha}$$

we have

$$\sigma_y^2 = \int_0^\infty S_{yy}(\omega)\, d\omega = \frac{\pi A}{2\alpha\omega_0^2} - \frac{3\pi A r\sigma_y^2}{2\alpha\omega_0^2}$$
$$= \sigma_{y0}^2 - \sigma_{y0}^2(3r\sigma_y^2)$$

where $\sigma_{y0}^2 = \pi A/2\alpha\omega_0^2 =$ response to a white noise spectrum A for a linear system.

Thus, we have the response variance

$$\sigma_y^2 = \sigma_{y0}^2 / (1 + 3r\sigma_{y0}^2) \qquad \blacksquare$$

If a system has nonlinearities in both the damping and restoring terms and if they are not separable, then the solution is rather complicated. Caughey (1963b) gives the solution for this case with a white noise input spectrum as follows:

We may write the equivalent damping and frequency from Eq. (15.8) under the condition $E[y\dot{y}] = 0$ as

$$\alpha_e = \alpha + \frac{E[\dot{y}g(y,\dot{y})]}{E[\dot{y}^2]}$$
$$\omega_e^2 = \omega_0^2 + \frac{E[yg(y,\dot{y})]}{E[y^2]} \qquad (15.26)$$

Assuming that the response is a narrow-band random process, we may write the response following Eq. (9.10) as

$$y(t) = a(t)\cos\{\omega_e t + \phi(t)\} = a\cos\psi \qquad (15.27)$$

and

$$\dot{y}(t) = -a\omega_e \sin\psi \qquad (15.28)$$

Here, the amplitude a and phase ψ are random variables. We first consider the expected value in Eq. (15.26) with respect to ψ. The first equation of Eq. (15.26) can be written as $\alpha_e =$

$$\alpha + \frac{\lim_{N\to\infty}(1/2N)\sum_{i=-N}^{N}\left[(1/2\pi)\int_{i2\pi}^{(i+1)2\pi} -a\sin\psi\, g\{a\cos\psi, -a\omega_e \sin\psi\}\,d\psi\right]}{\lim_{N\to\infty}(1/2N)\sum_{i=-N}^{N}\left[(1/2\pi)\int_{i2\pi}^{(i+1)2\pi} a^2\omega_e \sin^2\psi\,d\psi\right]}$$

$$(15.29)$$

Since a and ψ are slowly varying functions of time, the integration involved in the numerator of Eq. (15.29) becomes approximately

$$\frac{1}{2\pi}\int_{i2\pi}^{(i+1)2\pi} a\sin\psi\, g\{a\cos\psi, -a\omega_e\sin\psi\}\,d\psi$$

$$\sim \frac{1}{2\pi}\int_0^{2\pi} a_i\sin\psi\, g\{a_i\cos\psi, -a_i\omega_e\sin\psi\}\,d\psi = S(a_i) \quad (15.30)$$

Similarly, the integration involved in the denominator of Eq. (15.29) may be written as

$$\frac{1}{2\pi}\int_{i2\pi}^{(i+1)2\pi} a^2\omega_e\sin^2\psi\,d\psi \sim \frac{1}{2\pi}a_i^2\omega_e\int_0^{2\pi}\sin^2\psi\,d\psi = \frac{1}{2}a_i^2\omega_e \quad (15.31)$$

By accumulating the random variable a_i, we may write α_e as

$$\alpha_e = \alpha - \frac{2}{\omega_e}\frac{\lim_{N\to\infty}(1/2N)\sum_{i=-N}^{N}S(a_i)}{\lim_{N\to\infty}(1/2N)\sum_{i=-N}^{N}a_i^2} \quad (15.32)$$

By the same procedure shown in Eqs. (15.29) through (15.32), ω_e^2 given in Eq. (15.26) can be evaluated as follows:

$$\omega_e^2 = \omega_0^2 + 2\frac{\lim_{N\to\infty}(1/2N)\sum_{i=-N}^{N}C(a_i)}{\lim_{N\to\infty}(1/2N)\sum_{i=-N}^{N}a_i^2} \quad (15.33)$$

where

$$C(a_i) = \frac{1}{2\pi}\int_0^{2\pi} a_i\cos\psi\, g\{a_i\cos\psi, -a_i\omega_e\sin\psi\}\,d\psi \quad (15.34)$$

Using the narrow-band assumption, and in addition if we assume that the response is a Gaussian random process with mean value zero, then the amplitude a_i involved in Eqs. (15.32) and (15.33) obeys the following

Rayleigh distribution as discussed in Section 11.1.1:

$$f(a) = \frac{a}{\sigma_y^2} e^{-a^2/2\sigma_y^2}, \quad 0 \leq a < \infty \tag{15.35}$$

Here, the parameter σ_y^2 is equal to the variance of the response, although its magnitude is unknown at this stage.

By using the probability density function $f(a)$, we may write α_e and ω_e^2 in integral form as

$$\alpha_e = \alpha - \frac{2}{\omega_e} \frac{\int_0^\infty S(a)f(a)\,da}{\int_0^\infty a^2 f(a)\,da} = \alpha - \frac{1}{\omega_e \sigma_y^2} \int_0^\infty S(a)f(a)\,da \tag{15.36}$$

$$\omega_e^2 = \omega_0^2 + 2\frac{\int_0^\infty C(a)f(a)\,da}{\int_0^\infty a^2 f(a)\,da} = \omega_0^2 + \frac{1}{\sigma_y^2} \int_0^\infty C(a)f(a)\,da \tag{15.37}$$

Next, let us find the variance σ_y^2. The frequency response function, $H(\omega)$, of the system given in Eq. (15.3) can be written by neglecting the error term ε as follows:

$$H(\omega) = \frac{1}{(\omega_e^2 - \omega^2) + i\alpha_e \omega} \tag{15.38}$$

Then, the variance σ_y^2 for a white noise spectrum A for $0 \leq \omega < \infty$ can be obtained by

$$\sigma_y^2 = \int_0^\infty A|H(\omega)|^2\,d\omega = \int_0^\infty \frac{A}{(\omega_e^2 - \omega^2)^2 + (\alpha_e \omega)^2}\,d\omega$$

$$= \frac{\pi A}{2\alpha_e \omega_e^2} \quad \text{(see Example 15.1)} \tag{15.39}$$

Thus, the unknown variance σ_y^2 can be evaluated by applying the iteration method to Eqs. (15.36), (15.37), and (15.39), and thereby the statistical properties of the amplitude of the response can be evaluated from the probability density function given in Eq. (15.35).

As can be seen in Eqs. (15.36) and (15.37), the method for evaluating the equivalent α_e and ω_e^2 for a nonlinear system, in general, is through the probability density function of the response amplitude $f(a)$ in contrast with the method applicable for a system having nonlinearities in either the damping or restoring term. Evaluation of α_e and ω_e^2 for the latter case is made by Eq. (15.10) directly from the probability density function of the deviation from the mean $f(y)$. However, these two approaches yield the same answer for a system with nonlinearities only in one term (either damping or restoring). An example is given in Exercise 15.1.

15.3 PERTURBATION TECHNIQUE

Another approach for evaluating the response of a nonlinear system is to assume that the response may be expressed in a power series of the parameter associated with the nonlinear term. This approach is known as the *perturbation technique* (Crandall, 1963; Crandall et al., 1964, among others).

For convenience, let us write the nonlinear term involved in Eq. (15.2) as $\varepsilon g(y, \dot{y})$, where the parameter ε is assumed to be small. That is,

$$\ddot{y} + \alpha \dot{y} + \omega_0^2 y + \varepsilon g(y, \dot{y}) = f_x(t) \tag{15.40}$$

Assume that the solution of the above equation is given as a power series of ε. That is,

$$y(t) = y_0(t) + \varepsilon y_1(t) + \varepsilon^2 y_2(t) + \cdots \tag{15.41}$$

where $y_0(t)$ is the solution of the linear equation. By inserting Eq. (15.41) into Eq. (15.40) and by equating terms containing the same power of ε, we have the following set of linear equations for y_0, y_1, and y_2, etc.

$$\ddot{y}_0 + \alpha \dot{y}_0 + \omega_0^2 y_0 = f_x(t)$$

$$\ddot{y}_1 + \alpha \dot{y}_1 + \omega_0^2 y_1 = -g(y_0, \dot{y}_0)$$

$$\ddot{y}_2 + \alpha \dot{y}_2 + \omega_0^2 y_2 = -y_1 \frac{\partial}{\partial y} g(y_0, \dot{y}_0) - \dot{y}_1 \frac{\partial}{\partial \dot{y}} g(y_0, \dot{y}_0) \cdots \tag{15.42}$$

The third equation above is obtained by applying a Taylor series expansion to $\varepsilon g(y, \dot{y})$. That is,

$$\varepsilon g(y, \dot{y}) = \varepsilon g(y_0, \dot{y}_0) + \varepsilon \left(\varepsilon y_1 + \varepsilon^2 y_2 + \cdots \right) \frac{\partial}{\partial y} g(y_0, \dot{y}_0)$$

$$+ \varepsilon \left(\varepsilon \dot{y}_1 + \varepsilon^2 \dot{y}_2 + \cdots \right) \frac{\partial}{\partial \dot{y}} g(y_0, \dot{y}_0) + \cdots \tag{15.43}$$

PERTURBATION TECHNIQUE

As shown in Eq. (15.42), $y_{i+1}(t)$ can be obtained as the linear response to an excitation that is a nonlinear function of the previously determined $y_i(t)$. However, evaluation of $y_i(t)$ for $i \geq 2$ is extremely complicated; hence, only the first two terms are usually considered in practice. The following discussion, therefore, is limited to this case.

From the first equation of Eq. (15.42), the solution $y_0(t)$ can be expressed by a convolution integral given by

$$y_0(t) = \int_{-\infty}^{\infty} f_x(t - \tau) h(\tau)\, d\tau \tag{15.44}$$

where $h(\tau)$ is the impulse response function of a linear system, which is equal to the Fourier inverse transform of the frequency response function $H(\omega)$. Similarly, from the second equation in Eq. (15.42), we have

$$y_1(t) = -\int_{-\infty}^{\infty} g\{y_0(t-\tau), \dot{y}_0(t-\tau)\} h(\tau)\, d\tau \tag{15.45}$$

Thus, the response of a nonlinear system with an impulse function $h(\tau)$ can be obtained approximately in the time domain as follows:

$$\begin{aligned} y(t) &= y_0(t) + \varepsilon y_1(t) \\ &= \int_{-\infty}^{\infty} f_x(t-\tau) h(\tau)\, d\tau \\ &\quad - \varepsilon \int_{-\infty}^{\infty} g\{y_0(t-\tau), \dot{y}_0(t-\tau)\} h(\tau)\, d\tau \end{aligned} \tag{15.46}$$

Next, let us consider spectral analysis of the response $y(t)$ based on the series expansion given in Eq. (15.41). For this, we first evaluate the autocorrelation function. By taking the first order of the parameter ε in Eq. (15.41), the response autocorrelation function can be written as

$$\begin{aligned} R_{yy}(\tau) &= E[y(t) y(t+\tau)] \\ &= E[\{y_0(t) + \varepsilon y_1(t)\}\{y_0(t+\tau) + \varepsilon y_1(t+\tau)\}] \\ &\sim E[y_0(t) y_0(t+\tau)] \\ &\quad + \varepsilon \{E[y_0(t) y_1(t+\tau)] + E[y_1(t) y_0(t+\tau)]\} \\ &= R_{y_0 y_0}(\tau) + \varepsilon \{R_{y_0 y_1}(\tau) + R_{y_1 y_0}(\tau)\} \end{aligned} \tag{15.47}$$

where $R_{y_0y_0}(\tau)$ is the autocorrelation function of the linear part of the response, and $R_{y_0y_1}(\tau)$ and $R_{y_1y_0}(\tau)$ are the cross-correlation functions defined in Chapter 10. We have $R_{y_1y_0}(\tau) = R_{y_0y_1}(-\tau)$. The autocorrelation $R_{y_0y_0}(\tau)$ can be obtained as

$$R_{y_0y_0}(\tau) = E[y_0(t)y_0(t+\tau)]$$

$$= \int_{-\infty}^{\infty}\int_{-\infty}^{\infty} E[f_x(t-\tau_1)f_x(t+\tau-\tau_2)]h(\tau_1)h(\tau_2)\,d\tau_1\,d\tau_2$$

$$= \int_{-\infty}^{\infty}\int_{-\infty}^{\infty} R_{xx}(\tau+\tau_1-\tau_2)h(\tau_1)h(\tau_2)\,d\tau_1\,d\tau_2 \qquad (15.48)$$

where $R_{xx}(\tau+\tau_1-\tau_2)$ = input autocorrelation function.

By applying a formula similar to that for the autocorrelation function, the cross-correlation functions $R_{y_0y_1}(\tau)$ and $R_{y_1y_0}(\tau)$ can be evaluated by

$$R_{y_0y_1}(\tau) = E[y_0(t)y_1(t+\tau)]$$

$$= -\int_{-\infty}^{\infty}\int_{-\infty}^{\infty} E[f_x(t-\tau_1)g(t+\tau-\tau_2)]h(\tau_1)h(\tau_2)\,d\tau_1\,d\tau_2$$

$$\qquad (15.49)$$

where $g(\ )$ is the abbreviation of $g\{y(t), \dot{y}(t)\}$.

Similarly, we have

$$R_{y_1y_0}(\tau) = E[y_1(t)y_0(t+\tau)]$$

$$= -\int_{-\infty}^{\infty}\int_{-\infty}^{\infty} E[f_x(t+\tau-\tau_1)g(t-\tau_2)]h(\tau_1)h(\tau_2)\,d\tau_1\,d\tau_2$$

$$\qquad (15.50)$$

Thus, the response autocorrelation function $R_{yy}(\tau)$ can be evaluated from Eqs. (15.47) through (15.50).

The response spectral density function $S_{yy}(\omega)$ can then be obtained by taking the Fourier transform of the autocorrelation function. It can be written as

$$S_{yy}(\omega) = S_{y_0y_0}(\omega) + 2\varepsilon\, Co_{y_0y_1}(\omega) \qquad (15.51)$$

where $Co_{y_0y_1}(\omega)$ is the real part of the cross-spectral density function of $y_0(t)$ and $y_1(t)$ defined in Section 10.2.2.

As an application of the method developed above, let us consider the Duffing system defined in Eq. (15.19). By letting $\omega_0^2 r = \varepsilon$ in Eq. (15.19), the Duffing system may be written as

$$\ddot{y} + \alpha\dot{y} + \omega_0^2 y + \varepsilon y^3 = f_x(t) \tag{15.52}$$

where the excitation $f_x(t)$ is a steady-state Gaussian random process with the spectral density function $S_{xx}(\omega)$. Since the impulse response function, $h(t)$, of a linear system ($\varepsilon = 0$) is known, the autocorrelation function of the response, $R_{y_0y_0}(\tau)$, can be evaluated by applying Eq. (10.34). That is,

$$R_{y_0y_0}(\tau) = \frac{1}{2}\int_{-\infty}^{\infty} S_{xx}(\omega)|H(\omega)|^2 e^{i\omega\tau}\, d\omega \tag{15.53}$$

where

$$H(\omega) = \int_{-\infty}^{\infty} h(\tau) e^{-i\omega\tau}\, d\tau$$

In order to obtain the cross-correlation function $R_{y_0y_1}(\tau)$ given in Eq. (15.49), we first evaluate $E[f_x(t - \tau_1)g(t + \tau - \tau_2)]$. It becomes for the present nonlinear example,

$$E[f_x(t - \tau_1)g(t + \tau - \tau_2)] = E[f_x(t - \tau_1)y_0^3(t + \tau - \tau_2)]$$
$$= E[f_x(t - \tau_1)y_0(t + \tau - \tau_2)]$$
$$\times E[y_0^2(t + \tau - \tau_2)] \tag{15.54}$$

The last step in Eq. (15.54) is permissible because of the Gaussian random process assumption. Furthermore, from Eq. (15.44), we may write

$$E[f_x(t - \tau_1)y_0(t + \tau - \tau_2)]$$
$$= \int_{-\infty}^{\infty} E[f_x(t - \tau_1)f_x(t + \tau - \tau_2 - \tau_3)]h(\tau_3)\, d\tau_3$$
$$= \int_{-\infty}^{\infty} R_{xx}(\tau + \tau_1 - \tau_2 - \tau_3)h(\tau_3)\, d\tau_3 \tag{15.55}$$

and

$$E[y_0^2(t + \tau - \tau_2)] = \sigma_{y_0}^2 \qquad (15.56)$$

where $\sigma_{y_0}^2$ is the variance of the linear response. It should be noted, however, there are three different ways by which we can choose $E[y_0^2(t + \tau - \tau_2)]$ from $E[f_x(t - \tau_1)y_0^3(t + \tau - \tau_2)]$ in Eq. (15.54). Thus, from Eqs. (15.54) through (15.56), we have

$$E[f_x(t - \tau_1)g(t + \tau - \tau_2)]$$

$$= 3\sigma_{y_0}^2 \int_{-\infty}^{\infty} R_{xx}(\tau + \tau_1 - \tau_2 - \tau_3) h(\tau_3)\, d\tau_3$$

$$= \frac{3}{2}\sigma_{y_0}^2 \int_{-\infty}^{\infty}\int_{-\infty}^{\infty} S_{xx}(\omega) e^{i\omega(\tau+\tau_1-\tau_2-\tau_3)} h(\tau_3)\, d\tau_3\, d\omega \qquad (15.57)$$

and thereby, from Eq. (15.49), the cross-correlation function $R_{y_0 y_1}(\tau)$ becomes

$$R_{y_0 y_1}(\tau) = -\tfrac{3}{2}\sigma_{y_0}^2 \int_{-\infty}^{\infty}\int_{-\infty}^{\infty}\int_{-\infty}^{\infty}\int_{-\infty}^{\infty} S_{xx}(\omega) e^{i\omega(\tau+\tau_1-\tau_2-\tau_3)}$$

$$\times h(\tau_1) h(\tau_2) h(\tau_3)\, d\tau_1\, d\tau_2\, d\tau_3\, d\omega$$

$$= -\tfrac{3}{2}\sigma_{y_0}^2 \int_{-\infty}^{\infty} S_{xx}(\omega)|H(\omega)|^2 H^*(\omega) e^{i\omega\tau}\, d\omega \qquad (15.58)$$

The cross-correlation function $R_{y_1 y_0}(\tau)$ can then be obtained by replacing τ by $-\tau$ in the above equation. That is,

$$R_{y_1 y_0}(\tau) = -\tfrac{3}{2}\sigma_{y_0}^2 \int_{-\infty}^{\infty} S_{xx}(\omega)|H(\omega)|^2 H^*(\omega) e^{-i\omega\tau}\, d\omega \qquad (15.59)$$

Thus, the autocorrelation function of the nonlinear response, $R_{yy}(\tau)$, can be evaluated from Eqs. (15.53), (15.58), and (15.59) as

$$R_{yy}(\tau) = \tfrac{1}{2}\int_{-\infty}^{\infty} S_{xx}(\omega)|H(\omega)|^2 e^{i\omega\tau}\, d\omega$$

$$- 3\varepsilon\sigma_{y_0}^2 \int_{-\infty}^{\infty} S_{xx}(\omega)|H(\omega)|^2 H^*(\omega)\cos\omega\tau\, d\omega \qquad (15.60)$$

Then, the spectrum density function of the response becomes (see Table 10.1)

$$S_{yy}(\omega) = S_{xx}(\omega)|H(\omega)|^2 - 6\varepsilon\sigma_{y_0}^2 S_{xx}(\omega)|H(\omega)|^2 \operatorname{Re}\{H(\omega)\}$$
$$= S_{xx}(\omega)|H(\omega)|^2 \left[1 - 6\varepsilon\sigma_{y_0}^2 \operatorname{Re}\{H(\omega)\}\right] \quad (15.61)$$

Here, we have $\varepsilon = \omega_0^2 r$ for the Duffing system [see Eq. (15.52)]. Hence, Eq. (15.61) agrees with the result obtained in Eq. (15.25), which was derived by applying the equivalent linearization technique.

Next, let us evaluate the variance of the response of the Duffing system considered in Eq. (15.52). Although the variance can be evaluated from Eq. (15.60) by letting $\tau = 0$, it may be well to show here the computation method directly from Eq. (15.41). Since the mean response, $E[y]$, is zero, the variance is equal to the second moment $E[y^2]$, which may be approximately given from Eq. (15.41) as

$$E[y^2] \sim E[y_0^2] + 2\varepsilon E[y_0 y_1] \quad (15.62)$$

The first term can be evaluated from Eq. (15.48). That is, by letting $\tau = 0$, we have

$$E[y_0^2] = R_{y_0 y_0}(0) = \int_{-\infty}^{\infty}\int_{-\infty}^{\infty} R_{xx}(\tau_1 - \tau_2)h(\tau_1)h(\tau_2)\,d\tau_1\,d\tau_2 \quad (15.63)$$

$E[y_0 y_1]$ in the second term of Eq. (15.62) can be written for the Duffing system from Eq. (15.45) as

$$E[y_0 y_1] = -\int_{-\infty}^{\infty} h(\tau) E[y_0(t) y_0^3(t - \tau)]\,d\tau \quad (15.64)$$

Further, by applying Eq. (15.44), the above equation can be written as

$$E[y_0 y_1] = -\int_{-\infty}^{\infty} h(\tau)\,d\tau \int_{-\infty}^{\infty}\int_{-\infty}^{\infty}\int_{-\infty}^{\infty}\int_{-\infty}^{\infty} h(\tau_1)h(\tau_2)h(\tau_3)h(\tau_4)$$
$$\times E[f_x(t - \tau_1)$$
$$\times f_x(t - \tau - \tau_2) f_x(t - \tau - \tau_3) f_x(t - \tau - \tau_4)]\,d\tau_1\,d\tau_2\,d\tau_3\,d\tau_4$$
$$(15.65)$$

Since we assume the input is a Gaussian random process, the expected value involved in Eq. (15.65) can be written as follows (see Exercise 6.9):

$$E[f_x(t - \tau_1)f_x(t - \tau - \tau_2)f_x(t - \tau - \tau_3)f_x(t - \tau - \tau_4)]$$

$$= E[f_x(t - \tau_1)f_x(t - \tau - \tau_2)] \cdot E[f_x(t - \tau - \tau_3)f_x(t - \tau - \tau_4)]$$

$$+ E[f_x(t - \tau_1)f_x(t - \tau - \tau_3)] \cdot E[f_x(t - \tau - \tau_2)f_x(t - \tau - \tau_4)]$$

$$+ E[f_x(t - \tau_1)f_x(t - \tau - \tau_4)] \cdot E[f_x(t - \tau - \tau_2)f_x(t - \tau - \tau_3)]$$

$$= R_{xx}(\tau - \tau_1 + \tau_2) \cdot R_{xx}(\tau_3 - \tau_4) + R_{xx}(\tau - \tau_1 + \tau_3) \cdot R_{xx}(\tau_2 - \tau_4)$$

$$+ R_{xx}(\tau - \tau_1 + \tau_4) \cdot R_{xx}(\tau_2 - \tau_3) \qquad (15.66)$$

Thus, we can write

$$E[y_0 y_1] = -\int_{-\infty}^{\infty} h(\tau) \, d\tau \int_{-\infty}^{\infty} \int_{-\infty}^{\infty} \int_{-\infty}^{\infty} \int_{-\infty}^{\infty} h(\tau_1) h(\tau_2) h(\tau_3) h(\tau_4)$$

$$\times [R_{xx}(\tau - \tau_1 + \tau_2) \cdot R_{xx}(\tau_3 - \tau_4)$$

$$+ R_{xx}(\tau - \tau_1 + \tau_3) \cdot R_{xx}(\tau_2 - \tau_4)$$

$$+ R_{xx}(\tau - \tau_1 + \tau_4) \cdot R_{xx}(\tau_2 - \tau_3)] \, d\tau_1 \, d\tau_2 \, d\tau_3 \, d\tau_4 \qquad (15.67)$$

On the other hand, from Eq. (15.48) we have

$$R_{y_0 y_0}(0) = \int_{-\infty}^{\infty} \int_{-\infty}^{\infty} R_{xx}(\tau_1 - \tau_2) h(\tau_1) h(\tau_2) \, d\tau_1 \, d\tau_2 \qquad (15.68)$$

By applying the above equation together with Eq. (15.48), $E[y_0 y_1]$ can be written as

$$E[y_0 y_1] = -3 R_{y_0 y_0}(0) \int_{-\infty}^{\infty} R_{y_0 y_0}(\tau) h(\tau) \, d\tau \qquad (15.69)$$

Hence, by substituting Eqs. (15.63) and (15.69) into Eq. (15.62), the second moment (which is equal to the variance for zero mean) can be evaluated by the following formula:

$$E[y^2] = R_{y_0 y_0}(0) \left[1 - 6\varepsilon \int_{-\infty}^{\infty} R_{y_0 y_0}(\tau) h(\tau) \, d\tau \right] \qquad (15.70)$$

15.4 MARKOV PROCESS APPROACH—APPLICATION OF FOKKER–PLANCK EQUATION

The two approaches presented in the preceding sections for obtaining statistical properties of a nonlinear system are both essentially applicable for a system with weak nonlinearities. In these approaches the probability density function applicable for a nonlinear response random process is assumed to be a Gaussian distribution, which is not in conforming with the distribution obtained from analysis of responses of systems with strong nonlinearities. This shortcoming, however, is circumvented by the approach in which the Markov random process concept is applied.

The significant advantages of the Markov process approach are twofold: there is no restriction on the degree of nonlinearities, and the exact probability density function of the response $y(t)$ can be obtained as a solution of the Fokker–Planck equation. Although the Markov process approach offers these advantages, it does have one disadvantage. That is, the method is applicable for a system whose input spectrum has a constant value (white noise spectrum); hence, determination of the magnitude of a white noise spectrum equivalent to an arbitrarily given input spectrum is essential in practice.

The definition of the Fokker–Planck equation was given in Section 13.3 along with a discussion on its application to a nonlinear vibration system. For convenience of further discussion on this subject, it may be well to summarize the important results derived in Section 13.3 as follows:

For a system of single degree of freedom with both nonlinear damping, $g(\dot{y})$, and nonlinear restoring, $h(y)$, terms, the Fokker–Planck equation is given by

$$-\dot{y}\frac{\partial}{\partial y}f(y,\dot{y}) + \frac{\partial}{\partial \dot{y}}\big(\{g(\dot{y}) + h(y)\}f(y,\dot{y})\big) + \frac{\pi A}{2}\frac{\partial^2}{\partial \dot{y}^2}f(y,\dot{y}) = 0$$

where

$f(y,\dot{y})$ = joint probability density function of output displacement and velocity

A = constant spectral density function (white noise spectrum) of the excitation.

The solution of the Fokker–Planck equation yields the joint probability density function of displacement and velocity of the response. It is difficult, in general, to derive an analytical solution in closed form; however, if the damping of the system is linear, then the solution of the Fokker–Planck equation can be derived in closed form. That is, by letting $g(\dot{y}) = \alpha \dot{y}$, the probability density function of the response y becomes

$$f(y) = (\text{constant}) \exp\left\{ -\frac{2\alpha}{\pi A} \int_0^y h(y) \, dy \right\}$$

Example 15.2. Let us apply the Markov process approach to a Duffing system. By writing $h(y) = \omega_0^2(y + ry^3)$, the probability density function of the response $f(y)$ can be obtained as

$$f(y) = C \exp\left\{ -\frac{2\alpha \omega_0^2}{\pi A} \int_0^y (y + ry^3) \, dy \right\}$$

$$= C \exp\left\{ -\frac{\alpha \omega_0^2}{\pi A} \left(y^2 + \frac{ry^4}{2} \right) \right\}$$

where the constant C is a normalization factor to be determined from $\int_{-\infty}^{\infty} f(y) \, dy = 1$. It is noted that if $r = 0$ (linear system), then $f(y)$ becomes

$$f(y) = C \exp\left\{ -\frac{\alpha \omega_0^2 y^2}{\pi A} \right\}$$

which is in the form of a normal distribution with variance $\pi A / 2\alpha \omega_0^2$. ∎

It has been shown that the solution of the Fokker–Planck equation yields the probability density function of the response of a nonlinear system. The question arises as to how the white noise spectrum involved in the Fokker–Planck equation can be determined for a system that has an arbitrarily given input spectrum. Unfortunately, there is no complete method at present to answer this question. However, the following approach may be one way to determine the white noise spectrum for a system with strong nonlinearities:

The underlying assumption of the approach is that the response energy for a nonlinear system is equal to that for an equivalent linear system. With this assumption,

1. Establish a linear equation through the equivalent linearization technique, and obtain the coefficients α_e and ω_e of the linear system.

2. Consider the Markov process approach to this linear equation for a white noise spectrum A. Since the system is linear, the solution of the Fokker–Planck equation of the displacement is a normal distribution with zero mean and variance $\pi A/2\alpha_e \omega_e^2$, where A is unknown as this stage.

3. Evaluate the frequency response spectrum for the linear equation derived in Item (1).

4. Obtain the response spectrum for an arbitrarily given spectrum by applying Theorem 14.2.

5. Evaluate the area under the response spectrum, m_0, which is equal to the variance of the response.

6. By equating the variances obtained in Items (2) and (5), the unknown white noise spectrum can be obtained as $A = 2\alpha_e \omega_e^2 m_0/\pi$.

15.5 APPLICATION OF BISPECTRAL ANALYSIS

It was shown in Section 14.1 that the output of a linear system can be obtained in the time domain as the convolution integral of the output and frequency response function. The same concept can also be applied for evaluating the response of a nonlinear system (Tick, 1961). For this, let us write the output of a nonlinear system as

$$y(t) = y_1(t) + y_2(t)$$

$$= \int_{-\infty}^{\infty} h(\tau) x(t-\tau) \, d\tau$$

$$+ \int_{-\infty}^{\infty} \int_{-\infty}^{\infty} h(\tau_1, \tau_2) x(t-\tau_1) x(t-\tau_2) \, d\tau_1 \, d\tau_2 \quad (15.71)$$

Here, the second term represents the nonlinear effect on the response. $h(\tau_1, \tau_2)$ is the second-order impulse response function, and its Fourier transform is given by

$$H(\omega_1, \omega_2) = \int_{-\infty}^{\infty} \int_{-\infty}^{\infty} h(\tau_1, \tau_2) e^{-i(\omega_1 \tau_1 + \omega_2 \tau_2)} \, d\tau_1 \, d\tau_2 \quad (15.72)$$

Equation (15.72) indicates that the response of a nonlinear system can be obtained in the time domain if the second-order frequency response function $H(\omega_1, \omega_2)$ is known. The function $H(\omega_1, \omega_2)$ can be obtained in practice through experiments in the laboratory as is the case for the frequency response function $H(\omega)$. A significant difference between them, however, is that the input in the experiment for evaluating $H(\omega_1, \omega_2)$ is a random process in contrast to a series of sinusoidal inputs with different frequencies for evaluating $H(\omega)$. The principle for evaluating $H(\omega_1, \omega_2)$ is as follows:

Let the random processes $x(t)$ and $y(t)$ be the excitation and response, respectively, of a nonlinear system. We may consider the excitation as a Gaussian random process with zero mean; however, the mean of the response may not be zero. Analogous to the definition of the cross-spectral density function given in Theorem 10.3, the cross-bispectral density function of the output and input may be written as

$$B_{yxx}(\omega_1, \omega_2) = \frac{1}{\pi^2} \int_{-\infty}^{\infty} \int_{-\infty}^{\infty} M_{yxx}(\mu_1, \mu_2) e^{-i(\omega_1 \mu_1 + \omega_2 \mu_2)} d\mu_1 \, d\mu_2 \quad (15.73)$$

where

$$M_{yxx}(\mu_1, \mu_2) = \frac{1}{4} \int_{-\infty}^{\infty} \int_{-\infty}^{\infty} B_{yxx}(\omega_1, \omega_2) e^{i(\omega_1 \mu_1 + \omega_2 \mu_2)} d\omega_1 \, d\omega_2 \quad (15.74)$$

On the other hand, $M_{yxx}(\mu_1, \mu_2)$ can be evaluated as follows:

$$M_{yxx}(\mu_1, \mu_2) = E\big[\{y(t) - \bar{y}(t)\}x(t - \mu_1)x(t - \mu_2)\big] \quad (15.75)$$

where from Eq. (15.71) $\bar{y}(t)$ becomes

$$\bar{y}(t) = \int_{-\infty}^{\infty} \int_{-\infty}^{\infty} h(\tau_1, \tau_2) R_{xx}(\tau_1 - \tau_2) \, d\tau_1 \, d\tau_2 \quad (15.76)$$

By inserting Eqs. (15.71) and (15.76) into Eq. (15.75), we have

$$M_{yxx}(\mu_1, \mu_2) = \int_{-\infty}^{\infty} \int_{-\infty}^{\infty} h(\tau_1, \tau_2) E\big[x(t - \tau_1) x(t - \tau_2)$$

$$\times x(t - \mu_1) x(t - \mu_2)\big] \, d\tau_1 \, d\tau_2$$

$$- \int_{-\infty}^{\infty} \int_{-\infty}^{\infty} h(\tau_1, \tau_2) R_{xx}(\tau_2 - \tau_1) R_{xx}(\mu_2 - \mu_1) \, d\tau_1 \, d\tau_2$$

$$(15.77)$$

Since the input is a Gaussian random process, the result given in Exercise 6.9 can be applied to the first term of Eq. (15.77). Furthermore, by carrying out the Fourier transform on $h(\tau_1, \tau_2)$ and the Fourier inverse transform on the autocorrelation functions, Eq. (15.77) becomes

$$M_{yxx}(\mu_1, \mu_2) = \int_{-\infty}^{\infty} \int_{-\infty}^{\infty} h(\tau_1, \tau_2) R_{xx}(\mu_1 - \tau_1) R_{xx}(\mu_2 - \tau_2) \, d\tau_1 \, d\tau_2$$

$$+ \int_{-\infty}^{\infty} \int_{-\infty}^{\infty} h(\tau_1, \tau_2) R_{xx}(\mu_2 - \tau_1) R_{xx}(\mu_1 - \tau_2) \, d\tau_1 \, d\tau_2$$

$$= \frac{1}{2} \int_{-\infty}^{\infty} \int_{-\infty}^{\infty} H(\omega_1, \omega_2) S_{xx}(\omega_1) S_{xx}(\omega_2) e^{i(\omega_1 \mu_1 + \omega_2 \mu_2)} \, d\omega_1 \, d\omega_2$$

(15.78)

Thus, from a comparison between Eqs. (15.74) and (15.78), we can derive the following relationship:

$$H(\omega_1, \omega_2) = \frac{B_{yxx}(\omega_1, \omega_2)}{2 S_{xx}(\omega_1) S_{xx}(\omega_2)} \quad (15.79)$$

The above equation implies that the second-order frequency response function $H(\omega_1, \omega_2)$ of a nonlinear system can be evaluated from knowledge of the cross-bispectral density function $B_{yxx}(\omega_1, \omega_2)$, which can be obtained through experiments having a random process as an input. Then, by taking the inverse Fourier transform, the second-order impulse response function $h(\tau_1, \tau_2)$ can be obtained, and thereby the response of a nonlinear system to an arbitrarily given input can be evaluated from Eq. (15.71).

EXERCISES

15.1 Consider a Duffing system given by

$$\ddot{y} + \alpha \dot{y} + \omega_0^2 (y + r y^3) = f_x(t)$$

Let ω_e be the frequency obtained by the equivalent linearization technique. Derive the following formula by applying Eq. (15.37):

$$\omega_e^2 = \omega_0^2 + 3 r \omega_0^2 \sigma_y^2$$

15.2 The variance of the Duffing system with zero mean is given as follows (see Eq. 15.70):

$$\text{Var}[y] = R_{y_0 y_0}(0)\left[1 - 6\varepsilon \int_{-\infty}^{\infty} R_{y_0 y_0}(\tau) h(\tau)\, d\tau\right]$$

where

$$R_{y_0 y_0}(\tau) = \text{autocorrelation function of the linear system}$$

$$h(\tau) = \text{impulse response function}$$

$$\varepsilon = \omega_0^2 r$$

On the other hand, the autocorrelation function of the response $R_{yy}(\tau)$ is given in Eq. (15.60) as follows:

$$R_{yy}(\tau) = \frac{1}{2}\int_{-\infty}^{\infty} S_{xx}(\omega)|H(\omega)|^2 e^{i\omega\tau}\, d\omega$$

$$- 3\varepsilon\sigma_{y_0}^2 \int_{-\infty}^{\infty} S_{xx}(\omega)|H(\omega)|^2 H^*(\omega) \cos\omega\tau\, d\omega$$

where $S_{xx}(\omega) = $ input spectral density function. Derive $\text{Var}[y]$ from $R_{yy}(\tau)$.

15.3 The coupled equation of motion for a system with two degrees of freedom is given by

$$\ddot{x} + g_{xx}(\dot{x}) + h_{xx}(x) + m_{yx}\ddot{y} + g_{yx}(\dot{y}) + h_{yx}(y) = s_x(t)$$

$$\ddot{y} + g_{yy}(\dot{y}) + h_{yy}(y) + m_{xy}\ddot{x} + g_{xy}(\dot{x}) + h_{xy}(x) = s_y(t)$$

where $g(\ )$ and $h(\ )$ are nonlinear terms. The excitations $s_x(t)$ and $s_y(t)$ have a white noise spectrum A_x and A_y, respectively. Derive the Fokker–Planck equation.

CHAPTER 16

Non-Gaussian Stochastic Processes

Stochastic processes addressed in the preceding chapters are assumed to be weakly stationary, subject to the ergodic property, and the deviations from the mean obey a normal (Gaussian) distribution. This assumption has been found acceptable for predicting the statistical properties of the majority of phenomena observed in the diverse engineering fields and in the physical sciences. However, there are cases for which the Gaussian process assumption is not valid. As an example, Figure 16.1 shows the time history of wind-generated wave profiles in water of finite water depth. Because of nonlinear wave characteristics due to limited water depth, wave profiles show a definite excess of high crests and shallow troughs in contrast to those of ocean waves in deep water as illustrated in Figure 9.8.

Another example of a non-Gaussian stochastic process is the response of a system that has strong nonlinear characteristics. In this case, as stated in the previous chapter, the response is not a Gaussian process even though the input is Gaussian (Crandall, 1980).

This chapter presents the probability distribution for non-Gaussian stochastic processes. The probability density function applicable for a non-Gaussian stochastic process is given in the form of a series known as the Gram–Charlier series probability distribution. It was developed from two different approaches; one by applying the concept of polynomials orthogonal with respect to the normal distribution and the other by applying the cumulant generating function. However, the distributions can

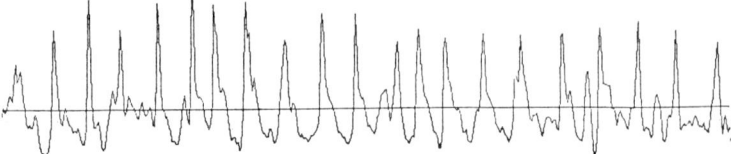

Figure 16.1 Example of non-Gaussian random process.

be reduced to the same formula. The derivation of each approach is given below.

16.1 PROBABILITY DISTRIBUTION BY APPLYING THE CONCEPT OF ORTHOGONAL POLYNOMIALS

Let us first give the definition of Hermite polynomials. The Hermite polynomial of degree n, denoted by $H_n(z)$ is defined as a function that satisfies the relationship given by

$$\frac{d^n}{dz^n}e^{-z^2/2} = (-1)^n H_n(z) \cdot e^{-z^2/2}, \qquad n = 0, 1, 2, \ldots \quad (16.1)$$

From the above equation, we have the following polynomials called the *Hermite polynomials*:

$$H_0(z) = 1$$

$$H_1(z) = z$$

$$H_2(z) = z^2 - 1$$

$$H_3(z) = z^3 - 3z$$

$$H_4(z) = z^4 - 6z^2 + 3$$

$$H_5(z) = z^5 - 10z^3 + 15z$$

$$H_6(z) = z^6 - 15z^4 + 45z^2 - 15, \ldots \quad (16.2)$$

APPLYING THE CONCEPT OF ORTHOGONAL POLYNOMIALS

Let $\alpha(z)$ be the standardized normal (Gaussian) probability density function given by

$$\alpha(z) = \frac{1}{\sqrt{2\pi}} e^{-z^2/2} \tag{16.3}$$

It can be shown that the polynomials $(1/\sqrt{n!})H_n(z)$ are orthogonal with respect to the standardized normal probability density function. To elaborate, let us consider the integration of the product of two Hermite polynomials and the standardized normal probability density function. That is,

$$\int_{-\infty}^{\infty} H_m(z) H_n(z) \alpha(z)\, dz = (-1)^n \int_{-\infty}^{\infty} H_m(z) \frac{d^n}{dz^n} \alpha(z)\, dz$$

$$= (-1)^{n-1} \left[H_m(z) \frac{d^{n-1}}{dz^{n-1}} \alpha(z) \right]_{-\infty}^{\infty}$$

$$+ (-1)^{n-1} \int_{-\infty}^{\infty} \frac{d}{dz} H_m(z) \frac{d^{n-1}}{dz^{n-1}} \alpha(z)\, dz \tag{16.4}$$

The first term becomes zero. For the integration of the second term, we may use the following property of the Hermite polynomials:

$$\frac{d}{dz} H_n(z) = n \cdot H_{n-1}(z) \tag{16.5}$$

By repeating the integration by parts and by repeatedly applying the property given in Eq. (16.5), we have

$$\int_{-\infty}^{\infty} H_m(z) \cdot H_n(z) \cdot \alpha(z)\, dz = (-1)^{n-m} m! \int_{-\infty}^{\infty} \frac{d^{n-m}}{dz^{n-m}} \alpha(z)\, dz$$

$$= m! \int_{-\infty}^{\infty} H_{n-m}(z) \cdot \alpha(z)\, dz$$

$$= \begin{cases} n! & \text{if } m = n \\ 0 & \text{if } m \neq n \end{cases} \tag{16.6}$$

Equation (16.6) may be written as

$$\int_{-\infty}^{\infty} \left\{ \frac{1}{\sqrt{m!}} H_m(z) \right\} \left\{ \frac{1}{\sqrt{n!}} H_n(z) \right\} \alpha(z) \, dz = \begin{cases} 1 & \text{if } m = n \\ 0 & \text{if } m \neq n \end{cases} \quad (16.7)$$

The relationship given in the above equation implies that $(1/\sqrt{n!})H_n(z)$ is a sequence of orthogonal polynomials with respect to $\alpha(z)$. With this in mind, let us express an arbitrarily given standardized probability density function, $f(z)$, in the following form:

$$f(z) = a_0 \alpha(z) + a_1 \alpha^{(1)}(z) + a_2 \alpha^{(2)}(z) + a_3 \alpha^{(3)}(z) + \cdots \quad (16.8)$$

where

$$a_n = \text{constants (unknown)}$$

$$\alpha^{(n)}(z) = \frac{d^n}{dz^n} \alpha(z)$$

$\alpha(z)$ = standardized normal probability density function.

From the definition of the Hermite polynomials given in Eq. (16.1), we have

$$f(z) = \alpha(z)\{a_0 H_0(z) - a_1 H_1(z) + a_2 H_2(z) - a_3 H_3(z) + \cdots\}$$

$$= \alpha(z) \left\{ c_0 \frac{H_0(z)}{\sqrt{0!}} - c_1 \frac{H_1(z)}{\sqrt{1!}} + c_2 \frac{H_2(z)}{\sqrt{2!}} - c_3 \frac{H_3(z)}{\sqrt{3!}} + \cdots \right\}$$

$$= \alpha(z) \sum_{n=0}^{\infty} (-1)^n c_n \frac{H_n(z)}{\sqrt{n!}} \quad (16.9)$$

where

$$c_n = \sqrt{n!} \, a_n \text{ (unknown)}.$$

In order to determine the unknown constant c_n, we may multiply $f(z)$ by $H_n(z)/\sqrt{n!}$, and integrate over the domain $(\infty, -\infty)$. Since $H_n(z)/\sqrt{n!}$ is orthogonal with respect to $\alpha(z)$, we have

$$\int_{-\infty}^{\infty} \frac{H_n(z)}{n!} f(z) \, dz = (-1)^n \frac{c_n}{n!} \int_{-\infty}^{\infty} H_n^2(z) \alpha(z) \, dz = (-1)^n c_n \quad (16.10)$$

Thus, the constant c_n can be determined as

$$c_n = (-1)^n \frac{1}{\sqrt{n!}} \int_{-\infty}^{\infty} H_n(z) f(z) \, dz \qquad (16.11)$$

By applying the Hermite polynomials given in Eq. (16.2), c_n can be evaluated as follows:

$$c_0 = \int_{-\infty}^{\infty} H_0(z) f(z) \, dz = 1$$

$$c_1 = -\int_{-\infty}^{\infty} H_1(z) f(z) \, dz = -m_1 = 0$$

$$c_2 = \frac{1}{\sqrt{2!}} \int_{-\infty}^{\infty} H_2(z) f(z) \, dz = \frac{1}{\sqrt{2!}} (m_2 - 1) = 0$$

$$c_3 = -\frac{1}{\sqrt{3!}} \int_{-\infty}^{\infty} H_3(z) f(z) \, dz = -\frac{m_3}{\sqrt{3!}}$$

$$c_4 = \frac{1}{\sqrt{4!}} \int_{-\infty}^{\infty} H_4(z) f(z) \, dz = \frac{1}{\sqrt{4!}} (m_4 - 3)$$

$$c_5 = \frac{1}{\sqrt{5!}} \int_{-\infty}^{\infty} H_5(z) f(z) \, dz = \frac{1}{\sqrt{5!}} (m_5 - 10 m_3), \dots \qquad (16.12)$$

where $m_j = j$th moment of the standardized random variable. Note that $m_1 = 0$ and $m_2 = 1$ since $f(z)$ is a standardized probability density function.

From Eqs. (16.9) and (16.12), the probability density function of a standardized random variable can be expressed as

$$f(z) = \frac{1}{\sqrt{2\pi}} e^{-z^2/2} \left[1 + \frac{m_3}{3!} H_3(z) + \frac{m_4 - 3}{4!} H_4(z) + \frac{m_5 - 10 m_3}{5!} \right.$$

$$\left. + \frac{m_6 - 15 m_4 + 30}{6!} H_6(z) + \cdots \right] \qquad (16.13)$$

This is called the *Gram–Charlier series of Type A* probability density function.

The parameters $m_3, m_4, m_5, \ldots$ in Eq. (16.13) are the moments of the standardized random variable. For a nonstandardized random variable X, which has the mean value μ and variance σ^2, the moment m_j may be written in the form of the central moments,

$$m_j = \frac{E\left[(x-\mu)^j\right]}{\sigma^j} = \frac{E\left[(x-\mu)^j\right]}{\left\{E\left[(x-\mu)^2\right]\right\}^{j/2}} \quad (16.14)$$

On the other hand, from the functional relationship given in Eq. (4.21), the central moments can be expressed in terms of the cumulants of the random variable as follows:

$$\mu_2 = k_2, \quad \mu_3 = k_3, \quad \mu_4 = k_4 + 3k_2^2,$$

$$\mu_5 = k_5 + 10k_3k_2$$

$$\mu_6 = k_6 + 15k_4k_2 + 10k_3^2 + 15k_2^3, \ldots \quad (16.15)$$

where k_j = cumulants.

Thus, from Eqs. (16.14) and (16.15), we have

$$m_3 = \mu_3/\left(\sqrt{\mu_2}\right)^3 = k_3/\left(\sqrt{k_2}\right)^3$$

$$m_4 = \mu_4/\mu_2^2 = \left(k_4/k_2^2\right) + 3$$

$$m_5 = \mu_5/\left(\sqrt{\mu_2}\right)^5 = k_5/\left(\sqrt{k_2}\right)^5 + 10k_3/\left(\sqrt{k_2}\right)^3 = k_5/\left(\sqrt{k_2}\right)^5 + 10m_3$$

$$m_6 = \mu_6/\mu_2^3 = k_6/k_2^3 + 15\left\{\left(k_4/k_2^2\right) + 3\right\} + 10\left(k_3^2/k_2^3\right) - 30$$

$$= k_6/k_2^3 + 15m_4 + 10m_3^2 - 30, \ldots \quad (16.16)$$

By writing $\lambda_j = k_j/(k_2)^{j/2}$, the moment m_j can be expressed as

$$m_3 = \lambda_3$$

$$m_4 - 3 = \lambda_4$$

$$m_5 - 10m_3 = \lambda_5$$

$$m_6 - 15m_4 + 30 = \lambda_6 + 10\lambda_3^2, \ldots \quad (16.17)$$

Hence, the probability density function of a random variable X with zero mean and variance σ^2 can be written from Eq. (16.13) with the aid of Eq. (16.17) as

$$f(x) = \frac{1}{\sigma\sqrt{2\pi}} e^{-x^2/2\sigma^2} \left[1 + \frac{\lambda_3}{3!} H_3\left(\frac{x}{\sigma}\right) + \frac{\lambda_4}{4!} H_4\left(\frac{x}{\sigma}\right) + \frac{\lambda_5}{5!} H_5\left(\frac{x}{\sigma}\right) \right.$$

$$\left. + \left(\frac{\lambda_6}{6!} + \frac{\lambda_3^2}{72}\right) H_6\left(\frac{x}{\sigma}\right) + \cdots \right] \quad (16.18)$$

where

$$\lambda_j = \frac{k_j}{(k_2)^{j/2}} = \frac{k_j}{\sigma^j}$$

Note that λ_3 is equal to the skewness, and λ_4 is equal to the kurtosis minus 3 as defined in Eq. (4.22).

16.2 PROBABILITY DISTRIBUTION BY APPLYING CUMULANT GENERATING FUNCTION

The probability density function given in Eq. (16.18) was also derived by Longuet-Higgins (1963) by applying the cumulant generating function. He further derived the joint non-Gaussian probability density function for two random variables. In the following, his derivation of the probability density function for a single random variable is outlined.

The cumulant generating function $\psi(t)$ is the logarithm of a characteristic function and it is defined in Eq. (4.19). That is,

$$\psi(t) = \ln \phi(t) = \sum_{j=1}^{\infty} \frac{(it)^j}{j!} k_j$$

where

$$\phi(t) = \text{characteristic function}$$

$$k_j = \text{cumulant}.$$

It was also shown in Eq. (4.10) that the probability density function of a random variable X can be evaluated from the characteristic function by

$$f(x) = \frac{1}{2\pi} \int_{-\infty}^{\infty} \phi(t) e^{-itx} dx$$

Hence, from the above two equations we can express the probability density function $f(x)$ in terms of the cumulants as follows:

$$f(x) = \frac{1}{2\pi} \int_{-\infty}^{\infty} \exp\{\psi(t) - itx\} dt$$

$$= \frac{1}{2\pi} \int_{-\infty}^{\infty} \exp\left\{(k_1 - x)it + \frac{k_2}{2!}(it)^2 + \frac{k_3}{3!}(it)^3 + \cdots\right\} dt \quad (16.19)$$

Next, the probability density function $f(x)$ given in Eq. (16.19) will be standardized. Since the cumulants k_1 and k_2 are the mean and variance, respectively, of the random variable X, we may use the following notation in the transformation.

$$z = \frac{x - k_1}{\sqrt{k_2}}, \qquad t = s/\sqrt{k_2}, \qquad \lambda_j = k_j / \left(\sqrt{k_2}\right)^j \quad (16.20)$$

Then, we can write

$$(k_1 - x)it = -izs$$

$$k_2(it)^2 = -s^2$$

$$k_3(it)^3 = \lambda_3(is)^3$$

$$k_4(it)^4 = \lambda_4(is)^4$$

$$k_5(is)^5 = \lambda_5(is)^5, \ldots \quad (16.21)$$

and thereby the probability density function of the standardized random variable Z becomes

$$f(z) = \frac{1}{2\pi} \int_{-\infty}^{\infty} \exp\left\{-\frac{1}{2}(s^2 + 2izs)\right\} \cdot \exp\left\{\sum_{j=3}^{\infty} \frac{1}{j!} \lambda_j (is)^j\right\} ds \quad (16.22)$$

We may expand the second exponential term of the integrand as follows:

$$\exp\left\{\sum_{j=3}^{\infty}\frac{1}{j!}\lambda_j(is)^j\right\} = 1 + \sum_{j=3}^{\infty}\frac{1}{j!}\lambda_j(is)^j + \frac{1}{2!}\left(\sum_{j=3}^{\infty}\frac{1}{j!}\lambda_j(is)^j\right)^2 + \cdots$$

$$= 1 + \frac{1}{3!}\lambda_3(is)^3 + \frac{1}{4!}\lambda_4(is)^4 + \frac{1}{5!}\lambda_5(is)^5$$

$$+ \left\{\frac{1}{6!}\lambda_6 + \frac{1}{2!(3!)^2}\lambda_3^2\right\}(is)^6$$

$$+ \left\{\frac{1}{7!}\lambda_7 + \frac{1}{3!4!}\lambda_3\lambda_4\right\}(is)^7 + \cdots \qquad (16.23)$$

Then, Eq. (16.22) can be written as

$$f(z) = \frac{1}{2\pi}\int_{-\infty}^{\infty}\exp\left\{-\frac{1}{2}(s^2 + izs)\right\}$$

$$\times \left[1 + \frac{\lambda_3}{3!}(is)^3 + \frac{\lambda_4}{4!}(is)^4 + \frac{\lambda_5}{5!}(is)^5\right.$$

$$\left. + \left\{\frac{\lambda_6}{6!} + \frac{\lambda_3^2}{2!(3!)^2}\right\}(is)^6 + \cdots\right] \qquad (16.24)$$

For the integration of the first term in the bracket, we may use the following formula associated with the normal probability distribution.

$$\frac{1}{\sqrt{2\pi}}\int_{-\infty}^{\infty}\exp\left\{-\frac{1}{2}(s^2 + 2isz)\right\} ds = \exp\left\{-\frac{z^2}{2}\right\} \qquad (16.25)$$

and for the integration of the second term and the rest we may write

$$\frac{1}{\sqrt{2\pi}}\int_{-\infty}^{\infty}\exp\left\{-\frac{1}{2}(s^2 + 2izs)\right\}(is)^n\, ds$$

$$= \frac{(-1)^n}{\sqrt{2\pi}}\frac{d^n}{dz^n}\int_{-\infty}^{\infty}\exp\left\{-\frac{1}{2}(s^2 + 2izs)\right\} ds$$

$$= (-1)^n \frac{d^n}{dz^n}e^{-z^2/2} = H_n(z)\cdot e^{-z^2/2} \qquad (16.26)$$

where $H_n(z)$ = Hermite polynomial of degree n.

By applying Eqs. (16.25) and (16.26), the standardized non-Gaussian probability density function given in Eq. (16.24) can be expressed as

$$f(z) = \frac{1}{2\pi} \int_{-\infty}^{\infty} \exp\left\{-\frac{1}{2}(s^2 + 2izs)\right\}\left[1 + \frac{\lambda_3}{3!}(is)^3 + \frac{\lambda_4}{4!}(is)^4 + \cdots\right] ds$$

$$= \frac{1}{\sqrt{2\pi}} e^{-z^2/2} \left[1 + \frac{\lambda_3}{3!} H_3(z) + \frac{\lambda_4}{4!} H_4(z) + \frac{\lambda_5}{5!} H_5(z)\right.$$

$$+ \left\{\frac{\lambda_6}{6!} + \frac{\lambda_3^2}{2!(3!)^2}\right\} H_6(z)$$

$$\left. + \left\{\frac{\lambda_7}{7!} + \frac{\lambda_3 \lambda_4}{3!4!}\right\} H_7(z) + \cdots\right] \qquad (16.27)$$

Thus, for a random variable with zero mean and variance σ^2, the probability density function is given by the same formula as derived in Eq. (16.18).

In applying the non-Gaussian probability density function given in Eqs. (16.18) and (16.27) in practice, the following remarks are given:

1. The probability density function at times becomes negative for large negative x depending on the number of terms considered as well as on the values of the parameters, $\lambda_3, \lambda_4, \ldots$. Hence, it is necessary to examine whether or not the negative density will cause any serious problem in analyzing the data.

2. The accuracy of a function that is expressed in the form of a series increases with an increase in higher order terms, in general. However, this is not the case of the non-Gaussian probability density function. That is, higher order terms do not necessarily yield better agreement with the histogram constructed from data as shown in the following example.

Example 16.1. An example of comparison between the probability density function given in Eq. (16.18) and observed data is shown in Figure 16.2. The histogram shown in the figure is that obtained of the wind-generated wave profile obtained in water of finite depth. As can be seen in the

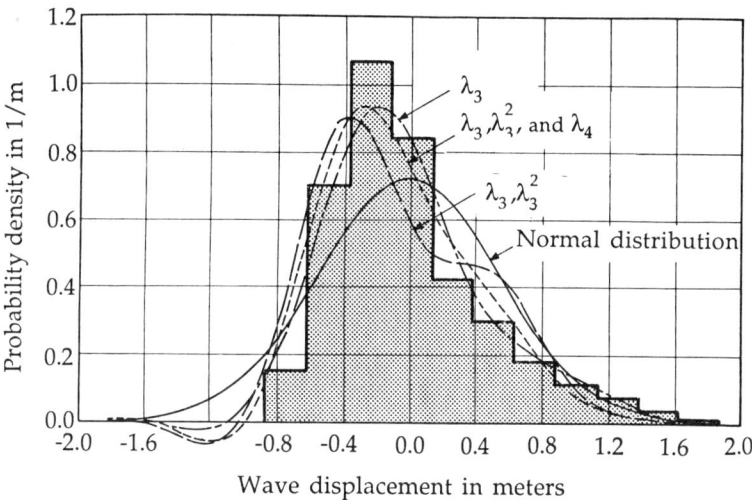

Figure 16.2 Comparison between histogram and non-Gaussian probability density function given in Eq. (16.18) for wind-generated wave profile in water of finite depth.

figure:

1. The histogram deviates substantially from the normal distribution.

2. The probability density function becomes negative for large negative x. However, the magnitude of the negative portion is relatively small, on the order of 2% or less. Furthermore, the negative density occurs outside the range of the histogram. Therefore, it will not cause any serious problem if we assume this negative probability density to be zero and, in turn, normalize the entire probability density function so that the area becomes unity.

3. The probability density function that includes the parameter λ_3 only agrees reasonably well with the histogram, but the agreement becomes poor if the term with the parameter λ_3^2 is included. However, the probability density function consisting of terms with the parameters λ_3, λ_3^2, and λ_4 agrees very well with the histogram. The addition of the higher terms such as λ_5, λ_6, ... does not yield any appreciable change in the shape of the density function. Thus, as far as the example shown in Figure 16.2 is concerned, the parameter λ_3 that represents the skewness of the random variable is the dominant parameter affecting the non-Gaussian characteristics and that combination of the parameters λ_3, λ_3^2, and λ_4 best represents the histogram constructed from observed data. ∎

EXERCISES

16.1 Derive the characteristic function of the Gram–Charlier series distribution with parameters λ_3 and λ_4.

16.2 Consider two random processes A and B. Sample moments $m_k = (1/n)\sum_{i=1}^{n} x_i^k$ are evaluated from 250 readings of the time history for each process. For the random process A, we have $m_1 = 2.0$, $m_2 = 13.0$, $m_3 = 52.0$, $m_4 = 380.0$, while for the random process B, we have $m_1 = 2.0$, $m_2 = 13.0$, $m_3 = 62.0$, and $m_4 = 475.0$. Identify whether these random processes are Gaussian or non-Gaussian.

16.3 Let X and Y be statistically independent non-Gaussian random processes each obeying the following Gram–Charlier series distribution:

X: Non-Gaussian with mean $= 0$, variance $= \sigma_x^2$, and λ_{4x}

Y: Non-Gaussian with mean $= 0$, variance $= \sigma_y^2$, and λ_{4y}

Consider the sum of two random processes $Z = aX + bY$, where $a > 0$, $b > 0$. Show that the random process Z approximately follows the Gram–Charlier series distribution with mean $= 0$, variance $= a^2\sigma_x^2 + b^2\sigma_y^2$ and the parameter given by

$$\lambda_{4z} = \frac{\lambda_{4x} a^4 \sigma_x^4 + \lambda_{4y} b^4 \sigma_y^4}{\left(a^2 \sigma_x^2 + b^2 \sigma_y^2\right)^2}$$

[Hint] Derive the characteristic functions $\phi_x(t)$ and $\phi_y(t)$, and apply $\phi_z(t) = \phi_x(at)\phi_y(bt)$.

CHAPTER 17

Counting Stochastic Processes

17.1 POISSON PROCESSES

17.1.1 Fundamentals of the Poisson Process

The definition of the Poisson process is given in Section 9.3.3. In this section the theoretical background of the Poisson process along with the assumptions involved are discussed in detail.

Let a random process $N(t)$ represent the total number of occurrences of an event in a time interval between 0 and t (see Figure 17.1). It is assumed that

(i) $N(t)$ has independent increments with $N(0) = 0$,

(ii) $N(t)$ has stationary increments, and

(iii) In a small time interval Δt, the number of occurrences of the event is at most one. The probability of one occurrence is proportional to Δt, namely $\nu(\Delta t)$.

Under these conditions, it can be proved that the counting process $N(t)$ becomes the Poisson process as presented in the following theorem:

Theorem 17.1. A counting process $N(t)$ that satisfies conditions (i) through (iii) obeys the Poisson distribution with mean νt, and the process $N(t)$ is called the *Poisson process*.

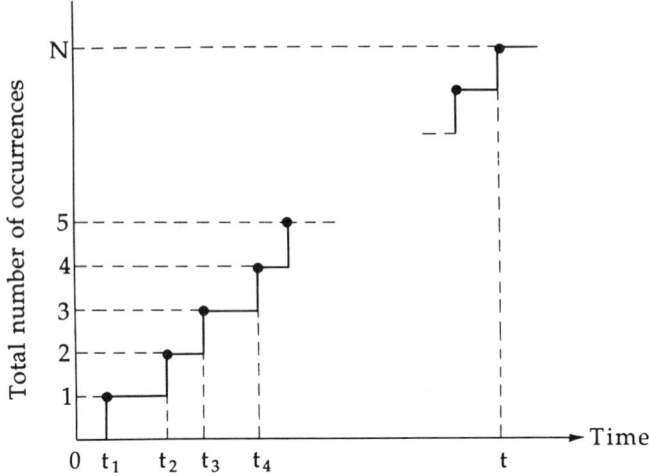

Figure 17.1 Counting random process $N(t)$.

Proof. Let us assume n occurrences of the event in the time interval $(t + \Delta t)$. Because of condition (iii), it is sufficient to consider the following two cases:

(a) n occurrences of the event in time t and no occurrence in Δt, and

(b) $(n - 1)$ occurrences of the event in time t and one occurrence in Δt.

For Case (a), we have

$$Pr\{N(t + \Delta t) = n\} = Pr\{N(t) = n\} \cdot Pr\{N(\Delta t) = 0 | N(t) = n\}$$
$$= p_n(t) \cdot p_0(\Delta t) = p_n(t)\{1 - \nu(\Delta t)\} \quad (17.1)$$

For Case (b), we have

$$Pr\{N(t + \Delta t) = n\} = Pr\{N(t) = n - 1\}$$
$$\cdot Pr\{N(\Delta t) = 1 | N(t) = n - 1\}$$
$$= p_{n-1}(t) p_1(\Delta t) = p_{n-1}(t) \cdot \nu(\Delta t) \quad (17.2)$$

POISSON PROCESSES

By adding these two cases, we can write

$$p_n(t + \Delta t) = p_n(t)\{1 - \nu(\Delta t)\} + p_{n-1}(t) \cdot \nu(\Delta t) \qquad (17.3)$$

Let Δt be small. Then, we can derive the following relationship:

$$\lim_{\Delta t \to 0} \frac{p_n(t + \Delta t) - p_n(t)}{\Delta t} = p_n'(t) = -\nu\{p_n(t) - p_{n-1}(t)\} \qquad (17.4)$$

Next, from the condition $N(0) = 0$, the above equation yields $p_0'(t) = -\nu p_0(t)$. As a solution of this equation, we have

$$p_0(t) = (\text{constant}) e^{-\nu t} \qquad (17.5)$$

Since $p_0(0) = 1$, the constant becomes unity. Then, we have $p_1'(t) = -\nu\{p_1(t) - e^{-\nu t}\}$, and its solution can be obtained as

$$p_1(t) = \nu t e^{-\nu t} \qquad (17.6)$$

By repeating this procedure, we can derive

$$p_2(t) = \frac{1}{2!}(\nu t)^2 e^{-\nu t}$$

$$p_3(t) = \frac{1}{3!}(\nu t)^3 e^{-\nu t}, \ldots \qquad (17.7)$$

In general, we may write

$$p_n(t) = \frac{1}{n!}(\nu t)^n e^{-\nu t} \qquad (17.8)$$

This is a Poisson distribution with mean νt. Thus, it can be proved that the random process $N(t)$ obeys the Poisson distribution.

It is stated in Chapter 9 that the parameter ν is called the intensity of the Poisson process, and it is considered constant in general. We will discuss a Poisson process for which the intensity is a function of time in Section 17.2.1.

The mean, variance, and autocorrelation function of the Poisson process are summarized below:

$$E[N(t)] = \nu t$$
$$\text{Var}[N(t)] = \nu t$$
$$\begin{aligned} R(\tau) &= E[N(t)N(t+\tau)] \\ &= E[N(t)\{\overline{N(t+\tau) - N(t)} + N(t)\}] \\ &= E[N(t)]E[N(t+\tau) - N(t)] + E[N^2(t)] \\ &= \nu^2 t\tau + (\nu t)^2 + \nu t \end{aligned} \qquad (17.9)$$

As can be seen in Eq. (17.9), the mean value depends on time t, hence the Poisson process $N(t)$ is not a stationary process. However, the increment Poisson process, $N(t+\tau) - N(t)$, is stationary as proved in Section 9.3.3.

17.1.2 *Some Properties of the Poisson Process*

(a) The sum of independent Poisson processes, $N_1(t), N_2(t), \ldots, N_n(t)$, with mean values $\nu_1 t, \nu_2 t, \ldots, \nu_n t$, respectively, is also a Poisson process with mean $(\nu_1 + \nu_2 + \cdots + \nu_n)t$.

Proof. Since the Poisson processes $N_1(t), N_2(t), \ldots$ are statistically independent, the characteristic function of the sum of these processes following Theorem 4.4 becomes

$$\phi(u) = \prod_{j=1}^{n} \exp\{\nu_j t(e^{iu} - 1)\} = \exp\left\{\sum_{j=1}^{n} \nu_j t(e^{iu} - 1)\right\}$$

This is the characteristic function of the Poisson distribution with mean $\sum_{j=1}^{n} \nu_j t$. Thus, the sum of the independent Poisson processes is also a Poisson process with a mean that is equal to the sum of means of the processes.

(b) The difference of two independent Poisson processes $N_1(t)$ and $N_2(t)$ with mean $\nu_1 t$ and $\nu_2 t$, respectively, is not a Poisson process; instead, it has the probability distribution given by

$$\Pr\{N_1(t) - N_2(t) = n\} = e^{-(\nu_1 + \nu_2)t}\left(\frac{\nu_1}{\nu_2}\right)^{n/2} I_n\left(2\sqrt{\nu_1 \nu_2}\, t\right) \qquad (17.10)$$

where $I_n(\ )$ is a modified Bessel function of order n.

Proof. Let us write $N(t) = N_1(t) - N_2(t)$. By applying the formula for evaluating the probability for the difference of two discrete-type random variables given in Exercise 7.2, we may write

$$Pr\{N(t) = n\} = \sum_{k=0}^{\infty} Pr\{N_1(t) = n + k\} Pr\{N_2(t) = k\}$$

$$= \sum_{k=0}^{\infty} \frac{e^{-\nu_1 t}(\nu_1 t)^{n+k}}{(n+k)!} \frac{e^{-\nu_2 t}(\nu_2 t)^k}{k!}$$

$$= e^{-(\nu_1 + \nu_2)t} \left(\frac{\nu_1}{\nu_2}\right)^{n/2} \cdot \sum_{k=0}^{\infty} \frac{\left(\sqrt{\nu_1 \nu_2}\, t\right)^{2k+n}}{k!(n+k)!}$$

On the other hand, we have the following expansion formula for the modified Bessel function of order n:

$$I_n(z) = \sum_{k=0}^{\infty} \frac{(z/2)^{2k+n}}{k!\,\Gamma(n+k+1)}$$

By letting $z = 2\sqrt{\nu_1 \nu_2}\, t$, we can write the probability as

$$Pr\{N(t) = n\} = e^{-(\nu_1 + \nu_2)} \left(\frac{\nu_1}{\nu_2}\right)^{n/2} I_n\left(2\sqrt{\nu_1 \nu_2}\, t\right)$$

(c) If the Poisson process $N(t)$ with mean νt is filtered such that every occurrence of the event is not counted, the process has a constant probability p of being counted. In this case, the resulting counting process is also a Poisson process with mean $p\nu t$.

Proof. Let $M(t)$ be the number of events actually counted in the time interval from 0 to t. It is assumed that each filtering is statistically independent. Let us consider the conditional probability that n events are counted in $(n + r)$ occurrences of the event in time t. This conditional probability can be evaluated by applying the binomial probability law as

$$Pr\{M(t) = n | N(t) = n + r\} = \binom{n+r}{n} p^n q^r, \quad \text{where } p + q = 1$$

Since $N(t)$ is a Poisson process, we have

$$Pr\{N(t) = n + r\} = e^{-vt}\frac{(vt)^{n+r}}{(n+r)!}$$

Considering that r can be any integer, $M(t)$ can be obtained as

$$Pr\{M(t) = n\} = \sum_{r=0}^{\infty}\binom{n+r}{n}p^n q^r \cdot e^{-vt}\frac{(vt)^{n+r}}{(n+r)!}$$

$$= e^{-vt}\frac{(pvt)^n}{n!}\sum_{r=0}^{\infty}\frac{(qvt)^r}{r!}$$

$$= e^{-vt}\frac{(pvt)^n}{n!}e^{qvt} = e^{-pvt}\frac{(pvt)^n}{n!}$$

Thus, it can be proved that the counting process $M(t)$ is a Poisson process with mean pvt.

Among the statistical properties of the Poisson process, those associated with time intervals between successive occurrences of events are extremely important. This subject will be discussed separately in the next section.

Example 17.1. Let X be the number of occurrences of an event that takes place in accordance with a Poisson process with intensity v. Find the number X that has the largest probability in a specified time t.

Consider the following ratio of probabilities:

$$\frac{Pr\{X = r + 1\}}{Pr\{X = r\}} = \frac{e^{-vt}(vt)^{r+1}/(r+1)!}{e^{-vt}(vt)^r/r!} = \frac{vt}{r+1}, \quad r = 0, 1, 2, \ldots$$

This implies that $Pr\{X = r + 1\} \gtreqless Pr\{X = r\}$ is equivalent to $vt \gtreqless r + 1$. Hence, by choosing the largest integer r that does not exceed vt, we can find X for which the probability becomes largest. That is,

$$Pr\{X = 0\} < Pr\{X = 1\} < \cdots < Pr\{X = r - 1\}$$
$$\leq Pr\{X = r\} > Pr\{X = r + 1\} > \cdots \blacksquare$$

A practical application of this example is given in Exercise 17.1.

Example 17.2. Let $N(t)$ be a Poisson process with intensity ν. Show that

$$Pr\{N(t) = m | N(t + \tau) = n\} = \binom{n}{m}\left(\frac{t}{t + \tau}\right)^m \left(\frac{\tau}{t + \tau}\right)^{n-m}$$

$$Pr\{N(t) = m | N(t + \tau) = n\}$$
$$= Pr\{N(t) = m, N(t + \tau) = n\} / Pr\{N(t + \tau) = n\}$$
$$= Pr\{N(t) = m, N(\tau) = n - m\} / Pr\{N(t + \tau) = n\}$$
$$= \frac{n!}{m!(n - m)!} \frac{t^m \tau^{n-m}}{(t + \tau)^n}$$
$$= \binom{n}{m}\left(\frac{t}{t + \tau}\right)^m \left(\frac{\tau}{t + \tau}\right)^{n-m} \blacksquare$$

17.1.3 *Interarrival Time and Waiting Time*

In this section we discuss the statistical properties of the time intervals between two successive occurrences of random events, called the *interarrival*

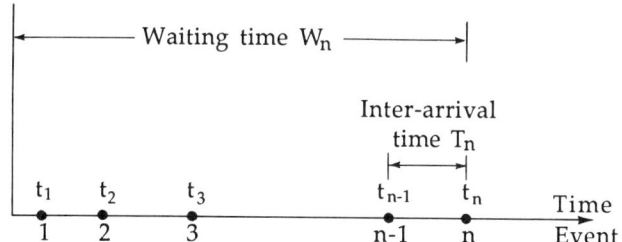

Figure 17.2 Definition of interarrival time, T_n, and waiting time, W_n.

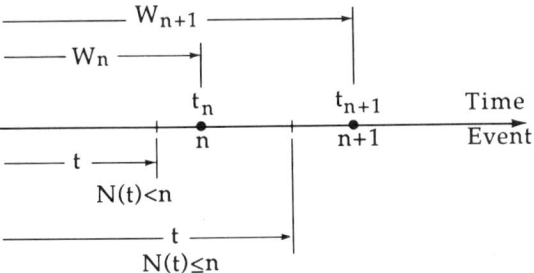

Figure 17.3 $Pr\{N(t) < n\}$ and $Pr\{N(t) \leqslant n\}$.

time. Also discussed is *waiting time*, which is referred to as the time up to a specific number of occurrences of the event from $t = 0$. Figure 17.2 shows the definition of the interarrival time, denoted by T_n, and the waiting time, denoted by W_n. As can be seen in the figure, we may write

$$W_n = T_1 + T_2 + T_3 + \cdots + T_n$$

$$T_n = W_n - W_{n-1} \tag{17.11}$$

We can now express the probability associated with the number of occurrences of a process $N(t)$ in terms of the waiting time as follows (see Figure 17.3):

$$Pr\{N(t) < n\} = Pr\{W_n > t\}$$
$$Pr\{N(t) \leqslant n\} = Pr\{W_{n+1} > t\}, \quad n = 0, 1, 2, \ldots \tag{17.12}$$

We can also write

$$Pr\{N(t) \geqslant n\} = Pr\{W_n \leqslant t\} = F_{W_n}(t)$$
$$Pr\{N(t) > n\} = Pr\{W_{n+1} \leqslant t\} = F_{W_{n+1}}(t) \tag{17.13}$$

Thus, the probability that there are exactly n occurrences of the event at time t can be written as

$$Pr\{N(t) = n\} = Pr\{N(t) \geqslant n\} - Pr\{N(t) > n\}$$
$$= F_{W_n}(t) - F_{W_{n+1}}(t), \quad n = 1, 2, 3, \ldots \tag{17.14}$$

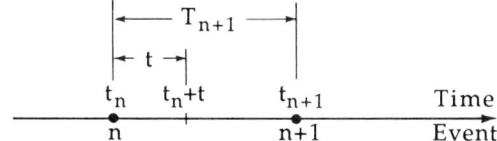

Figure 17.4 Explanatory figure representing $Pr\{T_{n+1} > t\}$.

In particular, for $n = 0$, we have

$$Pr\{N(t) = 0\} = Pr\{W_1 > t\} = 1 - F_{W_1}(t) \qquad (17.15)$$

It should be noted that the relationship between interarrival time, waiting time, and the probability of occurrence of an event discussed above is valid not only for a Poisson process, but also for any counting process. In particular, we have the following important theorem for a Poisson process:

Theorem 17.2. *The interarrival time of a Poisson process with intensity ν obeys an exponential probability law with mean $1/\nu$.*

Proof. The interarrival time T_{n+1} satisfies the following relationship (see Figure 17.4).

$$Pr\{T_{n+1} > t\} = Pr\{N(t_n + t) = n | N(t_n) = n\}$$
$$= Pr\{N(t) = 0 | N(t_n) = n\}$$

Since this relationship holds for any integer n, and since $N(t)$ obeys the Poisson distribution, we may write that the interarrival time T satisfies, in general:

$$Pr\{T > t\} = Pr\{N(t) = 0\} = e^{-\nu t}$$

and thereby,

$$Pr\{T \leq t\} = F(t) = 1 - e^{-\nu t}$$

This result implies that the interarrival time T obeys an exponential distribution with parameter ν.

It can also be proved that the interarrival times $T_1, T_2, \ldots$ are statistically independent. However, proof of this property is extremely complicated and is beyond the scope of our interest; hence, the proof is not given here. By using this independence property, the probability distribution of waiting time can be derived as follows:

Theorem 17.3. The waiting time to the nth occurrence of an event of a Poisson process with intensity ν follows the gamma probability distribution with parameters ν and n.

Proof. It was obtained in Theorem 17.2 that the interarrival time of a Poisson process obeys the exponential probability law. Since the waiting time W_n is the sum of n independent interarrival times, it can be proved from the definition of the gamma distribution (Definition 6.5) that W_n obeys the gamma probability law with parameters n and ν.

Example 17.3. Nuclear particles arrive at a counter in accord with a Poisson process with the intensity of one particle every 10 seconds. It is assumed that the counter records one in every three arrivals of the particles. Evaluate the mean and variance of the interrecording time.

Let T_n be the interarrival time between $(n-1)$th and nth arrivals of particles, and let $\overline{T}_n$ be the interrecording time between $(n-1)$th and nth recordings of the particles. Then, we have

$$\overline{T}_n = T_{3n} + T_{3n-1} + T_{3n-2}$$

Here, T_{3n}, T_{3n-1}, and T_{3n-2} are exponentially distributed. Since the sum of three exponential probability density functions is a gamma distribution, its probability density function becomes

$$f(\overline{T}_n) = e^{-\nu \overline{T}_n} \frac{\nu(\nu \overline{T}_n)^{m-1}}{\Gamma(m)} \quad \text{with } \nu \overline{T}_n = 0.1, \, m = 3$$

Hence, the mean and variance of $\overline{T}_n$ become

$$E[\overline{T}_n] = m/\nu \overline{T}_n = 30 \text{ (in seconds)}$$

and

$$\text{Var}[\overline{T}_n] = m/(\nu \overline{T}_n)^2 = 300 \text{ (in second}^2) \qquad \blacksquare$$

Example 17.4. Let U be the time until the next occurrence of an event after the last event has occurred in a Poisson process with a mean value νt.

Prove that U is exponentially distributed with mean $1/\nu$.

$$Pr\{U > u\} = Pr\{T_{n+1} > t - t_u + u \mid T_{n+1} > t - t_n\}$$

$$= \frac{Pr\{T_{n+1} > t - t_n + u\}}{Pr\{T_{n+1} > t - t_n\}}$$

$$= \frac{\exp\{-\nu(t - t_n + u)\}}{\exp\{-\nu(t - t_n)\}} = e^{-\nu u}$$

Thus, U is exponentially distributed with mean $1/\nu$. ∎

Theorem 17.4. *A total of n random events occurs in time T in accord with a Poisson process with intensity ν. Then, the waiting times $W_1, W_2, \ldots, W_n$ are equivalent to the ordered sample of a random variable that has a uniform distribution between 0 and T.*

Proof. Evaluate the joint probability that the waiting time W_1 is between t_1 and $t_1 + \tau_1$, and W_2 is between t_2 and $t_2 + \tau_2$, etc., given that n events have occurred in time T. This can be written as

$$Pr\{t_1 < W_1 < t_1 + \tau_1, t_2 < W_2 < t_2 + \tau_2, \ldots, t_n < W_n < t_n + \tau_n \mid N(t) = n\}$$

$$= \frac{[e^{-\nu\tau_1}(\nu\tau_1/1!)][e^{-\nu\tau_2}(\nu\tau_2/1!)] \cdots [e^{-\nu\tau_n}(\nu\tau_n/1!)][e^{-\nu\Delta}((\nu\Delta)^0/0!)]}{e^{-\nu T}[(\nu T)^n/n!]}$$

$$= \frac{n!}{T^n}(\tau_1, \tau_2, \ldots, \tau_n) \qquad (17.16)$$

where $\Delta = T - \sum_{j=1}^{n} \tau_j$.

The conditional joint probability obtained in Eq. (17.16) is equal to the joint probability of $(W_1, W_2, \ldots, W_n)$ multiplied by $(\tau_1, \tau_2, \ldots, \tau_n)$. Hence, the joint probability density function of the waiting times $W_1, W_2, \ldots, W_n$ becomes

$$f(W_1, W_2, \ldots, W_n) = \frac{n!}{T^n} \qquad (17.17)$$

Following the definition of the joint probability density function of the ordered sample given in Eq. (8.1), the right side of Eq. (17.17) is equal to

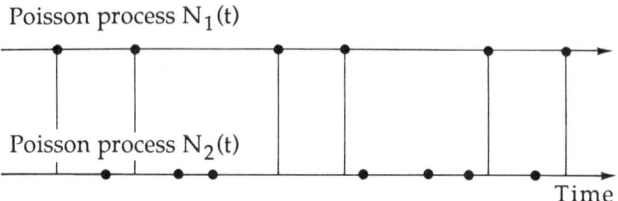

Figure 17.5 Poisson processes $N_1(t)$ and $N_2(t)$ in Example 17.5.

the joint probability density function of the ordered sample of a random variable, which has a uniform distribution between 0 and T. Hence, the waiting times $W_1, W_2, \ldots, W_n$, all of which are statistically independent, can be considered as an ordered sample of a random variable that obeys the uniform probability law between 0 and T.

Example 17.5. $N_1(t)$ and $N_2(t)$ are statistically independent Poisson processes with intensity ν_1 and ν_2, respectively. Let X be the number of occurrences of events of the process $N_2(t)$ during the interarrival time of the process $N_1(t)$ (see Figure 17.5). Find the probability density function of X.

The interarrival time, T, of the process $N_1(t)$ follows the exponential distribution with mean $1/\nu_1$. That is, $f(T) = \nu_1 e^{-\nu_1 T}$. On the other hand, for a given time interval T, $N_2(t)$ has the following conditional probability:

$$Pr\{X = x | T = T\} = e^{-\nu_2 T} \frac{(\nu_2 T)^x}{x!}$$

Hence, $Pr\{X = x\}$ can be evaluated by

$$Pr\{X = x\} = \int_0^\infty e^{-\nu_2 T} \frac{(\nu_2 T)^x}{x!} \cdot f(T) \, dT$$

$$= \frac{\nu_1 \nu_2^x}{x!} \int_0^\infty T^x e^{-(\nu_1 + \nu_2)T} \, dT$$

$$= \frac{\nu_1 \nu_2^x}{x!} \frac{x!}{(\nu_1 + \nu_2)^{x+1}} = \left(\frac{\nu_1}{\nu_1 + \nu_2}\right)\left(\frac{\nu_2}{\nu_1 + \nu_2}\right)^x \qquad \blacksquare$$

It is often necessary in practice to examine whether or not an observed phenomenon can be considered as a Poisson process. It is also often necessary to examine whether or not two independently observed sets of data are from the same Poisson process. Such types of statistical inference problems associated with a Poisson process can be solved through statistical tests on the waiting time of the process.

Let us examine whether or not an observed phenomenon can be considered as a Poisson process. The procedure is as follows: We first evaluate the parameter ν by counting the number of occurrences of the phenomenon for a preassigned sufficiently long observation time assuming that the observed random process follows the Poisson probability law. Next, obtain the waiting time W_n for a specified nth occurrence of the event. Here, the waiting time W_n follows the gamma probability law with parameter n and ν from Theorem 17.3. Hence, the random variable $2\nu W_n$ has a χ^2 distribution with $2n$ degrees of freedom (see transformation from the gamma distribution to χ^2 distribution given in Example 6.6). Then determine the values χ_1 and χ_2 of the χ^2 distribution such that the interval (between χ_1 and χ_2) provides a confidence interval for the parameter $2\nu W_n$ with confidence coefficient $1 - \alpha$ (see Chapter 12). That is,

$$Pr\{\chi_1 < 2\nu W_n < \chi_2\} = 1 - \alpha \qquad (17.18)$$

Hence, the confidence interval for the parameter ν becomes

$$Pr\{\chi_1/2W_n < \nu < \chi_2/2W_n\} = 1 - \alpha \qquad (17.19)$$

Thus, if the observed value of the parameter ν falls within this interval, we may conclude that the phenomenon can be considered as a Poisson process.

As another example of statistical inference associated with the Poisson process, let us consider the case of two sets of observed data, both of which are known to be Poisson processes, but we want to know if these Poisson processes are the same. For this problem, it is sufficient to examine whether or not the intensities of the two Poisson processes are the same. We evaluate the intensities ν_1 and ν_2 from each set of observations. We also evaluate the waiting times W_{n_1} and W_{n_2} for the n_1th and n_2th occurrences of the events, respectively, for each set. We showed that the random variables $2\nu_1 W_{n_1}$ and $2\nu_2 W_{n_2}$ follow χ^2 distributions with $2n_1$ and $2n_2$ degrees of freedom, respectively. By assuming that these two random variables are statistically independent, it can also be shown that the ratio $(2\nu_1 W_{n_1}/n_1)/(2\nu_2 W_{n_2}/n_2)$ has an F distribution with n_1 and n_2 degrees of freedom (see Theorem 12.1), and thereby we may carry out a statistical test on the hypothesis that $\nu_1 = \nu_2$ against the alternative $\nu_1 \neq \nu_2$. By choosing $F_1(\alpha/2)$ and $F_2(\alpha/2)$ as the lower and upper bounds, respectively, we may accept the hypothesis if

$$Pr\left\{F_1\left(\frac{\alpha}{2}\right) < \frac{W_{n_1} n_2}{W_{n_2} n_1} < F_2\left(\frac{\alpha}{2}\right)\right\} = 1 - \alpha \qquad (17.20)$$

Inversely, we may reject the hypothesis with a level of significance α if

$$\frac{W_{n_1} n_2}{W_{n_2} n_1} < F_1\left(\frac{\alpha}{2}\right) \quad \text{or} \quad \frac{W_{n_1} n_2}{W_{n_2} n_1} > F_2\left(\frac{\alpha}{2}\right) \qquad (17.21)$$

Example 17.6. City A has experienced 24 tropical storms in the past 15 years. The number of storms each year is counted and the observed frequency is tabulated as shown in Table 17.1. Included also in the table is the estimated probability of number of storms per year assuming that the tropical storms experienced by City A obeys a Poisson random process with mean $vt = 24/15 = 1.60$.

Since some discrepancy exists between the observed and estimated probabilities, we may examine the validity of the assumption through Eq. (17.18). The record shows that the waiting time to the tenth storm was 74 months (6.17 years). By taking the confidence coefficient $1 - \alpha = 0.95$, the χ^2 values with 20 degrees of freedom are 9.59 and 34.17. Hence, we have

Table 17.1

Observed Frequency of Occurrence of Storms Each Year and Computed Poisson Probability

Storms/year	Observed		Poisson probability
	Number	Frequency	
0	2	0.133	0.202
1	6	0.400	0.323
2	4	0.267	0.258
3	2	0.133	0.138
4	1	0.067	0.055
5	0	0	0.018
6	0	0	0.005
7	0	0	0.001

$$Pr\{9.59 < 2(6.17)\nu < 34.17\} = 0.95$$

which yields a confidence interval for the ν value of $(0.77, 2.77)$. Since the observed ν value of 1.60 falls within the confidence interval, we may safely assume that the occurrence of tropical storms observed at City A is a Poisson random process with the confidence coefficient 0.95. ∎

17.2 GENERALIZATION OF THE POISSON PROCESS

The Poisson process discussed heretofore was derived subject to the conditions presented in Section 17.1.1. By removing some of these conditions, the Poisson process can be generalized so that it may represent a variety of more realistic random phenomena observed in the natural sciences. There are a number of ways to generalize the Poisson process. To see how this can be done, let us examine the characteristic function of the Poisson process that is given by

$$\phi(u) = \exp\{\nu t(e^{iu} - 1)\} \tag{17.22}$$

In Section 17.1, the intensity ν is considered to be a constant and νt is the mean of the process. Thus, the mean value increases linearly with time. We now generalize the Poisson process such that the intensity is a function of time t. We may also generalize the process that the mean is a random variable. Furthermore, we may generalize the exponential function e^{iu} involved in the characteristic function such that it is an arbitrary function of u. We will discuss these generalizations of the Poisson process in the following:

17.2.1 Nonhomogeneous Poisson Process

Definition 17.1. A Poisson process with an intensity that is a function of time, $\nu(t)$, is defined as a *nonhomogeneous Poisson process*. $\nu(t)$ is called the *intensity function*.

The probability that n occurrences of the event in the time interval $(0, t)$ for a nonhomogeneous Poisson process is given by

$$Pr\{N(t) = n\} = \exp\left\{-\int_0^t \nu(t)\, dt\right\} \cdot \frac{\left\{\int_0^t \nu(t)\, dt\right\}^n}{n!} \tag{17.23}$$

Thus, the probability of no occurrence of the event in the time interval $(0, t)$ becomes

$$Pr\{N(t) = 0\} = \exp\left\{-\int_0^t \nu(t)\, dt\right\} \qquad (17.24)$$

The mean, variance, and autocorrelation function of the nonhomogeneous Poisson process are as follows:

$$E[N(t)] = \int_0^t \nu(t)\, dt$$

$$\mathrm{Var}[N(t)] = \int_0^t \nu(t)\, dt$$

$$R(\tau) = E[N(t)]E[N(t+\tau) - N(t)] + E[N^2(t)]$$

$$= \int_0^t \nu(t)\, dt \int_0^{t+\tau} \nu(t)\, dt + \int_0^t \nu(t)\, dt$$

$$= \int_0^t \nu(t)\, dt \left\{1 + \int_0^{t+\tau} \nu(t)\, dt\right\} \qquad (17.25)$$

It is noted that the increment process of a nonhomogeneous Poisson process is no longer stationary.

Next, let us consider the case where the intensity ν is a random variable with the probability density function $f(\nu)$. In this case, the probability of occurrence of the event in the time interval $(0, t)$ is given, in general, as

$$Pr\{N(t) = n\} = \int_0^\infty e^{-\nu t}\frac{(\nu t)^n}{n!} f(\nu)\, d\nu \qquad (17.26)$$

In particular, if the probability density function $f(\nu)$ is the following gamma distribution with parameters m and λ,

$$f(\nu) = \frac{1}{\Gamma(m)}\lambda^m \nu^{m-1} e^{-\lambda\nu}, \qquad 0 \leq \nu < \infty \qquad (17.27)$$

then it is shown in Section 5.2.2 that the probability is equal to a negative

GENERALIZATION OF THE POISSON PROCESS

binomial distribution. That is,

$$Pr\{N(t) = n\} = \binom{n+m-1}{n} p^m q^n \qquad (17.28)$$

where $n = 0, 1, 2, \ldots$, $p = \lambda/(t + \lambda)$, and $q = t/(t + \lambda)$.

17.2.2 Compound Poisson Process

As stated in Theorem 17.1, it is assumed that only one instantaneous event can occur for the Poisson process. We now remove this restriction and consider that several events may occur simultaneously as illustrated in Figure 17.6(a). That is, x_1 events occur simultaneously at time t_1, x_2 events at time t_2, x_3 events at time t_3, etc. Here, the number of events $x_1, x_2, x_3, \ldots$ are independent random variables with identical probability distributions. Under these conditions, the total number of occurrences in the time interval $(0, t)$, denoted by $Y(t)$, is called a compound Poisson process. The definition is given below.

Definition 17.2. A stochastic process $Y(t)$ is called a *compound Poisson process* (or *cluster Poisson process*) if it is the sum of random variables X_n given by

$$Y(t) = \sum_{n=1}^{N(t)} X_n \qquad (17.29)$$

where $N(t)$ is a Poisson process with intensity ν, and X_n are independent random variables with identical distribution.

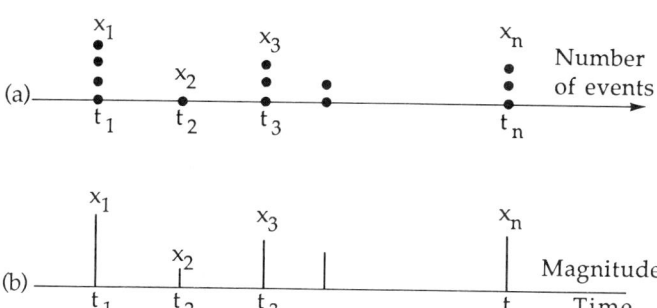

Figure 17.6 Definition of compound Poisson process $y = \sum_{i=1}^{N(t)} x_i$.

It is understood from the definition above that the random variables X_n of a compound Poisson process are not necessarily the number of events occurring simultaneously; instead, they can be the magnitudes of a random phenomenon as illustrated in Figure 17.6(b). The mathematical model given in Eq. (17.29) can be applied to a wide variety of practical problems in engineering and the physical sciences as will be shown in Examples 17.7 and 17.8.

Let us derive the characteristic function of a compound Poisson process. It is given by

$$\phi_Y(u) = E[e^{iuy}] = E\left[\exp\left\{iu \sum_{n=1}^{N(t)} X_n | N(t) = n\right\}\right]$$

$$= \sum_{n=0}^{\infty} \{\phi_x(u)\}^n \cdot e^{-vt} \frac{(vt)^n}{n!}$$

$$= e^{-vt} \sum_{n=0}^{\infty} \frac{\{vt\phi_x(u)\}^n}{n!} = e^{-vt} \cdot e^{vt\phi_x(u)}$$

$$= \exp\{vt(\phi_x(u) - 1)\} \qquad (17.30)$$

From comparison of Eq. (17.30) and the characteristic function of the Poisson process given in Eq. (17.22), it is understood that e^{iu} of the characteristic function of the Poisson distribution is generalized to the characteristic function of the random variable X that can be written as $\int \exp\{iux\} f(x) \, dx$.

The mean and variance of $Y(t)$ can be obtained from Eq. (17.30) as follows:

$$E[Y(t)] = \frac{1}{i}\left[\frac{d\phi_Y(u)}{du}\right]_{u=0} = \frac{1}{i} vt\phi'_x(0) = vtE[x] \qquad (17.31)$$

$$E[Y^2(t)] = \frac{1}{i^2}\left[\frac{d^2\phi_Y(u)}{du}\right]_{u=0} = \frac{1}{i^2}\left[vt\phi''_x(0) + (vt)^2\{\phi'_x(0)\}^2\right]$$

$$= vtE[x^2] + (vt)^2 (E[x])^2 \qquad (17.32)$$

GENERALIZATION OF THE POISSON PROCESS

Hence, we have

$$\text{Var}[Y(t)] = \nu t E[x^2] \tag{17.33}$$

In order to obtain $\text{Cov}[Y(s), Y(t)]$, we may use the following relationship:

$$\text{Var}[Y(t-s)] = \text{Var}[Y(t)] + \text{Var}[Y(s)] - 2\text{Cov}[Y(s), Y(t)] \tag{17.34}$$

By using the formula for variance given in Eq. (17.33), we have

$$\text{Cov}[Y(s), Y(t)] = \nu(\min s, t) E[X^2] \tag{17.35}$$

where $(\min s, t)$ implies either s or t, whichever is the smaller one.

We may recall that a random process $Y(t)$ that has the same form as that of a compound Poisson process was considered in Example 3.14 referring to the movement of nearshore sediment transport. $N(t)$ in that example, however, was not specified as a Poisson process; instead, the mean and variance of the process $Y(t)$ were derived through application of conditional moments. We may write here the mean and variance of that example again as follows:

$$E[y] = E[N]E[x]$$

$$\text{Var}[y] = E[N]\text{Var}[x] + \text{Var}[N](E[x])^2$$

If $N(t)$ is a Poisson process, then we have $E[N] = \text{Var}[N] = \nu t$, and thereby the mean and variance reduce to that given in Eqs. (17.31) and (17.33).

Example 17.7. Let x_i in Figure 17.6(b) be the amount of damage resulting from an impact implied at time t_i. It is assumed that x_i is exponentially distributed with parameter λ. Then, $Y(t) = \sum_{i=1}^{N(t)} x_i$ represents the total damage sustained at time t for an impact assumed to be a Poisson process with intensity ν. Let L be the critical value of $Y(t)$ when a failure takes place at time T. We will evaluate the expected failure time $E[T]$.

Since x_i is exponentially distributed with parameter λ, $Y(t) = \sum_{i=1}^{k} x_i$ follows the gamma distribution with parameters k and λ. That is,

$$f(y) = \frac{1}{\Gamma(k)} \lambda^k y^{k-1} e^{-\lambda y}$$

The expected failure time can be written as

$$E[T] = \int_0^\infty Pr\{T > t\}\, dt = \int_0^\infty Pr\{Y(t) > L\}\, dt$$

Note that

$$Pr\{\text{Failure}\} = Pr\{Y(t) > L\}$$

$$= \sum_{n=0}^\infty Pr\{Y(t) > L | N(t) = n\} Pr\{N(t) = n\}$$

From the probability density function $f(y)$, we have

$$Pr\{Y(t) > L | N(t) = n\} = \sum_{k=n}^\infty \frac{1}{\Gamma(k)} \lambda^k L^{k-1} e^{-\lambda L}$$

Since $N(t)$ is a Poisson process, we have

$$Pr\{N(t) = n\} = \frac{(\nu t)^n}{n!} e^{-\nu t}$$

Thus, we may write

$$E[T] = \sum_{n=0}^\infty \sum_{k=n}^\infty \left(\int_0^\infty \frac{(\nu t)^n}{n!} e^{-\nu t}\, dt \right) \frac{\lambda^k L^{k-1}}{\Gamma(k)} e^{-\lambda L}$$

$$= \frac{1}{\nu} \sum_{k=0}^\infty (1 + k) \frac{\lambda^k L^{k-1}}{\Gamma(k)} e^{-\lambda L} = \frac{1}{\nu}(1 + \lambda L) \quad \blacksquare$$

Example 17.8. Consider the Brownian motion of a particle on a line starting at $x = 0$ at time $t = 0$. Assume that random impacts with other particles occur following a Poisson process with intensity ν, and assume also that a particle moves either $+a$ or $-a$ at each impact with equal probability (see Figure 17.7). The location of the particle at time t can be expressed in the same form as given in Eq. (17.29) where X_n are independent identically distributed random variables with probability $Pr\{X = a\} = Pr\{X = -a\} = 1/2$. Since the characteristic function of x is given by

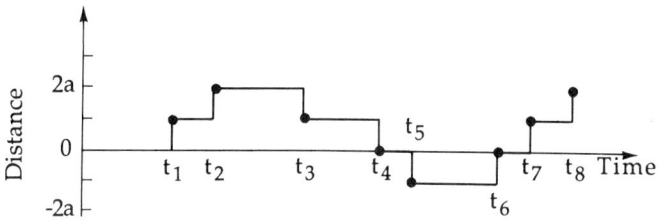

Figure 17.7 Brownian motion in Example 17.8

$\phi_x(u) = \cos au$, the characteristic function of $Y(t)$ from Eq. (17.30) becomes

$$\phi_Y(u) = \exp\{vt(\cos au - 1)\}$$

By expanding the cosine term into a series, it can be approximately written as $\phi_Y(u) = \exp\{-vta^2u^2/2\}$. If we assume that $v \to \infty$ and $a \to 0$ such that va^2 is a constant, σ^2, which represents the total mean square displacement of the particle per unit time. We may then write $\phi_Y(u) = \exp\{-\sigma^2 tu^2/2\}$. Thus, $Y(t)$ is a normal distribution that has stationary independent increments. ∎

17.2.3 Superimposed Poisson Process

We consider in this section a more generalized Poisson process in which events occur at times $\tau_1, \tau_2, \ldots, \tau_n$ following a Poisson process and each event occurs with random amplitude (intensity) $X_1, X_2, \ldots, X_n$ as shown in Figure 17.8. Here, $X_1, X_2, \ldots, X_n$ is a sequence of independent identically distributed random variables. Furthermore, the occurrence of the event induces a response, W_n, called the response function, which is a function of τ_n and X_n. If we write $t - \tau_n = s$, where $s > 0$, then the response $\Sigma w_n(s, X_n)$ constitutes a random process called the superimposed Poisson process.

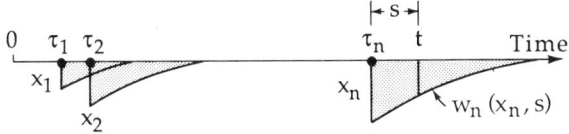

Figure 17.8 Superimposed Poisson process $y(t) = \sum_{i=1}^{N(t)} w_n$.

Definition 17.3. A stochastic process $Y(t)$ is called the *superimposed Poisson process* if it can be expressed by

$$Y(t) = \sum_{n=1}^{N(t)} w_n(t, \tau_n, X_n) = \sum_{n=1}^{N(t)} w_n(s, X_n) \qquad (17.36)$$

where $s = t - \tau_n$ and $s > 0$. $N(t)$ is a Poisson process with intensity ν, and τ_n is the time at which an event takes place. X_n represents the intensity (amplitude) of the event, which is a sequence of independent random variables with identical distribution $f(x)$.

The terminology "superimposed Poisson process" is not standard. Some call this process a *filtered Poisson process*. A number of examples of application of the process in the natural sciences and operation research are given in Parzen (1962).

The characteristic function of a superimposed Poisson process can be obtained in the same manner as that for a compound Poisson process; however, there is a difference in that the process is an accumulation of responses induced by random events with different intensities. We may write the characteristic function as

$$\phi_Y(u) = E[e^{iuY(t)}] = E[e^{iu\Sigma w_n}|N(t) = n]$$

$$= \exp\left\{-\nu\int_0^t d\tau\right\} \sum_{n=0}^{\infty} \frac{\left(\int_0^t \nu E[e^{iuw(t,\tau,x)}]\,d\tau\right)^n}{n!}$$

$$= \exp\left\{\nu\int_0^t E[e^{iuw(t,\tau,x)}] - 1]\,d\tau\right\} \qquad (17.37)$$

The mean and variance can be derived through the same procedure as shown in Eqs. (17.31) and (17.33). Those are,

$$E[Y(t)] = \nu\int_0^t E[w(t,\tau,x)]\,d\tau$$

$$\text{Var}[Y(t)] = \nu\int_0^t E[w^2(t,\tau,x)]\,d\tau \qquad (17.38)$$

$$\text{Cov}[Y(t_1), Y(t_2)] = \nu\int_0^{\min(t_1,t_2)} E[w(t_1,\tau,x)\cdot w(t_2,\tau,x)]\,d\tau$$

GENERALIZATION OF THE POISSON PROCESS

Here, $\text{Cov}[Y(t_1), Y(t_2)]$ can be derived by applying the relationship given in Eq. (17.34). The detailed derivation is left as an exercise for the reader.

Example 17.9. Let us consider a paralyzed situation of radioactive particles in a counter. Assume that the particles arrive at a counter following a Poisson process with intensity ν, and assume that the particle arriving at time τ_n paralyzes the counter for a random time X_n, as shown in Figure 17.9. X_n is a sequence of independent, identically distributed random variables with a cumulative distribution function $F_x(t)$. Then, the total number of particles that are paralyzing the counter at time t, denoted by $Y(t)$, can be expressed by Eq. (17.36).

Let us assume that the response function $w(s, X)$ is given for all X_n as

$$w(s, X) = \begin{cases} 1 & \text{for } 0 \leq s \leq X \\ 0 & \text{otherwise} \end{cases}$$

Let us write the characteristic function as

$$\phi_Y(u) = \exp\left\{ \nu \int_0^t E[e^{iuw(s, X)} - 1] \, ds \right\}$$

Since $w(s, X) = 0$ for $X < s$, $E[e^{iuw(s, X)} - 1]$ is equal to $(e^{iu} - 1) \times Pr\{X > s\}$. Hence, we have

$$\phi_Y(u) = \exp\left\{ \nu \int_0^t (e^{iu} - 1) \cdot Pr\{X > s\} \, ds \right\}$$

$$= \exp\left\{ \nu (e^{iu} - 1) \cdot \int_0^t \{1 - F_x(s)\} \, ds \right\}$$

By comparing the characteristic function derived above with that of the Poisson process given in Eq. (17.22), it can be seen that the superimposed Poisson process $Y(t)$ considered in this example is a Poisson process with

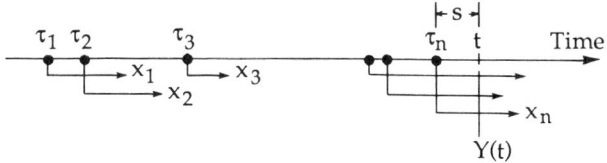

Figure 17.9 Schematic representation of $Y(t)$ in Example 17.9.

mean $\nu \int_0^t \{1 - F_x(s)\}\, ds$. It is noted that the counter is not paralyzed if $Y(t) = 0$. Since $Y(t)$ follows the Poisson process with mean $\nu \int_0^t \{1 - F_x(s)\}\, ds$, we can evaluate $Pr\{Y(t) = 0\}$ as follows:

$$Pr\{Y(t) = 0\} = \exp\left\{-\nu \int_0^t [1 - F_x(s)]\, ds\right\} \qquad \blacksquare$$

17.3 POISSON IMPULSE PROCESS AND RESPONSE

In this section we consider a stochastic process resulting from a superposition of random pulses arriving at random times. In particular, we assume that the times when impulses take place are regarded as a Poisson random process and thereby the sequence of random pulses is called a *Poisson impulse process*. It can be expressed as a sequence of delta functions that can be considered as time derivatives of a Poisson process $N(t)$ as shown in Figure 17.10.

We may write the process as

$$x(t) = \frac{d}{dt} N(t) = \sum_{n=1}^{N(t)} \delta(t - t_n) \qquad (17.39)$$

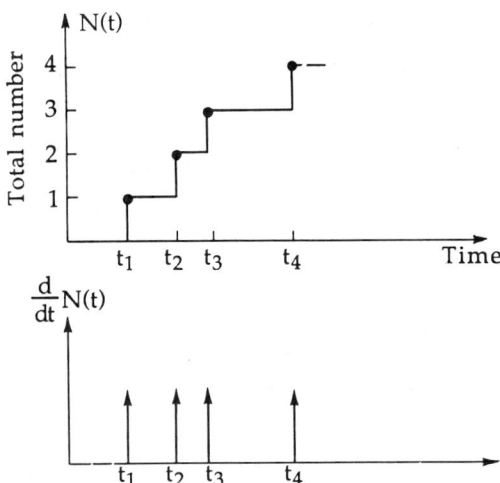

Figure 17.10 Poisson process and Poisson impulse process.

POISSON IMPULSE PROCESS AND RESPONSE

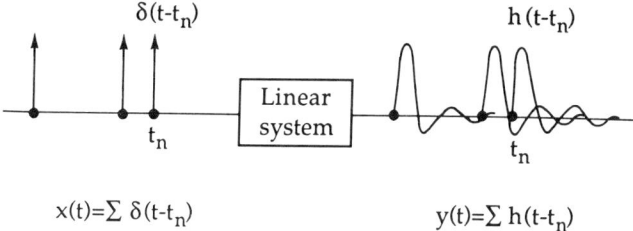

Figure 17.11 Poisson impulse process and response (shot noise process).

The response to an individual impulse applied to a linear system is defined in Section 14.1.2 as the impulse response function $h(t)$. Hence, we can write the response to a series of impulses, denoted by $Y(t)$, as

$$Y(t) = \sum_{n=1}^{N(t)} h(t - t_n) \qquad (17.40)$$

The input–output relationship of the system is shown in Figure 17.11. The response given in Eq. (17.40) is commonly known as the *shot noise process*.

We may generalize the Poisson impulse process by taking the severity of the impulse into consideration as shown in Figure 17.12. By letting a_n be a sequence of independent, identically distributed random variables representing the severity of impulses applied at time t_n, Eq. (17.40) can be generalized as

$$Y(t) = \sum_{n=1}^{N(t)} a_n h(t - t_n) \qquad (17.41)$$

Equation (17.41) is essentially the same equation defined as the superimposed Poisson process in Eq. (17.36). Therefore, the characteristic function of $Y(t)$ can be obtained from Eq. (17.37). That is, by letting $s = t - t_n$ and

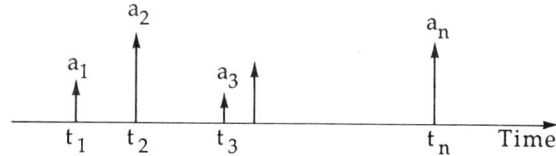

Figure 17.12 Poisson impulse process with various severities.

by taking the upper and lower limits to be infinite, we have

$$\phi_Y(u) = \exp\left\{\nu \int_{-\infty}^{\infty} E[e^{iuah(s)} - 1] \, ds\right\}$$

$$= \exp\left\{\nu \int_{-\infty}^{\infty} \{\phi_H(s) - 1\} \, ds\right\} \quad (17.42)$$

where $\phi_H(s)$ = characteristic function of $ah(s)$.

Let $E[a]$ and $E[a^2]$ be the first two moments of a sequence of random variable a_n. Then, the mean, variance, and covariance of $Y(t)$ can be obtained from Eq. (17.38) as follows:

$$E[y(t)] = \nu E[a] \int_{-\infty}^{\infty} h(s) \, ds = \nu E[a] H(0)$$

where $H(0)$ is defined in Eq. (14.11).

$$\mathrm{Var}[y(t)] = \nu E[a^2] \int_{-\infty}^{\infty} \{h(s)\}^2 \, ds$$

$$\mathrm{Cov}[y(t)] = \nu E[a^2] \int_{-\infty}^{\infty} h(s) h(s+\tau) \, ds \quad (17.43)$$

Formulas given in the above equation are known as *Campbell's theorem*.

The autocorrelation function of the response can be derived from Eq. (17.43) as

$$R_{yy}(\tau) = \{E[y(t)]\}^2 + \mathrm{Cov}[y(t), y(t+\tau)]$$

$$= \nu^2 (E[a])^2 H^2(0) + \nu E[a^2] \int_{-\infty}^{\infty} h(s) h(s+\tau) \, ds \quad (17.44)$$

Further, by taking the Fourier transform of the autocorrelation function, the spectral density function can be obtained. The second term of Eq. (17.44) is a convolution integral $h(s) * h(-s)$; hence, it can be written as $|H(\omega)|^2$ in the frequency domain. Then, following the definition of the transform given in Eq. (10.31), the spectral density function becomes

$$S_{yy}(\omega) = 2\nu^2 (E[a])^2 H^2(0) \delta(\omega) + \frac{\nu}{\pi} E[a^2] |H(\omega)|^2 \quad (17.45)$$

From the input–output relationship of a linear system, the spectral density function of the input, which is the Poisson impulse process in this case, can be derived as follows:

$$S_{xx}(\omega) = 2\nu^2(E[a])^2\delta(\omega) + \frac{\nu}{\pi}E[a^2] \qquad (17.46)$$

Then, the autocorrelation function becomes

$$R_{xx}(\tau) = \nu^2(E[a])^2 + \nu E[a^2]\delta(\tau) \qquad (17.47)$$

The direct derivation of the input autocorrelation function, $R_{xx}(\tau)$, and the spectral density function, $S_{xx}(\omega)$, is left as an exercise for the reader.

17.4 RENEWAL COUNTING PROCESSES

17.4.1 Renewal Counting Processes

One of the properties of the Poisson process is that the interarrival times of the process are a sequence of independent, identically distributed random variables that obey the exponential probability law. We now consider a more general case in which the interarrival times do not necessarily follow the exponential probability distribution. A counting process that possesses this property is called a renewal counting process defined as follows:

Definition 17.4. A counting process $N(t)$ that represents the total number of occurrences of an event in time between 0 and t is called a *renewal counting process* if the interarrival times are independent, identically distributed random variables.

In the following discussion of the renewal counting process, we will call, for convenience, the occurrence of an event as a *renewal*, the interarrival time as the *renewal period*, and the waiting time as the *renewal time*.

For a renewal counting process, the mean value function has several extremely important properties as given below.

Theorem 17.5. The mean value function of the renewal counting process, denoted by $m(t)$, is called the *renewal function* and is equal to the sum of the cumulative distribution function of all renewal times. That is,

$$m(t) = E[N(t)] = \sum_{n=1}^{\infty} F_n(t) \qquad (17.48)$$

where

$F_n(t)$ = cumulative distribution function of the nth renewal time.

Proof. The probability of exactly n renewals at time t is given in Eq. (17.14) as

$$Pr\{N(t) = n\} = F_n(t) - F_{n+1}(t), \qquad n = 1, 2, \ldots$$

Hence, the renewal function can be obtained as

$$m(t) = \sum_{n=0}^{\infty} n\{F_n(t) - F_{n+1}(t)\} = \sum_{n=1}^{\infty} F_n(t)$$

Let us write the relationship given in Eq. (17.48) in terms of the Laplace transform. We define $m^*(s)$ and $F_n^*(s)$ as the Laplace transform of $m(t)$ and $F_n(t)$, respectively. That is,

$$m^*(s) = \int_0^{\infty} m(t) e^{-st} \, dt$$
$$\qquad\qquad\qquad\qquad\qquad\qquad (17.49)$$
$$F_n^*(s) = \int_0^{\infty} F_n(t) e^{-st} \, dt$$

By letting the Laplace transform of the probability density function of the renewal time be $f_n^*(s)$, we have

$$F_n^*(s) = \frac{1}{s} f_n^*(s) \qquad (17.50)$$

Furthermore, because of the independent, identically distributed renewal period, we can write

$$f_n^*(s) = \{f_p^*(s)\}^n \qquad (17.51)$$

where

$f_p^*(s)$ = Laplace transform of the probability density function of the renewal period, $f_p(t)$.

Thus, Eq. (17.48) can be written in terms of the Laplace transform,

$$m^*(s) = \sum_{n=1}^{\infty} F_n^*(s) = \frac{1}{s} \sum_{n=1}^{\infty} \{f_p^*(s)\}^n$$

$$= \frac{f_p^*(s)}{s\{1 - f_p^*(s)\}} \qquad (17.52)$$

Equivalently, we may write

$$f_p^*(s) = \frac{s m^*(s)}{1 + s m^*(s)} \qquad (17.53)$$

The relationships given in Eqs. (17.52) and (17.53) are extremely useful in practice, since the probability distribution of the renewal period can be found from knowledge of the renewal function and vice versa.

Example 17.10. We may apply Eq. (17.53) to a Poisson process with the renewal function $m(t) = \nu t$. Its Laplace transform becomes $m^*(s) = \nu/s^2$, and thereby

$$f_p^*(s) = \nu/(s + \nu)$$

By taking the inverse Laplace transform, we have $f_p(t) = \nu e^{-\nu t}$, which is the probability density function of the interarrival time discussed in Theorem 17.2. ∎

17.4.2 Renewal Equation

Theorem 17.6. The renewal function $m(t)$ satisfies the following integral equation:

$$m(t) = F_p(t) + \int_0^t m(t - \tau) f_p(\tau) \, d\tau \qquad (17.54)$$

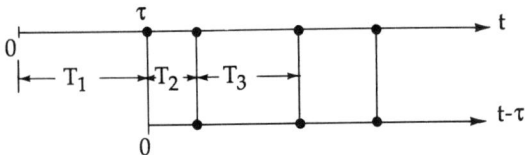

Figure 17.13 Renewal process in Theorem 17.6.

where

$F_p(t), f_p(t)$ = cumulative and probability density functions, respectively, of the renewal period.

Proof. Let us write the renewal function in the form of a conditional expectation given that the first renewal period $T_1 = \tau$. That is,

$$m(t) = \int_0^\infty E[N(t)|T_1 = \tau] f_p(\tau) \, d\tau$$

Note that if $t < \tau$, $E[N(t)|T_1 = \tau] = 0$, since there is no renewal for $t < \tau$. If $t > \tau$, we count the process based on the first renewal from $t = \tau$ as shown in Figure 17.13. Then, we have

$$E[N(t)|T_1 = \tau] = 1 + E[N(t - \tau)] = 1 + m(t - \tau)$$

Since $t > \tau$, the upper limit of the integral is t. Thus, we can write

$$m(t) = \int_0^t \{1 + m(t - \tau)\} f_p(\tau) \, d\tau$$

$$= F_p(t) + \int_0^t m(t - \tau) f_p(\tau) \, d\tau$$

Next, let us take the Laplace transform of the integral equation given in Theorem 17.6. By letting $t \to \infty$, we can write

$$\int_0^\infty m(t) e^{-st} \, dt = \int_0^\infty F_p(t) e^{-st} \, dt + \int_0^\infty \int_0^\infty m(t - \tau) f_p(\tau) e^{-st} \, d\tau \, dt$$

$$= \int_0^\infty F_p(t) e^{-st} \, dt + \int_0^\infty \int_0^\infty m(t - \tau) e^{-s(t-\tau)} f_p(\tau) e^{-s\tau} \, d\tau \, dt$$

Hence, we may write

$$m^*(s) = F_p^*(s) + m^*(s)f_p^*(s)$$
$$= (1/s)f_p^*(s) + m^*(s)f_p^*(s)$$

This is exactly the same relationship derived in Eq. (17.52).

The integral equation given in Theorem 17.6 is a special case of the following integral equation called the *renewal equation*:

$$x(t) = g(t) + \int_0^t x(t-\tau)f_p(\tau)\, d\tau \qquad (17.55)$$

where $x(t)$ is an unknown function to be evaluated, while $g(t)$ is any function associated with the renewal process. The function $g(t)$ must be nonnegative and its integral with respect to time must exist. $f_p(t)$ is the probability density function of the renewal period. The solution of this renewal equation is given in the following theorem.

Theorem 17.7. Let the renewal equation be

$$x(t) = g(t) + \int_0^t x(t-\tau)f_p(\tau)\, d\tau$$

Then, its solution is given by

$$x(t) = g(t) + \int_0^t g(t-\tau)\, dm(\tau)$$

where $m(t)$ = mean value function of the renewal process.

Proof. The Laplace transform of the renewal equation can be written as

$$x^*(s) = g^*(s) + x^*(s)f_p^*(s)$$

Hence, we have

$$x^*(s) = \frac{g^*(s)}{1 - f_p^*(s)} = g^*(s)\left\{1 + \frac{f_p^*(s)}{1 - f_p^*(s)}\right\}$$

From the result given in Eq. (17.52), we can write

$$x^*(s) = g^*(s)\{1 + sm^*(s)\}$$

Then, by taking the inverse Laplace transform, the desired result can be derived.

Example 17.11. Let $g(t) = k$, where k is a constant, in the renewal equation given in Eq. (17.55). In this case, the solution $x(t)$ becomes

$$x(t) = k + \int_0^t k\, dm(\tau) = k\{1 + E[N(t)]\}$$ ∎

17.4.3 Asymptotic Properties of Renewal Process

In this section some asymptotic properties of the renewal process $\{N(t), t \geq 0\}$ for large t are summarized without proof. For the proof of the following theorems the reader is referred to Prabhu (1965) and Medhi (1981).

Theorem 17.8. Let the mean and variance of the renewal period be $E[T] = \mu$ and $\text{Var}[T] = \sigma^2$, respectively. Then, we have the following asymptotic formula for the renewal function:

$$\lim_{t \to \infty} \frac{m(t)}{t} = \lim_{t \to \infty} \frac{E[N(t)]}{t} = 1/\mu \qquad (17.56)$$

This theorem is called the *elementary renewal theorem*. For the variance of the renewal process, we have the asymptotic formula

$$\lim_{t \to \infty} \frac{\text{Var}[N(t)]}{t} = \sigma^2/\mu^3 \qquad (17.57)$$

Theorem 17.9. The renewal process $N(t)$ is asymptotically normally distributed with mean t/μ and variance $t\sigma^2/\mu^3$ for large t. Hence, we have the following probability:

$$\lim_{t \to \infty} Pr\left\{\frac{N(t) - (t/\mu)}{\sqrt{t\sigma^2/\mu^3}} < a\right\} = \Phi(a) \qquad (17.58)$$

Theorem 17.10. Let $h(t)$ be a nonnegative function of t such that $\int_0^\infty h(t)\, dt < \infty$. Then, we have

$$\lim_{t \to \infty} \int_0^t h(t - \tau)\, dm(\tau) = \frac{1}{\mu} \int_0^t h(t)\, dt \qquad (17.59)$$

where μ is the mean renewal period. This theorem is called the *key renewal theorem*.

Example 17.12. Let the renewal period X have a gamma distribution given by

$$f(x) = \lambda^2 x e^{-\lambda x}$$

which has the mean value $2/\lambda$. Then, the renewal time to the nth event $W = X_1 + X_2 + \cdots + X_n$ also has a gamma distribution that can be written as

$$f(w) = \frac{1}{\Gamma(2n)} \lambda^{2n} w^{2n-1} e^{-\lambda w}$$

From the relationship given in Eq. (17.48), we can write

$$\frac{d}{dt} m(t) = \sum_{n=1}^{\infty} f(w) = \lambda e^{-\lambda w} \sum_{n=1}^{\infty} \frac{(\lambda w)^{2n-1}}{\Gamma(2n)}$$

$$= \frac{\lambda}{2} (1 - e^{-2\lambda w})$$

and thereby

$$m(t) = \int_0^t (\lambda/2)(1 - e^{-2\lambda w}) \, dw$$

$$= \frac{\lambda t}{2} - \frac{1}{4}(1 - e^{-2\lambda t})$$

Thus, for $t \to \infty$ we can write

$$\lim_{t \to \infty} \frac{m(t)}{t} = \lambda/2$$

which is equal to the inverse of the mean of the renewal period given in Theorem 17.8. ∎

17.4.4 Probability Distribution of Residual Time

An interesting subject associated with renewal theory is to evaluate the probability distribution of the time before the next occurrence of the event. Let us consider the renewal process in which n renewals have already

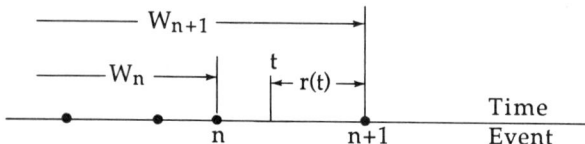

Figure 17.14 Definition of residual time $r(t)$ given in Eq. (17.60).

occurred, and let the present time t be in the interval between two renewal times, W_n and W_{n+1}, as shown in Figure 17.14.

The *residual time*, denoted by $r(t)$, is defined as the remaining time before the $(n + 1)$th renewal will take place. We may write the residual time as

$$r(t) = W_{n+1} - t \tag{17.60}$$

and evaluate the probability distribution of $r(t)$. For this, we first show that the residual time $r(t)$ satisfies the renewal equation given in the following theorem:

Theorem 17.11. Let $r(t)$ be the residual time of a renewal process. Then, the probability that $r(t) > a$ (where a is specified) satisfies the renewal equation

$$x(t, a) = 1 - F_p(t + a) + \int_0^t x(t - \tau, a) f_p(\tau) \, d\tau \tag{17.61}$$

where

$$x(t, a) = Pr\{r(t) > a\}$$

$f_p(t)$ = probability density of renewal period

$F_p(t)$ = cumulative distribution of renewal period.

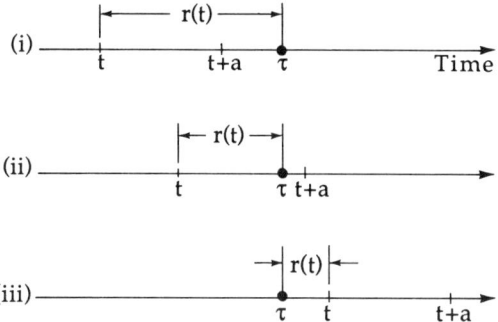

Figure 17.15 Three situations of time t relative to the first occurrence of the event $T_1 = \tau$ in Theorem 17.11.

Proof. We apply the same technique as used for the proof of Theorem 17.6. That is, by considering the conditional probability that the first renewal period, denoted by T_1, is equal to τ, we can write

$$Pr\{r(t) > a\} = Pr\{r(t) > a | T_1 = \tau\} f_p(\tau) \, d\tau$$

For the conditional probability in the above equation, we consider three possible situations of time t relative to the first occurrence of the event as illustrated in Figure 17.15.

(i) $t < t + a < \tau$.

In this case, we have $Pr\{r(t) > a | T_1 = \tau\} = 1$.

(ii) $t < \tau < t + a$.

As is clear from the figure, we have $Pr\{r(t) > a | T_1 = \tau\} = 0$.

(iii) $\tau < t < t + a$.

Since the event occurs at time τ, the probability distribution of the residual time must be considered to originate at τ. That is,

$$Pr\{r(t) > a | T_1 = \tau\} = Pr\{r(t - \tau) > a\}$$

Taking these three cases into consideration, we can write

$$Pr\{r(t) > a\} = \int \bigl(1 + Pr\{r(t - \tau) > a\}\bigr) f_p(\tau) \, d\tau$$

where the integral limits of the first term are $(t + a, \infty)$, while those of the second term are $(0, t)$. By writing $Pr\{r(t) > a\} = x(t, a)$, we can derive the following renewal equation:

$$x(t, a) = \int_{t+a}^{\infty} f_p(\tau) \, d\tau + \int_0^t x(t - \tau, a) f_p(\tau) \, d\tau$$

$$= 1 - F_p(t + a) + \int_0^t x(t - \tau, a) f_p(\tau) \, d\tau \qquad \blacksquare$$

By applying Theorem 17.7, the solution of the above renewal equation can be obtained. Thus, the probability that the residual time exceeds a specified value can be evaluated by the following formula:

$$Pr\{r(t) > a\} = 1 - F_p(t + a) + \int_0^t \{1 - F_p(t + a - \tau)\} \, dm(\tau)$$

(17.62)

This is equivalent to

$$Pr\{r(t) \leq a\} = F_p(t + a) - \int_0^t \{1 - F_p(t + a - \tau)\} \, dm(\tau) \quad (17.63)$$

Example 17.13. Assume that the renewal period follows the exponential probability law for which the cumulative distribution function can be written as $F_p(t) = 1 - e^{-\nu t}$. Since this is a Poisson process, we have $m(t) = \nu t$. Then, the cumulative distribution function of the residual time can be obtained from Eq. (17.63) as

$$Pr\{r(t) \leq a\} = \{1 - e^{-\nu(t+a)}\} - \nu \int_0^t e^{-\nu(t+a-\tau)} \, d\tau$$

$$= 1 - e^{-\nu a}$$

Thus, it can be proved that the residual time is also exponentially distributed in this case. $\blacksquare$

EXERCISES

17.1 Analysis of records obtained in the Gulf of Mexico indicates that tropical storms come to the Gulf in accordance with a Poisson process with intensity 0.68 per year. Obtain the number of storms having the highest probability in a 5-year period.

17.2 Let $Y(t)$ be a superimposed Poisson process with intensity ν. That is,

$$Y(t) = \sum_{n=1}^{N(t)} w_n(t, \tau_n, X_n)$$

Prove that the covariance of $Y(t)$ is given by

$$\text{Cov}[Y(t_1), Y(t_2)] = \nu \int_0^{\min(t_1, t_2)} E[w(t_1, \tau, x) w(t_2, \tau, x)] \, d\tau$$

17.3 Consider a compound Poisson process $Y(t) = \sum_{n=1}^{N(t)} X_n$, where $N(t)$ is a Poisson process with intensity ν and X_n are independent random variables having an exponential distribution with parameter λ. Show that $Y(t)$ is approximately normally distributed with mean $\nu t/\lambda$ and variance $2\nu t/\lambda^2$.

17.4. Let W_n be the waiting time of a Poisson process with intensity ν. Equation (17.13) gives the probability that the nth event occurs before time t as follows:

$$Pr\{W_n < t\} = F_{W_n}(t)$$

Show that the cumulative distribution function $F_{W_n}(t)$ can be expressed as

$$F_{W_n}(t) = 1 - \sum_{k=0}^{n-1} e^{-\nu t} \frac{(\nu t)^k}{k!}$$

Show also that the probability density function of the waiting time is given by

$$f_{W_n}(t) = \nu e^{-\nu t} \frac{(\nu t)^{n-1}}{(n-1)!}$$

17.5 Let $X(t)$ be a Poisson process with various impact severities as illustrated in Figure 17.12. The severity a_n is a sequence of independent, identically distributed random variables. Let its first two moments be $E[a]$ and $E[a^2]$, respectively. Prove that the autocorrelation function $R_{xx}(\tau)$ and the spectral density function $S_{xx}(\omega)$ are given as follows:

$$R_{xx}(\tau) = \nu^2(E[a])^2 + \nu E[a^2]\delta(\tau)$$

$$S_{xx}(\omega) = 2\nu^2(E[a])^2 \delta(\omega) + \frac{\nu}{\pi}E[a^2]$$

17.6 Let the renewal period of a renewal process following a gamma probability distribution given by

$$f_p(t) = \frac{1}{\Gamma(m)}\nu^m t^{m-1} e^{-\nu t}$$

Show that

$$Pr\{N(t) = n\} = e^{-\nu t} \sum_{r=nm}^{(n+1)m-1} \frac{(\nu t)^r}{r!}$$

17.7 Show that the probability generating function of a renewal process, denoted by $G(z)$, is given by

$$G(z) = \sum_{n=0}^{\infty} z^n Pr\{N(t) = n\} = 1 + \sum_{n=1}^{\infty} z^{n-1}(z-1)F_n(t)$$

Show also that the Laplace transform of $G(z)$ can be written as

$$G^*(s) = \frac{1 - f_p^*(s)}{s\{1 - zf_p^*(s)\}}$$

where

$f_p^*(s) = $ Laplace transform of the probability density function of the renewal period $f_p(t)$.

APPENDIX A

Fourier Transform

A.1 DERIVATION OF THE FOURIER TRANSFORM

Let us consider a periodic function $f(x)$ with period 2π expressed by the following Fourier series:

$$f(x) = \frac{a_0}{2} + \sum_{n=1}^{\infty} (a_n \cos nx + b_n \sin nx) \tag{A.1}$$

where

$$a_n = \frac{1}{\pi} \int_{-\pi}^{\pi} f(x) \cos nx\, dx, \quad n = 0, 1, 2, \ldots$$

$$b_n = \frac{1}{\pi} \int_{-\pi}^{\pi} f(x) \sin nx\, dx, \quad n = 1, 2, \ldots$$

We transform $f(x)$ to a function of time $f(t)$ with period T by using the relationship given by $x = (2\pi/T)t = \omega t$, where ω is called the frequency. That is,

$$f(t) = \frac{a_0}{2} + \sum_{n=1}^{\infty} (a_n \cos n\omega t + b_n \sin n\omega t) \tag{A.2}$$

where

$$a_n = \frac{2}{T}\int_{-T/2}^{T/2} f(t)\cos n\omega t\, dt$$

$$b_n = \frac{2}{T}\int_{-T/2}^{T/2} f(t)\sin n\omega t\, dt$$

We may write Eq. (A.2) as follows:

$$f(t) = \frac{a_0}{2} + \sum_{n=1}^{\infty}\left\{\frac{1}{2}(a_n - ib_n)e^{in\omega t} + \frac{1}{2}(a_n + ib_n)e^{-in\omega t}\right\} \quad (A.3)$$

By defining

$$\frac{a_0}{2} = C_0$$

$$\frac{1}{2}(a_n - ib_n) = C_n \quad (A.4)$$

$$\frac{1}{2}(a_n + ib_n) = C_{-n}$$

Eq. (A.3) can be written as

$$f(t) = \sum_{n=-\infty}^{\infty} C_n e^{in\omega t} \quad (A.5)$$

where

$$C_n = \frac{1}{T}\int_{-T/2}^{T/2} f(t)(\cos n\omega t - i\sin n\omega t)\, dt$$

$$= \frac{1}{T}\int_{-T/2}^{T/2} f(t)e^{-in\omega t}\, dt \quad (A.6)$$

Equation (A.5) is called the complex form of a Fourier series expansion of a periodic function $f(t)$, and this expression will be used for the Fourier transform of a nonperiodic function. By combining Eqs. (A.5) and (A.6), we can write $f(t)$ as

$$f(t) = \sum_{n=-\infty}^{\infty} \left[\frac{1}{T} \int_{-T/2}^{T/2} f(t) e^{-in\omega t} \, dt \right] e^{in\omega t} \quad (A.7)$$

In order to express a nonperiodic function in terms of frequency, let us write $n\omega = \omega_n$ in Eq. (A.7) and the increment of the frequency as $\Delta\omega_n$. Since $1/T = \Delta\omega_n/2\pi$, Eq. (A.7) becomes

$$f(t) = \sum_{n=-\infty}^{\infty} \left[\frac{\Delta\omega_n}{2\pi} \int_{-T/2}^{T/2} f(t) e^{-i\omega_n t} \, dt \right] e^{i\omega_n t}$$

$$= \frac{1}{2\pi} \sum_{n=-\infty}^{\infty} \left[\int_{-T/2}^{T/2} f(t) e^{-i\omega_n t} \, dt \right] e^{i\omega_n t} \cdot \Delta\omega_n \quad (A.8)$$

For a nonperiodic function, let $T \to \infty$ and $\Delta\omega_n \to 0$, and thereby Eq. (A.8) can be expressed in integrable form. By writing ω_n as ω, we have

$$f(t) = \frac{1}{2\pi} \int_{-\infty}^{\infty} \left[\int_{-\infty}^{\infty} f(t) e^{-i\omega t} \, dt \right] e^{i\omega t} \, d\omega$$

$$= \frac{1}{2\pi} \int_{-\infty}^{\infty} F(\omega) \cdot e^{i\omega t} \, d\omega \quad (A.9)$$

where

$$F(\omega) = \int_{-\infty}^{\infty} f(t) \cdot e^{-i\omega t} \, dt \quad (A.10)$$

Equations (A.9) and (A.10) are called the *Fourier transform pair*, denoted by $f(t) \leftrightarrow F(\omega)$. Here, $F(\omega)$ is called the Fourier transform of $f(t)$, which may be denoted by $F(\omega) = \mathscr{F}\{f(t)\}$, while $f(t)$ is the Fourier inverse transform of $F(\omega)$ denoted by $f(t) = \mathscr{F}^{-1}\{F(\omega)\}$.

The condition for existence of the Fourier transform of $f(t)$ is given, in general, by

$$\int_{-\infty}^{\infty} |f(t)| \, dt < \infty \quad (A.11)$$

It is noted that the Fourier transform pair defined in Eqs. (A.9) and (A.10) is not unique; instead, the following is also defined as the Fourier transform pair:

$$F(\omega) = \frac{1}{\pi} \int_{-\infty}^{\infty} f(t) e^{-i\omega t} \, dt$$

and

$$f(t) = \frac{1}{2} \int_{-\infty}^{\infty} F(\omega) e^{i\omega t} \, d\omega \tag{A.12a}$$

$$F(\omega) = \frac{1}{\sqrt{2\pi}} \int_{-\infty}^{\infty} f(t) e^{-i\omega t} \, dt$$

and

$$f(t) = \frac{1}{\sqrt{2\pi}} \int_{-\infty}^{\infty} F(\omega) e^{i\omega t} \, d\omega \tag{A.12b}$$

The difference in definition of the Fourier transform pair stems from the definition of $F(\omega)$, which is given in the parentheses of Eq. (A.9). If $1/\pi$ is included in the parentheses, then the Fourier transform pair is defined by Eq. (A.12a).

The Fourier transform pair defined in Eqs. (A.9) and (A.10) is given in terms of the frequency ω in cps. The Fourier transform pairs in terms of the frequency f in Hz is given as follows [the Fourier transform of a function $x(t)$ is considered here in order to avoid possible confusion]:

$$X(f) = \int_{-\infty}^{\infty} x(t) e^{-i2\pi f t} \, dt$$

$$x(t) = \int_{-\infty}^{\infty} X(f) e^{i2\pi f t} \, df \tag{A.13}$$

Since the Fourier transform of a function $f(t)$ is a complex function of ω, we may write $F(\omega)$ as

$$F(\omega) = P(\omega) + iQ(\omega) = A(\omega) \cdot e^{i\theta(\omega)} \tag{A.14}$$

where

$$A(\omega) = \sqrt{\{P(\omega)\}^2 + \{Q(\omega)\}^2}$$

$$\theta(\omega) = \tan^{-1} \frac{Q(\omega)}{P(\omega)}$$

APPENDIX A

Then, from Eqs. (A.10) and (A.14), we have

$$P(\omega) = \int_{-\infty}^{\infty} f(t) \cdot \cos \omega t \, dt$$

$$Q(\omega) = -\int_{-\infty}^{\infty} f(t) \cdot \sin \omega t \, dt$$

(A.15)

It can easily be proved that

$$P(-\omega) = P(\omega)$$

$$Q(-\omega) = -Q(\omega)$$

(A.16)

and thereby we have

$$F(-\omega) = P(\omega) - iQ(\omega) = \text{conjugate of } F(\omega) = F^*(\omega) \quad (A.17)$$

If $f(t)$ is a real-valued even function, then from Eq. (A.15), we can derive the following Fourier transform pair:

$$F(\omega) = P(\omega) = 2\int_0^{\infty} f(t) \cos \omega t \, dt$$

and

$$f(t) = \frac{1}{\pi} \int_0^{\infty} P(\omega) \cos \omega t \, d\omega \quad (A.18)$$

Similarly, if $f(t)$ is a real-valued odd function, the following Fourier transform pair can be derived:

$$F(\omega) = Q(\omega) = -2\int_0^{\infty} f(t) \sin \omega t \, dt$$

and

$$f(t) = -\frac{1}{\pi} \int_0^{\infty} Q(\omega) \sin \omega t \, d\omega \quad (A.19)$$

Examples of the Fourier transform are given below:

Example A.1. $f(t) = e^{-a|t|}$ (see Figure A.1).

$$\mathscr{F}\{e^{-a|t|}\} = \int_{-\infty}^{\infty} e^{-a|t|-i\omega t}\, dt$$

$$= \int_{-\infty}^{0} e^{at-i\omega t}\, dt + \int_{0}^{\infty} e^{-at-i\omega t}\, dt$$

$$= \frac{1}{a-i\omega} + \frac{1}{a-i\omega} = \frac{2a}{a^2+\omega^2} \qquad \blacksquare$$

Example A.2. $f(t) = e^{-at^2}$ (see Figure A.2).

$$\mathscr{F}\{e^{-at^2}\} = \int_{-\infty}^{\infty} e^{-at^2-i\omega t}\, dt$$

$$= \int_{-\infty}^{\infty} \exp\left\{-a\left(t+\frac{i\omega}{2a}\right)^2 - \frac{\omega^2}{4a}\right\} dt$$

$$= \sqrt{\pi/a}\, \exp\left\{-\frac{\omega^2}{4a}\right\} \qquad \blacksquare$$

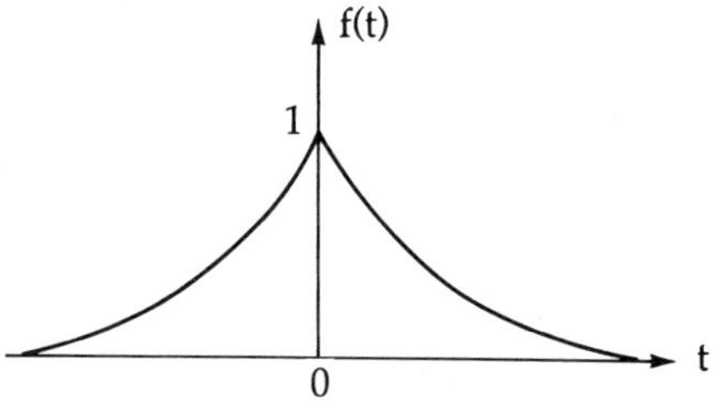

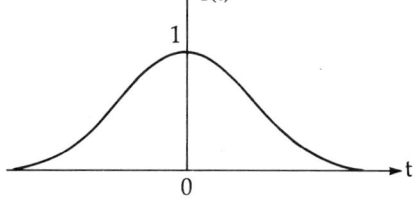

Figure A.1 Function $f(t)$ in Example A.1.

Figure A.2 Function $f(t)$ in Example A.2.

APPENDIX A

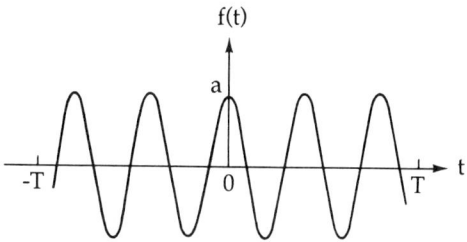

Figure A.3 Function $f(t)$ in Example A.3.

Example A.3. $f(t) = a \cos \omega_0 t$, $-T \leqslant t \leqslant T$ (see Figure A.3). Since $f(t)$ is a real-valued even function, we can write

$$\mathscr{F}\{f(t)\} = \int_{-T}^{T} a \cos \omega_0 t \, e^{-i\omega t} \, dt$$

$$= 2a \int_{0}^{T} \cos \omega_0 t \cos \omega t \, dt$$

$$= a \int_{0}^{T} \{\cos(\omega + \omega_0)t + \cos(\omega - \omega_0)t\} \, dt$$

$$= a \left\{ \frac{\sin(\omega + \omega_0)T}{\omega + \omega_0} + \frac{\sin(\omega - \omega_0)T}{\omega - \omega_0} \right\} \qquad \blacksquare$$

Example A.4.

$$f(t) = \begin{cases} 1 & \text{for } t > 0 \\ -1 & \text{for } t < 0 \end{cases} \text{(see Figure A.4)}$$

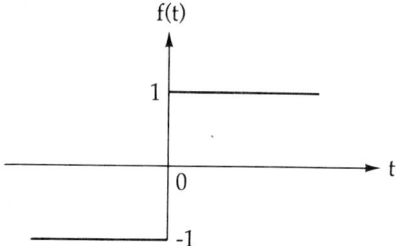

Figure A.4 Function $f(t)$ in Example A.4.

The function $f(t)$ is called a *sign function*, denoted by sgn t (signum t). The Fourier transform of sgn t cannot be obtained directly; instead, it can be evaluated by taking the Fourier transform $e^{-\tau|t|}$ sgn t and by letting $\tau \to 0$. That is,

$$\mathscr{F}\{e^{-\tau|t|} \operatorname{sgn} t\} = \int_{-\infty}^{\infty} e^{-i\omega t - \tau|t|} \operatorname{sgn} t \, dt$$

$$= \int_{-\infty}^{0} e^{-(\tau - i\omega)t} \, dt + \int_{0}^{\infty} e^{-(\tau + i\omega)t} \, dt$$

$$= -\frac{1}{\tau - i\omega} + \frac{1}{\tau + i\omega}$$

By letting $\tau \to 0$, we have $\mathscr{F}\{\operatorname{sgn} t\} = 2/i\omega$. ∎

A.2 PROPERTIES OF THE FOURIER TRANSFORM

Some useful properties associated with the Fourier transform are summarized below without proof. The reader may refer to Champeney (1973) and Papoulis (1962), among others, for details of these properties

1. *Linearity* (Superposition Theorem)
   Let
   $$f_1(t) \leftrightarrow F_1(\omega)$$
   $$f_2(t) \leftrightarrow F_2(\omega)$$
   $$\vdots$$
   $$f_n(t) \leftrightarrow F_n(\omega)$$
   then
   $$\{a_1 f_1(t) + a_2 f_2(t) + \cdots + a_n f_n(t)\}$$
   $$\leftrightarrow \{a_1 F_1(\omega) + a_2 F_2(\omega) + \cdots + a_n F_n(\omega)\} \quad \text{(A.20)}$$
   where $a_1, a_2, \ldots$ are constants.

APPENDIX A

11. *Frequency Differentiation*
   If
   $$f(t) \leftrightarrow F(\omega)$$
   then
   $$(-it)^n f(t) \leftrightarrow \frac{d^n F(\omega)}{d\omega^n} \qquad (A.30)$$
   if it exists.

Example A.5. Fourier transform of $f(t) = 1/t$.

The direct derivation of the Fourier transform is extremely complicated since $f(t) = \infty$ at $t = 0$. However, the Fourier transform can be simply obtained by applying Property (6) to the result shown in Example A.4. That is, from Example A.4, we have

$$\operatorname{sgn} t \leftrightarrow \frac{2}{i\omega}$$

Then, by applying Property (6), the following relationship can be derived:

$$\frac{2}{it} \leftrightarrow -2\pi \operatorname{sgn} \omega$$

Hence,

$$\frac{1}{t} \leftrightarrow -i\pi \operatorname{sgn} \omega \qquad \blacksquare$$

APPENDIX B

Hilbert Transform

Let $x(t)$ be a real-valued function in the interval $-\infty < x < \infty$. The *Hilbert transform* of the function $x(t)$, denoted by $\mathcal{H}\{x(t)\}$ or $\tilde{x}(t)$, is defined as

$$\tilde{x}(t) = \frac{1}{\pi} \int_{-\infty}^{\infty} \frac{x(\tau)}{t - \tau} d\tau \qquad (B.1)$$

Equation (B.1) is a convolution integral of $x(t)$ and $1/\pi t$. Hence we may write the Hilbert transform of $x(t)$ as

$$\tilde{x}(t) = x(t) * (1/\pi t) \qquad (B.2)$$

As given in Eq. (A.28), the convolution integral of two functions in the time domain can be expressed as the product of their Fourier transforms in the frequency domain. Hence, by taking the Fourier transform of each term, the above equation can be expressed in the frequency domain as

$$\tilde{X}(\omega) = X(\omega) \cdot (-i \operatorname{sgn} \omega) \qquad (B.3)$$

Note that $\operatorname{sgn} \omega$ is the sign function shown in Figure A.4 and $\mathcal{F}\{1/\pi t\} = -i \operatorname{sgn} \omega$ from Example A.5.

APPENDIX B

Example B.1. $\mathcal{H}\{\sin t\} = -\cos t$, $\mathcal{H}\{\cos t\} = \sin t$.
By letting $t - \tau = u$, we have

$$\mathcal{H}\{\sin t\} = \frac{1}{\pi}\int_{-\infty}^{\infty}\frac{\sin \tau}{t - \tau}d\tau = \frac{1}{\pi}\int_{-\infty}^{\infty}\frac{\sin(t - u)}{u}du$$

$$= \frac{1}{\pi}\sin t\int_{-\infty}^{\infty}\frac{\cos u}{u}du - \frac{1}{\pi}\cos t\int_{-\infty}^{\infty}\frac{\sin u}{u} = -\cos t$$

$\mathcal{H}\{\cos t\}$ can be proved in the same fashion. ∎

Example B.2.

$$x(t) = \frac{1}{a^2 + t^2} \quad (\text{see Figure B.1})$$

The Hilbert transform $\tilde{x}(t)$ will be obtained by applying Eq. (B.3). That is, first obtain the Fourier transform of $x(t)$ as

$$X(\omega) = \mathcal{F}\{1/(a^2 + t^2)\} = \frac{\pi}{a}\exp\{-a|\omega|\}$$

where $X(\omega)$ is an even function. Then, by applying Eq. (B.3),

$$\tilde{x}(t) = \mathcal{F}^{-1}\{\tilde{X}(\omega)\} = \frac{1}{a}\int_{0}^{\infty}e^{-a\omega}\sin \omega t \, d\omega = \frac{t}{a(a^2 + t^2)}$$

Note that $x(t)$ is an even function of t, but $\tilde{x}(t)$ is an odd function. The reverse is also true; if $x(t)$ is an odd function, then $\tilde{x}(t)$ is an even function. ∎

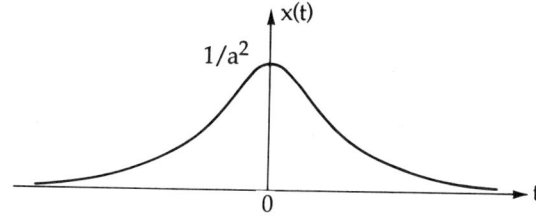

Figure B.1 Function $x(t)$ in Example B.2.

Next, consider a function $x(t)$, which is defined as $x(t) = 0$ for $t < 0$. This function is called the *causal function*. Let us write the Fourier transform of $x(t)$ as $X(\omega) = P(\omega) + iQ(\omega)$. Then, it can be shown that $P(\omega)$ and $Q(\omega)$ are uniquely determined in terms of the other by the following Hilbert transform:

$$P(\omega) = \frac{1}{\pi} \int_{-\infty}^{\infty} \frac{Q(\lambda)}{\omega - \lambda} d\lambda$$

$$Q(\omega) = -\frac{1}{\pi} \int_{-\infty}^{\infty} \frac{P(\lambda)}{\omega - \lambda} d\lambda$$

(B.4)

In order to prove this important property, the causal function $x(t)$ is expressed as the sum of even and odd functions, denoted as $x_1(t)$ and $x_2(t)$, respectively, as shown in Figure B.2. Since $x_1(t) = x_2(t)$ for $t > 0$,

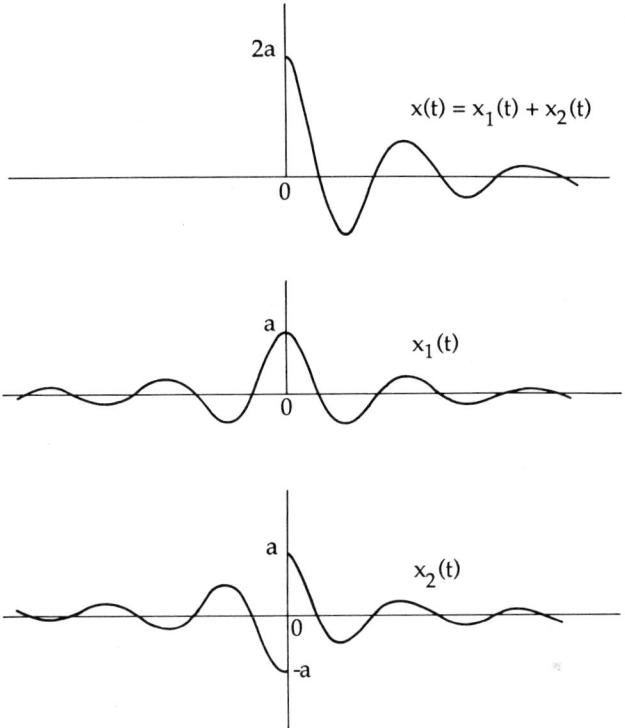

Figure B.2 Causal function $x(t)$ and its decomposition into even and odd functions.

APPENDIX B

and $x_2(t) = -x_1(t)$ for $t < 0$, we may write

$$x_1(t) = x_2(t) \operatorname{sgn} t \quad \text{and} \quad x_2(t) = x_1(t) \operatorname{sgn} t$$

By applying Eq. (A.27) as well as the result obtained in Example A.4, the above equations can be written in the frequency domain as

$$X_1(\omega) = (1/2\pi) X_2(\omega) * (2/i\omega)$$

$$X_2(\omega) = (1/2\pi) X_1(\omega) * (2/i\omega)$$

Since $x_1(t)$ is a real-valued even function, $X_1(\omega)$ is equal to $P(\omega)$ [see Eq. (A.18)]. Likewise, $x_2(t)$ is a real-valued odd function and thereby from Eq. (A.19) $X_2(\omega) = iQ(\omega)$. Thus, we have

$$P(\omega) = \frac{1}{\pi} Q(\omega) * (1/\omega)$$

$$Q(\omega) = -\frac{1}{\pi} P(\omega) * (1/\omega)$$

from which Eq. (B.4) can be derived.

Some useful properties of the Hilbert transform are summarized below without proof.

1. *Linearity*

$$\mathcal{H}\{a_1 x_1(t) + a_2 x_2(t)\} = a_1 \tilde{x}_1(t) + a_2 \tilde{x}_2(t) \quad (\text{B.5})$$

2. *Modulation*

$$\mathcal{H}\{x(t) \cos(\omega t + \varepsilon)\} = x(t) \sin(\omega t + \varepsilon) \quad (\text{B.6})$$

3. *Successive Hilbert transforms*

$$\mathcal{H}\{\tilde{x}(t)\} = -x(t) \quad (\text{B.7})$$

4. *Orthogonality*

$$\int_{-\infty}^{\infty} x(t) \tilde{x}(t) \, dt = 0 \quad (\text{B.8})$$

5. *Convolution*

$$\mathcal{H}\{x(t) * y(t)\} = \tilde{x}(t) * y(t) = x(t) * \tilde{y}(t) \qquad (B.9)$$

6. *Fourier transform of* $\tilde{x}(t)$

$$\mathcal{F}\{\tilde{x}(t)\} = \begin{cases} -iX(\omega) & \text{for } \omega < 0 \\ 0 & \omega = 0 \\ iX(\omega) & \omega > 0 \end{cases} \qquad (B.10)$$

APPENDIX C

Unit Impulse Function (Delta Function)

The *unit impulse function*, which is also known as the *delta function* (*δ-function*), is defined as follows:

$$\delta(t - t_0) = \begin{cases} \infty & \text{for } t = t_0 \\ 0 & \text{for } t \neq t_0 \end{cases}$$

and

$$\int_{-\infty}^{\infty} \delta(t - t_0)\, dt = \lim_{\varepsilon \to 0} \int_{t_0 - \varepsilon}^{t_0 + \varepsilon} \delta(t - t_0)\, dt = 1 \qquad (C.1)$$

Following the above definition, the delta function $\delta(t)$ can be interpreted as $\delta(t - 0)$. Figure C.1 shows the delta functions $\delta(t - t_0)$ and $\delta(t)$.

The delta function $\delta(t - t_0)$ is equal to the derivative of the *unit step function* $\mathsf{u}(t - t_0)$ which is defined as

$$\mathsf{u}(t - t_0) = \begin{cases} 1 & \text{for } t > t_0 \\ 0 & \text{for } t < t_0 \end{cases}$$

The delta function has several important properties that are often applied to stochastic process analysis. These properties are summarized below.

1. Let $h(t)$ be a continuous function at $t = t_0$. Then, we have

$$\int_{-\infty}^{\infty} h(t)\delta(t - t_0)\, dt = h(t_0) \qquad \text{(C.2)}$$

Proof. From the definition of $\delta(t - t_0)$ we may write

$$\int_{-\infty}^{\infty} h(t)\delta(t - t_0)\, dt = \lim_{\varepsilon \to 0} \int_{t_0-\varepsilon}^{t_0+\varepsilon} h(t)\delta(t - t_0)\, dt$$

$$= h(t_0) \lim_{\varepsilon \to 0} \int_{t_0-\varepsilon}^{t_0+\varepsilon} \delta(t - t_0)\, dt = h(t_0)$$

2. $\delta(at) = (1/|a|)\,\delta(t).$ \qquad (C.3)

Proof. Let us write $at = \tau$. Then, for $a > 0$, we have

$$\int_{-\infty}^{\infty} \delta(at)\, dt = \frac{1}{a}\int_{-\infty}^{\infty} \delta(\tau)\, d\tau$$

For $a < 0$,

$$\int_{-\infty}^{\infty} \delta(at)\, dt = \frac{1}{a}\int_{\infty}^{-\infty} \delta(\tau)\, d\tau = -\frac{1}{a}\int_{-\infty}^{\infty} \delta(\tau)\, d\tau$$

Thus, it can be proved that $\delta(at) = (1/|a|)\,\delta(t)$.

3. $\delta(-t) = \delta(t).$ \qquad (C.4)

This can be proved by letting $a = -1$ in Eq. (C.3).

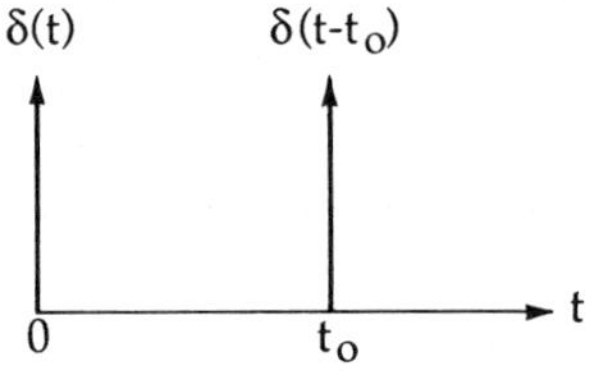

Figure C.1 Definition of delta functions $\delta(t)$ and $\delta(t - t_0)$.

APPENDIX C   479

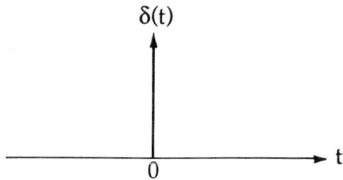

Figure C.2 Delta function $\delta(t)$ and its Fourier transform.

4. $\mathscr{F}\{\delta(t - t_0)\} = e^{-i\omega t_0}$. (C.5)

The proof follows from the result obtained in Eq. (C.2) by letting $h(t) = \exp\{-i\omega t\}$. Note that for $t_0 = 0$, we have

$$\mathscr{F}\{\delta(t)\} = 1 \quad (C.6)$$

In other words, the Fourier transform of $\delta(t)$ in the time domain is unity in the frequency domain as illustrated in Figure C.2.

5. $\mathscr{F}^{-1}\{\delta(\omega - \omega_0)\} = (1/2\pi)e^{i\omega_0 t}$. (C.7)

Proof. $\mathscr{F}^{-1}\{\delta(\omega - \omega_0)\} = (1/2\pi)\int_{-\infty}^{\infty} \delta(\omega - \omega_0)e^{i\omega t}d\omega = (1/2\pi)e^{i\omega_0 t}$.

By letting $\omega_0 = 0$ in Eq. (C.7), we can derive the following formula:

$$\mathscr{F}^{-1}\{2\pi\delta(\omega)\} = 1 \quad (C.8)$$

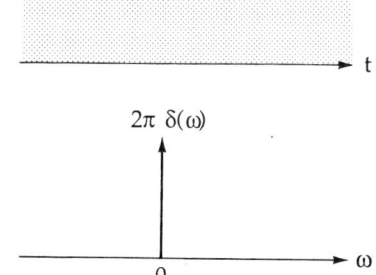

Figure C.3 $f(t) = 1$ and its Fourier transform.

In other words, the Fourier transform of unity in the time domain is equal to $2\pi\delta(\omega)$ in the frequency domain as shown in Figure C.3.

In applying the results given in Eqs. (C.6) and (C.8) to the Wiener–Khintchen theorem, care must be taken regarding the definition of the spectral density function shown in Table 10.1. If the spectral density function defined in Eq. (10.15a) is used, then we have the following:

(i) If the autocorrelation function $R_{xx}(\tau)$ is $\pi\delta(\tau)$ in the time domain, the spectral density function $S_{xx}(\omega)$ becomes unity in the frequency domain.

(ii) If the autocorrelation function $R_{xx}(\tau)$ is unity in the time domain, the spectral density function $S_{xx}(\omega)$ becomes $2\delta(\omega)$ in the frequency domain.

APPENDIX D

Statistical Goodness of Fit Tests

D.1 CHI-SQUARE (χ^2) TEST

This test establishes the confidence with which samples of observed occurrences of particular events can be assumed to belong to a hypothesized distribution. The test is valid for a large number of observations (at least 120 or so is necessary). The observed data are classified in k mutually exclusive groups with increasing magnitude, and the test is performed by comparing the observed occurrences in the various groups with those that would occur according to the hypothesized theoretical distribution. The underlying principle of the test is the multinomial distribution defined in Eq. (5.15). That is, the observed number of outcomes in the ith group, denoted by x_i, may be interpreted as the product of the probability of occurrence of the event belonging to the ith group, p_i, and the total number n of independent trials.

Let us assume that the probability of occurrence of the event belonging to the ith group is p_i, then the observed numbers $x_1, x_2, x_3, \ldots, x_k$ are random variables that have the multinomial distribution given by

$$p\{X_1 = x_1, X_2 = x_2, \ldots, X_k = x_k\} = \frac{n!}{x_1! x_2! \cdots x_k!} p_1^{x_1} p_2^{x_2} \cdots p_k^{x_k}$$

(D.1)

where $\sum_{i=1}^{k} p_i = 1$ and $\sum_{i=1}^{k} x_i = n$.

We now test by setting up the hypothesis that the observed data come from a multinomial distribution with probabilities p_i for the ith group where p_i are known. The hypothesis may be written as $x_i = np_i$. It can be shown that the quantity

$$\chi^2 = \sum_{i=1}^{k} \frac{(x_i - np_i)^2}{np_i} \tag{D.2}$$

is distributed approximately as the χ^2 distribution with $(k-1)$ degrees of freedom. Hence, χ^2 given in Eq. (D.2) is used as the test statistic for making the decision to accept or reject the hypothesis. That is, a level of significance α is assigned for the test and the critical value $\chi^2_{k-1}(\alpha)$ is determined.

Note that if the observed data perfectly fit the hypothesized distribution then $\chi^2 = 0$. This implies that the larger the χ^2 value, the more the hypothesis should be discredited. Therefore, if the χ^2 value evaluated by Eq. (D.2) is greater than the critical value we may reject the hypothesis; otherwise we may accept it. The most usual value to assign α is 0.05, that is, the test is conducted with 5% risk.

In carrying out the χ^2 test in practice it is recommended that np_i should be at least as large as 5. If difficulty is experienced in meeting this requirement, then several groups with smaller probabilities np_i may be combined so that this requirement may be satisfied for the new group.

Example D.1. Measurements are made on crest-to-trough excursions of the random vibration of a mechanical system. The histogram constructed from a total of 140 observations is shown in Figure D.1. The random vibration is assumed to be a narrow-banded Gaussian random process, and thereby the probability distribution of the crest-to-trough excursions should follow the Rayleigh probability law. Spectral analysis is also carried out, from which the variance of the process is obtained as $\sigma^2 = 0.55$. The theoretical Rayleigh distribution is then evaluated by applying Eq. (11.25), and the result is included in Figure D.1. As can be seen in the figure, some discrepancy is observed between the histogram and the Rayleigh distribution, and therefore it is necessary to examine whether the data are from a random process considered to be a narrow-banded Gaussian process. The χ^2 test is carried out with 9 degrees of freedom, since the last 4 groups in a total of 12 groups in the histogram are combined in order to meet the

APPENDIX D 483

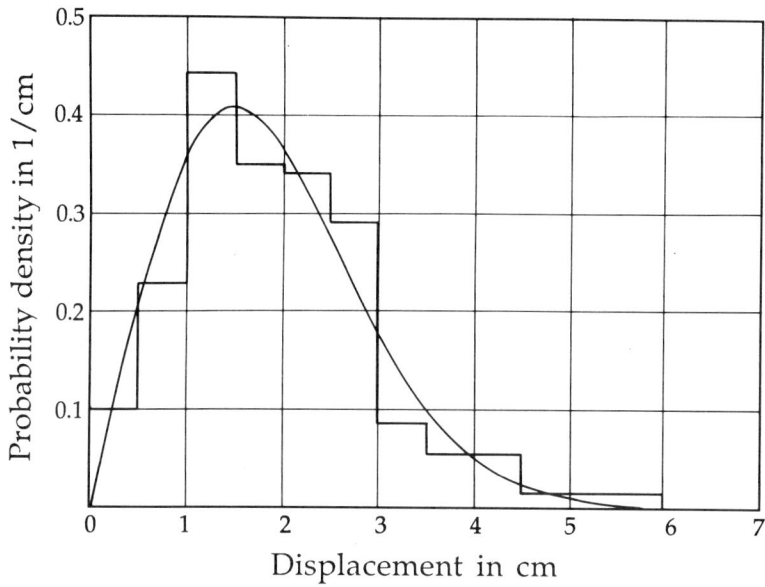

Figure D.1 Comparison of histogram constructed from observed data and theoretical probability density function in Example D.1.

requirement that $np_i \geq 5$. By choosing $\alpha = 0.05$, we find $\chi^2(0.05) = 16.92$. Since the computed value of $\chi^2 = 5.02$ is less than 16.92, we may conclude that the observed data are from a narrow-band Gaussian random process at the 5% level of significance. ∎

In the foregoing discussion it is assumed that p_i are known. However, this may not always be the case. If p_i are unknown but can be evaluated through parameters estimated from observed data $(x_1, x_2, \ldots, x_n)$, then the χ^2 test can be carried out with $(k - c - 1)$ degrees of freedom, where c is the number of parameters used for estimating p_i.

Example D.2. Let us consider the observed data of the random vibration of the mechanical system given in Example D.1. Suppose spectral analysis is not carried out on the data, then the parameter of the theoretical distribution (Rayleigh distribution) is unknown. In this case, the value of the parameter R of the Rayleigh distribution must be estimated through the maximum likelihood method as presented in Example 12.1. We have the estimated parameter of the Rayleigh distribution $\hat{R} = 4.78$ in this example.

Then, the χ^2 test is carried out with 8 degrees of freedom, since the estimation of the parameter R results in an additional reduction of 1 degree of freedom. By choosing $\alpha = 0.05$, we have $\chi_8^2(0.05) = 15.51$. The computed χ^2 value is 5.62, hence we may accept the hypothesized distribution. ∎

D.2 KOLMOGOROV–SMIRNOV TEST

This test does not consider the difference between the observed frequency in a classified group and the hypothesized probability distribution for that group; instead, it is concerned with the deviation of individual outcomes from the hypothesized distribution. The observed sample, therefore, is not classified into groups, but is rearranged in an ordered sequence for this test.

Let $x_1 < x_2 < \cdots < x_n$ be the ordered sample of size n, and let its cumulative distribution be $F_n(x)$ given by

$$F_n(x) = \begin{cases} 0 & \text{for } x < x_1 \\ r/n & \text{for } x_r \leq x \leq x_{r+1} \quad r = 1, 2, \ldots (n-1) \\ 1 & \text{for } x > x_n \end{cases} \quad (D.3)$$

$F_n(x)$ is a step function as illustrated in Figure D.2. In the case where several values are observed to be the same for $x = x_j$, then $F_n(x_j)$ increases significantly. If $F(x)$ is the cumulative distribution function of the hypothesized distribution, then the Kolmogorov–Smirnov test is based on the statistic:

$$D_n = \sup |F_n(x) - F(x)| \qquad (D.4)$$

where "sup" implies the maximum value of the entire range of x values.

Kolmogorov shows that if the sample of size n is large (on the order of 35 or greater) the probability of D_n exceeding a value ε is approximately given by

$$\Pr\{D_n > \varepsilon\} = 2 \sum_{r=1}^{\infty} (-1)^{r-1} \exp\{-2r^2 n \varepsilon^2\} \qquad (D.5)$$

Since the series converges rapidly, we may take the first term only, namely $2\exp\{-2n\varepsilon^2\}$. By letting this value be equal to the level of

APPENDIX D 485

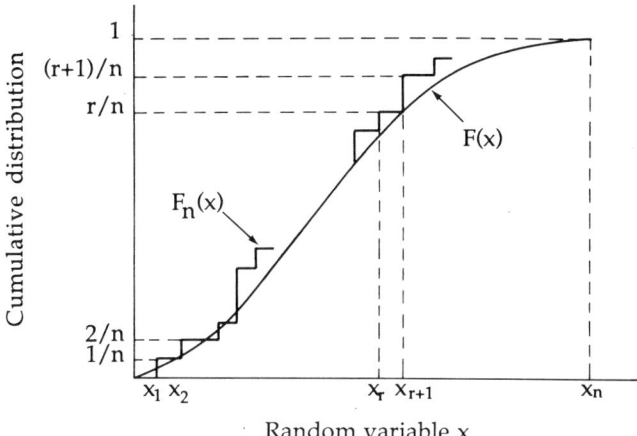

Figure D.2 Step function $F_n(x)$ and cumulative distribution function $F(x)$ for Kolmogorov–Smirnov test.

significance α, we have

$$\varepsilon = \sqrt{-\ln(\alpha/2)/2n}$$

$$= \begin{cases} 1.63/\sqrt{n} & \text{for } \alpha = 0.01 \\ 1.48/\sqrt{n} & \text{for } \alpha = 0.025 \\ 1.36/\sqrt{n} & \text{for } \alpha = 0.05 \end{cases} \quad (D.6)$$

The test is then carried out by which we may reject the hypothesis with the level of significance α if the statistic D_n exceeds ε.

The Kolmogorov–Smirnov test has the following features: (1) the sample size is taken into consideration in the test, (2) it can be applied to data obtained from a small number of observations, (3) the observed individual outcomes are considered rather than a group of observed values, and (4) only the information on the maximum deviation of the cumulative distribution functions is required. It is noted that the parameter of the distribution must be known in evaluating the statistic D_n, and that the test can be applied only to data for continuous-type random variables.

Example D.3. In the measured crest-to-trough excursions of the random vibration presented in Example D.1, the largest discrepancy between the cumulative distribution of the ordered sample and the hypothesized Rayleigh distribution is observed at $r = 38$ where $x = 1.25$. The cumulative distribution of the Rayleigh distribution with $R = 4.40$ is 0.299 for $x = 1.25$, and the cumulative distribution for $r = 38$ in a total of 140 ordered samples becomes $F_n(x) = 38/140 = 0.271$, and thereby we have $D_n = 0.028$. On the other hand, the ε value is 0.115 for $\alpha = 0.05$ and $n = 140$. Thus, we may accept the hypothesized probability distribution with a level of significance of $\alpha = 0.05$. ∎

References

Aitchison, J., and Brown, J. A. G. (1957). *The Lognormal Distribution.* Cambridge Univ. Press, Cambridge.

Anderson, T. W. (1966). *An Introduction to Multivariate Statistical Analysis.* Wiley, New York.

Arhan, M. K., Cavanié, A., and Ezraty, R. (1976). *Etude Théorique et Expérimentale de la Relation Hauteur-Période des Vagues de Tempête.* I. F. P. 24191 Centre National pour l'Exploitation des Océans.

Ariaratnam, S. T. (1960). "Random Vibrations of Non-Linear Suspensions." *J. Mech. Eng. Sci.*, Vol. 2, No. 3, pp. 195–201.

Bhat, U. N. (1972). *Elements of Applied Stochastic Processes.* Wiley, New York.

Bouws, E. (1978). "Wind and Wave Climate in the Netherlands Sector of the North Sea." The Netherlands Meteorol. Inst. Sci. Rep. WR 78-9.

Bury, K. (1976). *Statistical Models in Applied Science.* Wiley, New York.

Cartwright, D. E., and Longuet-Higgins, M. S. (1956). "The Statistical Distribution of the Maxima of a Random Function." *Proc. Roy. Soc. London, Ser. A.*, Vol. 237, pp. 212–232.

Cartwright, D. E., and Smith, N. D. (1964). "Buoy Techniques for Obtaining Directional Wave Spectra." *Buoy Technol., Mar. Tech. Soc.*, pp. 173–182.

Caughey, T. K. (1959). "Response of a Nonlinear String to Random Loading." *J. Appl. Mech., Am. Soc. Mech. Eng.*, pp. 341–344.

Caughey, T. K. (1963a). "Deviation and Application of the Fokker-Planck Equation to Discrete Nonlinear Dynamic Systems Subject to White

Random Excitation." *J. Acoust. Soc. Am.*, Vol. 35, No. 11, pp. 1683–1692.

Caughey, T. K. (1963b). "Equivalent Linearization Techniques." *J. Acoust. Soc. Am.*, Vol. 35, No. 11, pp. 1706–1711.

Cavanié, A., Arhan, M. K., and Ezraty, E. (1976). "A Statistical Relationship between Individual Heights and Periods of Storm Waves." *Proc. Conf. Behav. Offshore Struct.*, Vol. 2, pp. 354–360.

Champeney, D. C. (1973). *Fourier Transforms and Their Physical Applications*. Academic Press, London.

Chang, K., and Kazda, L. F. (1959). "A Study of Nonlinear Systems with Random Inputs." *Trans. Am. Inst. Elect. Eng.*, Vol. 78, Part II, pp. 100–105.

Cox, D. R., and Miller, H. D. (1965). *The Theory of Stochastic Processes*. Wiley, New York.

Cramér, H. (1966). *Mathematical Methods of Statistics*. Princeton Univ. Press, Princeton.

Crandall, S. H. (1962). "Random Vibration of a Nonlinear System with a Set-Up Spring." *J. Appl. Mech., Am. Soc. Mech. Eng.*, pp. 477–482.

Crandall, S. H. (1963). "Perturbation Techniques for Random Vibration of Nonlinear Systems." *J. Acoust. Soc. Am.*, Vol. 35, No. 11, pp. 1700–1705.

Crandall, S. H. (1964). "The Spectrum of Random Vibration of a Nonlinear Oscillator." *Trans. 11th Cong. Appl. Mech.*, pp. 239–243.

Crandall, S. H. (1980). "Non-Gaussian Closure for Random Vibration of Nonlinear Oscillators." *J. Non-Linear Mech.*, Vol. 15, pp. 303–313.

Crandall, S. H., Khabbaz, G. R., and Manning, J. E. (1964). "Random Vibration of an Oscillator with Nonlinear Damping." *J. Acoust. Soc. Am.*, Vol. 36, No. 7, pp. 1330–1334.

Davis, M. C., and Zarnick, E. E. (1964). "Testing Ship Models in Transient Waves." *Proc. 5th Symp. Naval Hydro.*, pp. 507–543.

Fisher, R. A., and Tippett, L. H. C. (1928). "Limiting Forms of the Frequency Distribution of the Largest or Smallest Number of a Sample." *Cambridge Philos. Soc.*, Vol. 24, Part 2, pp. 180–190.

Fisz, M. (1963). *Probability Theory and Mathematical Statistics*. Wiley, New York.

Fréchet, M. (1927). "Sur la loi de probabilité de l'écart maximum." *Ann. Soc. Polonaise Math.*, Vol. 6, p. 93.

Freeman, H. (1963). *Introduction of Statistical Inference*. Addison-Wesley, Palo Alto.

Gnedenko, B. V. (1943). "Sur la distribution limite du terme maximum d'une série aléatoire." *Ann. Math.*, Vol. 44, p. 423.

Goda, Y. (1970). "Numerical Experiments on Wave Statistics with Spectral Simulation." *Port Harbor Res. Inst. Rep.*, Vol. 9, No. 3, pp. 3–57.

Godfrey, M. D. (1965). An Exploratory Study of the Bispectrum of Economic Time Series. *Appl. Statist. (London)*, Vol. 14, pp. 48–69.

Gumbel, E. J. (1958). *Statistics of Extremes*. Columbia Univ. Press, New York.

Hasselmann, K., Munk, W., and McDonald, G. (1962). "Bispectra of Ocean Waves." *Proc. Symp. Time Series Anal.*, Chapter 8, pp. 125–139.

Kendall, M. G., and Stuart, A. (1963). *The Advanced Theory of Statistics*. Hafner, New York.

Kim, Y. C., and Powers, E. J. (1978). "Digital Bispectral Analysis of Self-Excited Fluctuation Spectra." *Phys. Fluids*, Vol. 21, pp. 1452–1453.

Kryloff, N., and Bogoliuboff, N. (1947). *Introduction to Non-Linear Mechanics* (Translated by S. Lefschetz). Princeton Univ. Press, Princeton.

Lii, K. S., Rosenblatt, M., and Van Atta, C. (1976). "Bispectral Measurements in Turbulence." *J. Fluid Mech.*, Vol. 77, pp. 45–62.

Liu, P. C., and Green, A. W. (1978). "Higher Order Wave Spectra." *Proc. 16th Coastal Eng. Conf.*, Vol. 1, pp. 360–371.

Longuet-Higgens, M. S. (1957). "The Statistical Analysis of a Random Moving Surface." *Philos. Trans. Roy. Soc. London, Ser. A.*, Vol. 249, pp. 321–387.

Longuet-Higgins, M. S. (1975). "On the Joint Distribution of the Periods and Amplitudes of Sea Waves." *J. Geophys. Res.*, Vol. 80, No. 18, pp. 2688–2694.

Longuet-Higgins, M. S. (1963). "The Effect of Non-Linearities on Statistical Distributions in the Theory of Sea Waves." *J. Fluid Mech.*, Vol. 17, Part 3, pp. 459–480.

Longuet-Higgins, M. S. (1983). "On the Joint Distribution of Wave Periods and Amplitudes in a Random Wave Field." *Proc. Roy. Soc. London, Ser. A.*, Vol. 389, pp. 241–258.

Longuet-Higgins, M. S. Cartwright, D. E., and Smith, N. D. (1961). "Observations of the Directional Spectrum of Sea Waves Using the Motions of a Floating Buoy." *Proc. Conf. Ocean Wave Spectra*, pp. 111–132.

Lyon, R. H. (1960a). "On the Vibration Statistics of a Randomly Excited Hard-Spring Oscillator." *J. Acoust. Soc. Am.*, Vol. 32, No. 6, pp. 716–719.

Lyon, R. H. (1960b). "Equivalent Linearization of the Hard Spring Oscillator." *J. Acoust. Soc. Am.*, Vol. 32, pp. 1161–1162.

McComas, C. H., and Briscoe, M. G. (1980). "Bispectra of Internal Waves." *J. Fluid Mech.*, Vol. 97, pp. 205–213.

Medhi, J. (1981). *Stochastic Processes*. Wiley, New York.

Middleton, D. (1960). *An Introduction to Statistical Communication Theory*. McGraw-Hill, New York.

Ochi, M. K. (1973). "On Prediction of Extreme Values." *J. Ship Res.*, Vol. 17, No. 1, pp. 29–37.

Ochi, M. K., and Whalen, J. E. (1980). "Prediction of the Severest Significant Wave Height." *Proc. 17th Conf. Coastal Eng.*, Vol. 1, pp. 587–599.

Panicker, N. N. (1974). "Review of Techniques for Directional Wave Spectra." *Proc. Ocean Wave Meas. Anal.*, Vol. 1, pp. 669–688.

Panicker, N. N., and Borgman, L. E. (1971). "Directional Spectra from Wave-Gage Arrays." *Proc. 12th Coastal Eng. Conf.*, Vol. 1, pp. 117–136.

Papoulis, A. (1962). *The Fourier Integral and Its Applications*. McGraw-Hill, New York.

Parzen, E. (1958). "Conditions that a Stochastic Processes Be Ergodic." *Ann. Math. Stat.*, Vol. 29, pp. 299–301.

Parzen, E. (1962). *Stochastic Processes*. Holden-Day, San Francisco.

Prabhu, N. U. (1965). *Stochastic Processes, Basic Theory and Its Applications*. Macmillan, New York.

Rice, O. (1944). "Mathematical Analysis of Random Noise." *Bell Syst. Tech. J.*, Vol. 23, pp. 282–332.

Rice, O. (1945). "Mathematical Analysis of Random Noise." *Bell Syst. Tech. J.*, Vol. 24, pp. 46–156.

Bohatgi, V. K. (1976). *An Introduction to Probability Theory and Mathematical Statistics*. Wiley, New York.

Sato, T., Sasaki, K., and Taketani, M. (1980). "Bispectral Passive Velocimeter of a Moving Noisy Machine." *J. Acoust. Soc. Am.*, pp. 1729–1735.

Tayfun, M. A. (1981). "Distribution of Crest-to-Trough Wave Heights." *Proc. Am. Soc. Civ. Eng.*, Vol. 107, WW3, pp. 149–158.

Tick, L. J. (1961). "The Estimation of Transfer Functions of Quadratic Systems." *Technometrics*, Vol. 3, No. 4, pp. 563–567.

Wang, M. C., and Uhlenbeck, G. E. (1945). "On the Theory of the Brownian Motion II." *Rev. Modern Phys.*, Vol. 17, Nos. 2 and 3, pp. 323–342.

Weibull, W. (1939). "A Statistical Theory of the Strength of Materials." *Ing. Vetenskaps Akad. Handl.*, Stockholm, No. 151.

Weibull, W. (1951). "A Statistical Distribution Function of Wide Applicability." *J. Appl. Mech., Am. Soc. Mech. Eng.*, Vol. 18, pp. 293–297.

Yamanouchi, Y., and Ohtsu, K. (1972). "On Nonlinear Ship Response and Higher Order Response Spectrum." *Trans. Soc. Nav. Arch. Jpn.*, Vol. 131, pp. 115–135.

Yeh, T. T., and Van Atta, C. W. (1973). "Spectral Transfer of Scales and Velocity Fields in Heated-Grid Turbulence." *J. Fluid Mech.*, Vol. 58, pp. 233–261.

Bibliography

Ang, A. H. S., and Tang, W. H. (1984). *Probability Concepts in Engineering Planning and Design* (2 volumes). Wiley, New York.

Bendat, J. S. (1958). *Principles and Applications of Random Noise Theory.* Wiley, New York.

Bhat, U. N. (1972). *Elements of Applied Stochastic Processes.* Wiley, New York.

Cox, D. R., and Miller, H. D. (1965). *The Theory of Stochastic Processes.* Wiley, New York.

Cramér, H. (1966). *Mathematical Methods of Statistics.* Princeton Univ. Press, Princeton.

Fisz, M. (1963). *Probability Theory and Mathematical Statistics.* Wiley, New York.

Gumbel, E. J. (1958). *Statistics of Extremes.* Columbia Univ. Press, New York.

Kendall, M. G., and Stuart, A. (1963). *The Advanced Theory of Statistics.* Hafner, New York. (3 volumes).

Lin, Y. K. (1967). *Probabilistic Theory of Structural Dynamics.* McGraw-Hill, New York.

Medhi, J. (1981). *Stochastic Processes.* Wiley, New York.

Papoulis, A. (1984). *Probability, Random Variables, and Stochastic Processes.* McGraw Hill, New York.

Parzen, E. (1962). *Stochastic Processes.* Holden-Day, San Francisco.

Solodovnikov, V. V. (1960). *Introduction to the Statistical Dynamics of Automatic Control Systems* (Translated by J. B. Thomas and L. A. Zadeh). Dover, New York.

Taylor, H. M. and Karlin, S. (1984). *An Introduction to Stochastic Modeling.* Academic Press, New York.

Index

Asymptotic extreme value distribution, 178
   Type I distribution, 179, 324
   Type II distribution, 186
   Type III distribution, 190
Autocorrelation function, 219
   two-dimensional, 260
Auto-covariance function, 204
Average frequency, 306
Average period, 304

Band-width parameter, 230
Bayes formula, 39
Bernoulli distribution, 92
   characteristic function of, 77
Bernoulli process, 217
Bernoulli trials, 91
Beta distribution:
   definition, 131
   generalized beta distribution, 132
   relationship with binomial distribution, 108, 132
Binomial distribution:
   approximation to hypergeometric distribution, 98
   characteristic function of, 92
   definition, 91
   limiting form of, 101
   mean of, 48, 93
   moment generating function of, 74
   probability generating function of, 88
   relationship with beta distribution, 108
Bispectral analysis, 259
   application to nonlinear system, 407
Bispectrum, 261

Bi-variate normal distribution, 63, 118
   characteristic function of, 83
   conditional distribution of, 64
   conditional mean and variance of, 64
   definition, 118
   expected value $E[xy]$ of, 68
   expected value $E[x^2y^2]$ of, 69
Borel field (σ-field), 9
Brownian motion, 215, 335
Buffon's needle problem, 41

Campbell's theorem, 448
Cauchy distribution, 49
   characteristic function of, 78
   mean of, 49
   moment generating function of, 75
   relationship with normal distribution, 160
Cauchy-type distribution, 186
Causal function, 474
Central limit theorem, 115
Central moments, 44
   of two random variables, 54
Chapman–Kolmogorov equation, 341
Characteristic function, 76
   conditional, 84
   joint, 83
Characteristic largest value, 172
Chi-square (χ^2) distribution, 125
   limiting form of, 127
   relationship with gamma distribution, 125
   relationship with normal distribution, 126, 133, 142
Chi-square (χ^2) test, 481
Coefficient of variation, 51

495

Coherency function, 245, 382
Conditional probability, 30
Conditional random variable:
　density function of, 32
　distribution function of, 32
　mean of, 62
　moments of, 62
　variance of, 62
Confidence coefficient, 325
Confidence interval, 325
Convergence in the mean square, 252
Convolution integral, 369
Convolution theorem, 470
Correlation coefficient, 55
Co-spectrum, 244
Counting random process, 101, 211, 423
Covariance, 54
Covariance matrix, 58
Critical damping, 373
Critical region, 332
Cross-correlation function, 239
Cross-spectral analysis, 239
Cross-spectral density function, 242
Cumulant generating function, 86
Cumulants, 84
Cumulative distribution function, 16
　joint, 26
　marginal, 29

Damping factor, 373
Delta function, 477
De Morgan's law, 8
Differentiable in mean square, 253
Differentiated random process, 249
Diffusion equation, 340
Diffusion process, 210
Directional spectrum, 245
Domain, 14
Duffing system, 392, 401, 406

Elementary renewal theorem, 454
Ensemble, 202
Envelope process, 279
Equivalent linearization technique, 389
Error function, 113
Estimator, 318
　best unbiased, 318
　maximum likelihood, 318
　minimum variance, 318
　unbiased, 318
Expected number of maxima, 284

Expected value (expectation), 43
　of random vector, 53
Exponential distribution:
　definition, 122
　moment generating function of, 74
　relationship with Rayleigh distribution, 142
Exponential-type distribution, 181
Extreme value, 168
　asymptotic distributions of, 178
　design extreme value, 176
　estimation from observed data, 194
　probable extreme value, 172
　risk parameter of, 176
Extreme value (largest):
　cumulative distribution function of, 171
　probability density function of, 171
Extreme value (smallest):
　cumulative distribution function of, 171
　probability density function of, 171

F-distribution, 331
Field, 8
Fokker–Planck equation, 353, 405
Fourier transform, 461
　properties of, 468
Frequency response function, 370
　linear system, 372

Gamma distribution:
　definition, 122
　generalized gamma distribution, 123
　maximum likelihood estimator of, 324
　relationship with χ^2-distribution, 124
　relationship with Poisson distribution, 108
Gaussian distribution, see Normal distribution
Gaussian process, see Normal process
Gram–Charlier series of Type A distribution, 415

Hazard function (or rate), 38, 175
Hermite polynomial, 412
Hilbert transform, 472
　properties of, 475
Hypergeometric distribution, 96
　limiting form of, 98

Impulse response function, 368, 372
Independent increment process, 207
Initial probability density function, 168
Integrated random process, 249
Interarrival time, 211, 429

INDEX   **497**

Poisson process, 431
Interquartile range, 22, 311
Inverse image, 14

Jacobian of transformation of random variables, 139
Joint probability:
amplitude and period of, 295
density function, 27
distribution function, 26
mass function, 26
two amplitudes, 278

Key renewal theorem, 455
Kolmogorov forward equation, 360
Kolmogorov–Smirnov test, 484
Kurtosis, 51

Leptokurtic distribution, 51
Level of significance, 322
Level (or threshold) crossing, 304
Likelihood function, 318
Lindberg–Lévy theorem, 115
Linear system, 367
time invariant, 367
Log-normal distribution, 117
relationship with normal distribution, 117, 133
Lower and upper quartile, 21

Marginal distribution, 28
probability density function of, 29
probability mass function of, 29
Markov chain, 210
two state, 345
Markov process, 210, 341
homogeneous in time, 211
two-state, 350
Maximum likelihood estimator, 318
Maximum point process, 284
Mean, 44
Mean frequency, 229
Mean period, 304
Mean value function, 204
Measurable space, 9
Median, 21
Minimum point process, 284
Modal frequency, 229
Mode, 24
Moment generating function, 73
Moments, 44

about the mean of spectrum, 230
of order $k + \ell$, 53
of random vector, 53
of spectral density function, 229
Multinomial distribution, 98
Multivariate normal distribution, 120
Mutually exclusive events, 33

Narrow-band process, 212
Negative binomial distribution, 95
limiting form of, 104, 135
Negative hypergeometric distribution, 109
Negative maxima, 284
Negative minima, 284
Non-Gaussian process, 411
probability density function of, 415, 420
Nonlinear system, 387
application of bispectral analysis, 407
application of Fokker–Planck equation, 405
equivalent linearization, 389
perturbation technique, 398
Non-narrow-band process, 214
distribution of maxima, 283
Normal approximation:
for χ^2 distribution, 127
for Poisson distribution, 114
Normal distribution:
characteristic function of, 79
confidence interval for variance, 329
definition, 111
kurtosis of, 52
maximum likelihood estimator of, 320
mean and variance of, 112
standardized, 50, 112
sum of distributions, 115, 163
three-sigma rule, 112
Normal process:
amplitude distribution, 267
crest-to-trough distribution, 273
definition, 214

One-step transition probability, 345
Order statistics, 169
j-th order statistics, 169
ordered sample, 169

Parseval theorem, 226
Period:
average, 304
distribution of, 309
mean, 304

Period *(Continued)*:
  zero-crossing, 304, 306
Perturbation technique, 398
Platykurtic distribution, 51
Poisson distribution:
  approximation to binomial distribution, 101
  approximation to negative binomial distribution, 104
  characteristic function of, 103
  confidence interval of, 332
  cumulant generating function of, 86
  definition, 102
  limiting form of, 114
  mean and variance of, 47, 103
  moment generating function of, 73
  probability generating function of, 88
  relationship with gamma distribution, 108
  sum of distributions, 149, 162
Poisson impulse process, 446
Poisson process:
  autocorrelation function of, 426
  compound (or cluster), 439
  definition, 216, 423
  filtered, 444
  generalization of, 437
  increment process, 216
  intensity function, 437
  non-homogenous, 437
  properties of, 426
  superimposed, 444
Positive maxima, 284
Positive minima, 284
Posterior probability, 39
Power spectrum, 227
Priori probability, 39
Probability, 2
  posterior, 39
  priori, 39
Probability density function, 19
  joint, 27
  marginal, 29
Probability distribution function, 16
Probability of failure, 155, 160
Probability generating function, 87
Probability mass function, 16
Probability measure, 9
Probability space, 10
Probable extreme amplitude, 313

Quadrature spectrum, 244

Quantile of order p, 21
Quartile, 21

Random event, 2
Random phenomena, 1
Random process, *see* Stochastic process
Random telegraph signal, 110, 224
Random variable, 14
  continuous type, 19
  discrete type, 16
Random vector, 25
  continuous type, 27
  discrete type, 26
Random walk, 336
Range, 14
Rayleigh distribution:
  amplitude of normal random process, 271
  confidence interval of, 326, 330
  definition, 129
  extreme value distribution of, 172
  maximum likelihood estimator of, 319
  mean and variance of, 48, 130
  probable extreme value of, 175
  relationship with normal distribution, 130, 145
  skewness, kurtosis of, 52
Reliability function, 37
Renewal, 449
Renewal counting process, 211, 449
  asymptotic properties of, 454
  elementary renewal theorem, 454
  key renewal theorem, 455
Renewal equation, 451
Renewal function, 449
Renewal period, 449
Renewal time, 449
Residual time, 456
Response amplitude operator, 370, 379
Response of linear system, 370
  to dual inputs, 379
  input and output mean levels, 371
  input and output spectral density functions, 378
Return period, 38, 174
Risk parameter, 176
Root-mean-square (rms) value, 320, 321

Sample function, 202
Sample space, 2
Schwarz's inequality, 56
Semi-invariants, 87
Set, 4

INDEX **499**

complement, 6
difference, 6
disjoint, 6
element of, 4
empty (null), 4
equal, 4
intersection, 5
mutually exclusive, 6
subset, 4
symmetric difference, 6
union, 4
Shot noise process, 217, 447
Sign function, 468
Significant amplitude, 281
Skewness, 51
Spectral analysis:
  auto-, 219
  cross-, 239
  higher order, 259
Spectral density function, 226
Spectral peakedness parameter, 231
Spectral width parameter, 230
Squared random process, 258
Square-law detector, 257
Standard deviation, 45
Standardized random variable, 50
Statistical goodness of fit tests, 481
Statistical independence, 33, 34
Stochastic process:
  definition, 201
  differentiated, 249
  envelope, 279
  ergodic process, 205
  evolutionary, 204
  integrated, 249
  narrow-band, 212
  non-Gaussian, 411
  squared, 258
  stationary (or steady state), 204
  strictly stationary, 204
  weakly (or covariance) stationary, 205
  wide-band, 213

T-distribution, 148
Threshold crossing, 304
Transfer function, 370
Transformation of random variable(s), 138, 144
  through characteristic function, 161
  difference of two random variables, 154
  product of two random variables, 156
  ratio of two random variables, 159
  sum of two random variables, 151
Transition probability density function, 211, 343
Transition probability matrix, 345
  limiting value of, 351
  n-step, 345
Truncated distribution, 35
  probability density function of, 36
  probability mass function of, 36
Two-state Markov chain, 345
Type I asymptotic extreme values, 179
  probability distribution for largest value, 181
  probability distribution for smallest value, 185
Type II asymptotic extreme values, 186
  probability distribution for largest value, 187
  probability distribution for smallest value, 189
  relationship with Type I distribution, 187
Type III asymptotic extreme values, 190
  probability distribution for largest value, 190
  probability distribution for smallest value, 194
  relationship with Type I distribution, 190

Uncorrelated random variables, 55
Uniform distribution:
  definition, 131
  maximum likelihood estimator of, 322
  relationship with beta distribution, 131
Unit impulse function, 477
Unit step function, 286, 477

Variance, 44
Variance spectrum, 234
Venn diagram, 4

Waiting time, 430
  Poisson process, 433
Weibull distribution, 128
  extreme value of, 176
  maximum likelihood estimator of, 322
  three parameter Weibull distribution, 129
White noise spectrum, 237
  band-limited, 237
Wide-band process, *see* Non-narrow-band process
Wiener–Khintchine theorem, 232
Wiener–Lévy process, 215, 339

Zero-crossing, 304
  average number of, 313
Zero-one distribution, 92

Applied Probability and Statistics (Continued)

GUPTA and PANCHAPAKESAN • Multiple Decision Procedures: Theory and Methodology of Selecting and Ranking Populations
GUTTMAN, WILKS, and HUNTER • Introductory Engineering Statistics, *Third Edition*
HAHN and SHAPIRO • Statistical Models in Engineering
HALD • Statistical Tables and Formulas
HALD • Statistical Theory with Engineering Applications
HAND • Discrimination and Classification
HEIBERGER • Computation for the Analysis of Designed Experiments
HOAGLIN, MOSTELLER and TUKEY • Exploring Data Tables, Trends and Shapes
HOAGLIN, MOSTELLER, and TUKEY • Understanding Robust and Exploratory Data Analysis
HOCHBERG and TAMHANE • Multiple Comparison Procedures
HOEL • Elementary Statistics, *Fourth Edition*
HOEL and JESSEN • Basic Statistics for Business and Economics, *Third Edition*
HOGG and KLUGMAN • Loss Distributions
HOLLANDER and WOLFE • Nonparametric Statistical Methods
HOSMER and LEMESHOW • Applied Logistic Regression
IMAN and CONOVER • Modern Business Statistics
JESSEN • Statistical Survey Techniques
JOHNSON • Multivariate Statistical Simulation
JOHNSON and KOTZ • Distributions in Statistics
        Discrete Distributions
        Continuous Univariate Distributions—1
        Continuous Univariate Distributions—2
        Continuous Multivariate Distributions
JUDGE, GRIFFITHS, HILL, LÜTKEPOHL and LEE • The Theory and Practice of Econometrics, *Second Edition*
JUDGE, HILL, GRIFFITHS, LÜTKEPOHL and LEE • Introduction to the Theory and Practice of Econometrics, *Second Edition*
KALBFLEISCH and PRENTICE • The Statistical Analysis of Failure Time Data
KASPRZYK, DUNCAN, KALTON, and SINGH • Panel Surveys
KAUFMAN and ROUSSEEUW • Finding Groups in Data: An Introduction to Cluster Analysis
KEENEY and RAIFFA • Decisions with Multiple Objectives
KISH • Statistical Design for Research
KISH • Survey Sampling
KUH, NEESE, and HOLLINGER • Structural Sensitivity in Econometric Models
LAWLESS • Statistical Models and Methods for Lifetime Data
LEAMER • Specification Searches: Ad Hoc Inference with Nonexperimental Data
LEBART, MORINEAU, and WARWICK • Multivariate Descriptive Statistical Analysis: Correspondence Analysis and Related Techniques for Large Matrices
LINHART and ZUCCHINI • Model Selection
LITTLE and RUBIN • Statistical Analysis with Missing Data
McNEIL • Interactive Data Analysis
MAGNUS and NEUDECKER • Matrix Differential Calculus with Applications in Statistics and Econometrics
MAINDONALD • Statistical Computation
MALLOWS • Design, Data, and Analysis by Some Friends of Cuthbert Daniel
MANN, SCHAFER and SINGPURWALLA • Methods for Statistical Analysis of Reliability and Life Data
MARTZ and WALLER • Bayesian Reliability Analysis
MASON, GUNST, and HESS • Statistical Design and Analysis of Experiments with Applications to Engineering and Science
MIKE and STANLEY • Statistics in Medical Research: Methods and Issues with Applications in Cancer Research

Applied Probability and Statistics (Continued)

MILLER • Beyond ANOVA, Basics of Applied Statistics
MILLER • Survival Analysis
MILLER, EFRON, BROWN, and MOSES • Biostatistics Casebook
MONTGOMERY and PECK • Introduction to Linear Regression Analysis
NELSON • Applied Life Data Analysis
NELSON • Accelerated Testing: Statistical Models, Testing Plans, and Data Analysis
OCHI • Applied Probability and Stochastic Processes in Engineering and Physical Sciences
OSBORNE • Finite Algorithms in Optimization and Data Analysis
OTNES and ENOCHSON • Applied Time Series Analysis: Volume I, Basic Techniques
OTNES and ENOCHSON • Digital Time Series Analysis
PANKRATZ • Forecasting with Univariate Box-Jenkins Models: Concepts and Cases
PLATEK, RAO, SARNDAL and SINGH • Small Area Statistics: An International Symposium
POLLOCK • The Algebra of Econometrics
RAO and MITRA • Generalized Inverse of Matrices and Its Applications
RÉNYI • A Diary on Information Theory
RIPLEY • Spatial Statistics
RIPLEY • Stochastic Simulation
ROSS • Introduction to Probability and Statistics for Engineers and Scientists
ROUSSEEUW and LEROY • Robust Regression and Outlier Detection
RUBIN • Multiple Imputation for Nonresponse in Surveys
RUBINSTEIN • Monte Carlo Optimization, Simulation, and Sensitivity of Queueing Networks
RYAN • Statistical Methods for Quality Improvement
SCHUSS • Theory and Applications of Stochastic Differential Equations
SEARLE • Linear Models
SEARLE • Linear Models for Unbalanced Data
SEARLE • Matrix Algebra Useful for Statistics
SKINNER, HOLT, and SMITH • Analysis of Complex Surveys
SPRINGER • The Algebra of Random Variables
STAUDTE and SHEATHER • Robust Estimation and Testing
STEUER • Multiple Criteria Optimization
STOYAN • Comparison Methods for Queues and Other Stochastic Models
STOYAN, KENDALL and MECKE • Stochastic Geometry and Its Applications
THOMPSON • Empirical Model Building
TIJMS • Stochastic Modeling and Analysis: A Computational Approach
TITTERINGTON, SMITH, and MAKOV • Statistical Analysis of Finite Mixture Distributions
UPTON and FINGLETON • Spatial Data Analysis by Example, Volume I: Point Pattern and Quantitative Data
UPTON and FINGLETON • Spatial Data Analysis by Example, Volume II: Categorical and Directional Data
VAN RIJCKEVORSEL and DE LEEUW • Component and Correspondence Analysis
WEISBERG • Applied Linear Regression, *Second Edition*
WHITTLE • Optimization Over Time: Dynamic Programming and Stochastic Control, Volume I and Volume II
WHITTLE • Systems in Stochastic Equilibrium
WILLIAMS • A Sampler on Sampling
WONNACOTT and WONNACOTT • Econometrics, *Second Edition*
WONNACOTT and WONNACOTT • Introductory Statistics, *Fourth Edition*
WONNACOTT and WONNACOTT • Introductory Statistics for Business and Economics, *Third Edition*
WOOLSON • Statistical Methods for The Analysis of Biomedical Data